Fault-Tolerant Control
of Deterministic Input/Output Automata

Dissertation zur Erlangung des Grades eines
Doktor-Ingenieurs
der Fakultät für Elektrotechnik und Informationstechnik
an der Ruhr-Universität Bochum

von

Melanie Schuh, geb. Schmidt
geboren in Bochum

Bochum, 2017

1. Gutachter: Prof. Dr.-Ing. Jan Lunze

                 Ruhr-Universität Bochum, Deutschland

2. Gutachter: Prof. Dr. Jean-Jacques Lesage

                 École normale supérieure de Cachan, Frankreich

Eingereicht am: 12.10.2016

Tag der mündlichen Prüfung: 23.03.2017

Bibliografische Information der Deutschen Nationalbibliothek

Die Deutsche Nationalbibliothek verzeichnet diese Publikation in der
Deutschen Nationalbibliografie; detaillierte bibliografische Daten sind
im Internet über http://dnb.d-nb.de abrufbar.

ISBN 978-3-8325-4539-0

Logos Verlag Berlin GmbH
Comeniushof, Gubener Str. 47,
10243 Berlin
Tel.: +49 (0)30 42 85 10 90
Fax: +49 (0)30 42 85 10 92
INTERNET: http://www.logos-verlag.de

Contents

Acknowledgements

This thesis summarizes the results of my five years as a research associate at the Institute of Automation and Computer Control at Ruhr-Universität Bochum, Germany. I would like to thank Prof. Dr.-Ing. Jan Lunze for his constant support during these past years. He believed that I could accomplish a doctorate degree even before I had thought of it and supported me in all my decisions. I would also like to thank Prof. Jean-Jacques Lesage for accepting to review this thesis. During meetings at various conferences he was always interested in my work and encouraged me with his support.

Special thanks go to my fellow PhD-students Sven Bodenburg, Ozan Demir, Fabian Just, Yannick Nke, Tobias Noeßelt, Andrej Mosebach, Sebastian Pröll, Kai Schenk, Christian Stöcker, Michael Ungermann, Philipp Welz, Christian Wölfel, Daniel Vey and Markus Zgorzelski. They were always willing to discussing the science of "drawing circles" and, finally, proof-reading this manuscript. Thanks also go to Dr.-Ing. Johannes Dastych, Kerstin Funke, Susanne Malow, Andrea Marschall, Rolf Pura and Udo Wieser for their constant technical and administrative support.

I would like to thank my family, who was always interested in my work, tried to understand my research and supports me through my life. Finally, I would like to express my gratitude to René, who became the perfect union of husband, colleague and friend for me. Thank you for sharing your life and bringing up our daughter Marie with me.

Lengerich, July 2017   Melanie Schuh

Abstract

This thesis deals with active fault-tolerant control of discrete event systems modeled by deterministic Input/Output (I/O) automata. Fault-tolerant control aims at enabling technical systems to fulfill their given task despite the presence of a fault while complying with safety constraints. Active fault-tolerant control realizes three operating modes – nominal control, fault diagnosis and controller reconfiguration. Thereby it differs from passive fault-tolerant control, where these explicit operating modes are abandoned and a single robust controller is used, both, for the faultless and the faulty case.

A new fault-tolerant controller which autonomously ensures the fulfillment of the control aim, both, in the faultless and the faulty case is developed. The control aim is to steer the plant into a desired final state while guaranteeing the avoidance of illegal transitions. The fault-tolerant controller is defined in a systematic way based on models of the faultless and the faulty plant. Corresponding to the three operating modes, the proposed integrated fault-tolerant controller consists of a tracking controller, a diagnostic unit and a reconfiguration unit.

As long as no fault is present, the newly defined tracking controller controls the plant in a feedback loop in order to guarantee the fulfillment of the control aim. It consists of a trajectory planning unit and a controller. The trajectory planning unit plans a reference trajectory based on the model of the plant. Then the controller, which is a deterministic I/O automaton, steers the plant along this reference trajectory. At the same time the diagnostic unit detects whether a fault occurred.

If a fault is detected, a novel active diagnosis method is used in order to identify the present fault as well as the current state of the faulty plant. That is, the diagnostic unit generates a diagnostic trajectory as an input for the controller. This input sequence is chosen such that the diagnostic result is improved until it ideally becomes unambiguous.

The developed reconfiguration unit uses the ambiguous or unambiguous diagnostic result provided by the diagnostic unit to reconfigure the tracking controller. That is, it provides the model and the current state of the faulty plant to the trajectory planning unit. Furthermore, the transitions in the controller are adapted to the fault. Consequently, no complete redefinition of the tracking controller is necessary.

As a main result, it is proved that the plant in the fault-tolerant control loop fulfills the control aim in the faultless as well as in the faulty case if the control loop is recoverable. The applicability of the fault-tolerant control method is demonstrated by means of a handling process at the Handling System HANS.

1 Introduction

1.1 Fault-tolerant control

1.1.1 Motivation for fault-tolerant control

Fault-tolerant control aims at controlling technical systems, both, in the absence and presence of faults. Faults may influence the actuators, sensors or system dynamics of the technical system in a negative way. For example, an actuator like a valve might get stuck or a sensor might provide a corrupted output signal due to deterioration. Control units are often designed such as to ensure the fulfillment of a given task by a *faultless* plant. Consequently, the given task can usually not be completed by the controlled plant any more if a fault occurs. Even worse, the technical system might get damaged or people might get harmed. Therefore, for all kinds of technical systems an effective fault handling is necessary.

There are various established approaches to deal with faults. In the simplest case, the violation of a threshold by a physical signal might be detected and reported to a human operator, who then takes necessary actions to counteract the influence of the fault or shut down the entire system to prevent damages. Another example is the use of physical redundancy, which is mainly used in safety-critical systems, where after the detection of a fault instead of the faulty component an identical substitute is used.

In this thesis the effects of faults are counteracted by using analytical redundancy, which relies on the evaluation of mathematical models of the controlled system. The main aim of fault-tolerant control is to keep the system in operation after the occurrence of a fault such that safety requirements are not violated and, if possible, the given task is still fulfilled by the closed-loop system. The necessity for intervention by a human operator shall be avoided. In an active fault-tolerant control loop, usually three operating modes are present:

1. **Nominal operating mode**, where the faultless plant is controlled by a nominal controller. Simultaneously, diagnosis is executed in order to detect whether a fault occurred.

2. **Diagnostic mode**, where a fault is present in the plant and it is tried to identify the fault.

3. **Reconfiguration mode**, where information about the fault is used to reconfigure the controller. Afterwards the reconfigured controller controls the faulty plant, preserving the

specifications of the nominal operating mode.

Often these operating modes are considered independently of each other. However, in order to achieve fault-tolerance without the need for intervention of a human operator, a fault-tolerant controller which realizes all three operating modes autonomously needs to be specified.

The main aim of this thesis is to develop a method for designing an integrated fault-tolerant controller that realizes the three operating modes of nominal operation, fault diagnosis and controller reconfiguration autonomously.

1.1.2 Structure of the fault-tolerant control loop

This thesis deals with the fault-tolerant control of discrete event systems within the framework shown in Fig. 1.1. A technical system called plant $\mathcal{P}$ is to be controlled such that it reaches a desired final state, which represents the completion of a given task. At the same time it has to be guaranteed that the plant does not violate given safety constraints. Since the plant $\mathcal{P}$ might be affected by a fault f, the use of a fault-tolerant controller $\mathcal{C}_{\mathrm{FTC}}$ is necessary. It consists of a *tracking controller* $\mathcal{C}_{\mathcal{T}}$, a *diagnostic unit* $\mathcal{D}$ and a *reconfiguration unit* $\mathcal{R}$. Roughly speaking, these three components of the fault-tolerant controller $\mathcal{C}_{\mathrm{FTC}} = (\mathcal{C}_{\mathcal{T}}, \mathcal{D}, \mathcal{R})$ correspond to the three operating modes of the fault-tolerant control loop mentioned above. In Fig. 1.1 normal arrows represent the transmission of scalar signals, while double arrows correspond to the transmission of more complex information like sets or algorithms.

In the execution layer of the fault-tolerant control loop in Fig. 1.1 the tracking controller $\mathcal{C}_{\mathcal{T}}$ controls the plant $\mathcal{P}$ in a common feedback loop by generating control signals for the actuators of the plant based on the evaluation of the sensor measurements of the plant, the desired final state and the safety constraints for the plant $\mathcal{P}$. The desired final state for the plant $\mathcal{P}$ is given

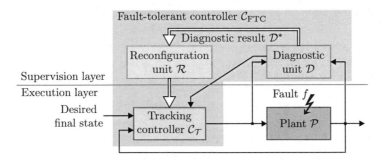

Figure 1.1: Structure of the fault-tolerant control loop.

at runtime by some external unit, for example, a higher-level controller or a human operator. In contrast, the safety constraints do not change over time and are therefore already considered during the definition of the tracking controller C_T.

In the supervision layer of the fault-tolerant control loop in Fig. 1.1 the diagnostic unit D uses active diagnosis in order to detect and identify a present fault f. Active diagnosis means that the diagnostic unit D generates test signals for the closed-loop system when a fault is detected (diagonal arrow in Fig. 1.1) and evaluates the input and output signals of the faulty plant P. Then it provides the diagnostic result D^* to the reconfiguration unit R. The reconfiguration unit R uses the information contained in the diagnostic result D^* in order to reconfigure the tracking controller C_T. The resulting reconfigured tracking controller C_T^r controls the faulty plant P such that it again reaches the desired final state while obeying the safety constraints.

1.1.3 Fundamental questions

This section states fundamental questions that are answered in this thesis in order to solve the proposed fault-tolerant control problem. They can be assigned to the three different modes of the fault-tolerant control loop and the realization of these modes by one fault-tolerant controller as follows:

- **Nominal operating mode** (Chapter 3)
    - How should the tracking controller C_T be designed such that the controlled plant P reaches a desired final state given at runtime when no fault is present?

- **Diagnostic mode** (Chapter 4)
    - How can the diagnostic unit D detect and identify the present fault f?
    - Under which conditions is the fault diagnosis possible?
    - Which information does the diagnostic unit D have to provide to the reconfiguration unit R in form of the diagnostic result D^*?

- **Reconfiguration mode** (Chapter 5)
    - How must the reconfiguration unit R modify the tracking controller C_T based on the diagnostic result D^*?
    - Under which conditions is the reconfiguration possible?

- **Integrated fault-tolerant controller** (Chapter 6)
    - Under which conditions is it possible to find a fault-tolerant controller such that a desired final state is reached by the plant P while obeying the safety constraints even when a fault f is present?

– How do the tracking controller $\mathcal{C}_T$, the diagnostic unit $\mathcal{D}$ and the reconfiguration unit $\mathcal{R}$ have to interact in order to achieve fault-tolerance in the closed-loop system?

1.2 Literature on fault-tolerant control of discrete event systems modeled by automata

This section gives a brief literature review on fault-tolerant control methods for discrete event systems modeled by different kinds of automata. A concise definition of basic terms and techniques for fault-tolerant control, both, of discrete event and continuous systems is given in [29]. There exist very little integrated contributions that address all three problems of controller design, fault diagnosis and controller reconfiguration. Therefore, at first literature dealing with each of these problems individually is given.

1.2.1 Literature on controller design

There exist various methods for designing controllers for discrete event systems modeled by automata. They differ from each other because of addressing different system classes, control aims and/or specification models. Some system classes to which different controller design methods correspond are, for example,

- finite state machines [35],

- I/O automata [63],

- asynchronous sequential machines [114].

Possible control aims are, for example,

- limiting the plant's behavior to a legal behavior [88],

- state attraction [30, 54],

- matching the behavior of the closed-loop system with a desired behavior [44].

Formally, a control aim is expressed by a specification, which can again be given in different forms, for example by

- finite state machines/regular languages [88],

- sets of desired or forbidden states, outputs or transitions [77].

This thesis addresses the control of deterministic I/O automata with the aim of steering a plant into a desired final state (state attraction) while avoiding the violation of safety constraints.

The state attraction problem has been considered previously within the supervisory control theory (SCT) framework [30, 54]. Within this framework a plant is seen as an event generator, where a supervisor can *prevent* events from occurring but can not *enforce* their occurrence. However, the enforcement of events is an important feature in the control of real-world systems. To overcome this problem, the following extensions to SCT have been proposed. In order to steer the plant into a marked state, in [26] a controller that actively enforces actions is obtained from the parallel composition of the supervisor and the plant model. In [31, 94] a controller which can be represented by a Mealy automaton and implemented on a programmable logic controller is extracted from the supervisor by selecting only one controllable event for every state. [85] starts with an I/O model of the plant, which is first transformed into a standard automaton. Afterwards a supervisor is derived using SCT and implemented as a Moore automaton. In contrast, this thesis presents a controller definition tailored to I/O automata, which naturally respect the causality relation between the controller and the plant.

In [25] the state attraction problem is considered in the context of biological processes, which is solved by searching for a path to the desired final state and defining a supervisor based on this path and sets of events that can be enabled, forced or preempted. Besides of the different modeling formalism, the main difference to the method in this thesis lies in the fact that the path to the desired final state is hard-coded into the controller. Therefore, it is not possible to select different desired final states at runtime like in this thesis.

An approach for the control of nondeterministic I/O automata based on the knowledge of a current state estimate for the plant is given in [77]. Due to the presence of nondeterminism, the resulting controller structure is more complicated and less suitable for a rigorous analysis than the one presented in this thesis.

1.2.2 Literature on fault diagnosis

An overview of fault diagnosis methods for discrete event systems has been given recently in [125]. Some of these methods for model-based fault diagnosis are summarized in the following. It can be distinguished between language-based approaches and state-based approaches. The main difference between these two classes of approaches is the different interpretation of faults. In language-based approaches a fault is seen as a single event after which the plant might return to its normal operation. For example, this type of fault typically occurs in computer systems. In state-based approaches a fault changes the dynamics of the system, e.g., the state transition function. Therefore, state-based approaches are more suited for dealing with faults in dynamical systems. In this thesis, a state-based fault diagnosis approach is developed.

Language-based approaches. The most common approach for fault diagnosis in discrete event systems is the so-called diagnoser approach [97, 98]. It is a language-based approach which considers a fault to be an unobservable event in the plant. Therefore, fault diagnosis means to infer from a trace of observed events to the occurrence of a fault event. The diagnoser is an observer-like automaton which gives, in addition to the state estimate, information on occurred faults whenever a new event is observed. Similar to when an observer is explicitly constructed, the diagnoser is of exponential complexity in the number of states in the plant. A means to reduce the exponential complexity of the diagnoser is to construct a so-called verifier or twin plant structure [51, 122]. Compared to the diagnoser its states do not contain sets of states but pairs of non-faulty and faulty states to be distinguished.

The diagnoser concept is extended to safe diagnosability in [79]. There it is claimed that the fault has to be identified before any safety constraints are violated by the plant. Various further extensions of the diagnoser, e.g., towards the diagnosis of intermittent faults [33, 41, 50, 123, 124], diagnosis for timed discrete event systems [21, 39, 49, 115, 117], diagnosis for stochastic discrete event systems [37, 38, 61, 112, 113, 118] or robust diagnosis [27, 32, 34, 58, 110] have been proposed as well.

All aforementioned methods are passive fault diagnosis methods, which means that the plant either produces outputs autonomously or is excited by some external source (e.g., A controller). In contrast, in active fault diagnosis the plant is actively influenced in order to enhance the fault diagnosis. In [96] SCT is used to synthesize a supervisor which guarantees that the controlled plant is diagnosable. Similarly, in [80] a supervisor ensuring safe diagnosability of the plant is synthesized. [36] uses a different approach by utilizing the nominal controller instead of designing a new one. The nominal controller continues its nominal operation unless it receives an input from the diagnostic unit. Such an input is generated only if it is thereby possible to decide whether the plant is faultless or faulty. In [116] an I/O automaton is converted into a finite state machine such that a diagnoser can be computed, which is converted back into an I/O automaton. When a fault is detected by the diagnoser, an input sequence called homing sequence is computed using testing theory (cf. [56]). The homing sequence, which guarantees the identification of the fault, is constructed based on the diagnoser.

The fault diagnosis method in this thesis does not belong to the language-based approaches, but to the state-based approaches discussed in the following paragraph. With respect to active diagnosis, the method in this thesis is most closely related to the ones in [116], because homing sequences will be used for the fault diagnosis once a fault has been detected as well. However, no diagnoser is computed such that no conversion of the I/O automaton is necessary.

State-based approaches. In contrast to language-based approaches, state-based diagnosis approaches assume that the fault is reflected in the states of the system. Therefore, in this case fault diagnosis means to determine the current state of the discrete event system, hence it is an observation problem. One of the first contributions on this topic is [60], which focuses on analyzing the diagnosability of a system by partitioning the state space. In [46–48], the diagnoser approach is modified in order to make it applicable to state-based diagnosis. In contrast to the language-based approach, this allows to solve the fault diagnosis problem even if the initial state of the system is unknown. Furthermore, the computational complexity is reduced by using a model reduction scheme. The results are extended to temporary failures in [28].

In [103, 104, 106] the problem of determining the current state of a plant in order to identify a present fault is considered as well. To perform the identification, an automaton whose states represent sets of possible current states of the plant and its reduction to a detector whose states contain pairs of plant states are proposed. Extensions to sensor activation problems and an application to stochastic systems are presented in [102, 107].

There also exist some state-based approaches for the fault diagnosis of timed discrete event systems [20, 69, 70, 99, 100, 108] or stochastic discrete event systems [57, 67, 68]. By incorporating this additional information level, a more accurate plant model can be obtained.

[90–93] follow the opposite direction by using only the nominal model of the plant and thereby reducing the necessary amount of model information. During the fault diagnosis I/O-pairs of the controlled plant are compared with the nominal model in order to find unexpected or missed events. The resulting so-called residuals are used to generate an estimate about the current state of the system and the faulty component. The method is enriched with timing information in [101], where additionally early and late events are considered.

In [40] an active state-based fault diagnosis method is presented. An observer automaton is computed based on which a preset input sequence can be found (if the plant is active diagnosable) such that from the outputs of the plant the fault can be determined. Both, the construction of the observer automaton and the computation of the preset input sequences is of exponential complexity.

The fault diagnosis method in this thesis belongs to the class of state-based approaches. The main difference to existing approaches is that an active diagnosis method is presented, such that the identification of the present fault can be enforced. Compared to the active diagnosis method in [40], the fault diagnosis method in this thesis does not rely on an explicitly constructed observer such that its computational complexity is reduced to linear.

1.2.3 Literature on controller reconfiguration

There mainly exist three kinds of approaches for controller reconfiguration based on a given diagnostic result: switching control, controller modification and restart states. The main difference between these different kinds of approaches is the time at which the reconfigured controller is computed. While in switching control all computations are executed offline, controller modification is performed at runtime. When restart states shall be used, the computations are also executed offline. However, in contrast to the two other kinds of approaches, the manual intervention is required.

Switching control. In switching control a bank of controllers corresponding to different fault cases is designed offline. Then, once a fault is identified, the respective controller is connected to the plant. In [71] a fault-tolerant control architecture that switches among a bank of controllers and objectives once the systems capabilities changes, e.g., when a fault occurs, is presented for the application to highway systems. In [43], a so-called megacontroller decides, which of a number of precomputed control laws is appropriate in case of a sensor failure.

There also exist some approaches which consider explicitly the switching procedure itself. [55] deals with the problem that a nominal specification has to be ensured until a fault occurs and afterwards a degraded specification has to be fulfilled for all times. Consequently, the nominal controller is not only responsible for the control in the faultless case, but also has to ensure the degraded specification between the occurrence and the detection of the fault. After the fault detection it is switched to a precomputed controller as usual. In contrast, [78] proposes a method to change from one configuration (i.e., nominal controller and faultless plant) to another (i.e., reconfigured controller and faulty plant) within a bounded number of steps. Therefore, for each configuration one coordination automaton is computed which guarantees the fulfillment of a configuration-specific specification prior to the change (for example switching to a different tool for a new product type).

The reconfiguration method in this thesis considers the online modification of the nominal controller in case of a fault. Consequently, there is no need to predefine a bank of controllers and switch to one of them.

Modification of nominal controller. Another approach for controller reconfiguration is the modification of the nominal controller. For the case of actuator faults, [73] proposes to remove all transitions using the faulty actuator from the nominal controller. If afterwards the specification can not be fulfilled by the controlled plant, the controller is completely redesigned. [75] focuses on the online recovery from a fault. If the fault happens some time before it is detected and identified, the controller is adapted such that it steers the plant from the current (faulty) state to the last nominal state. If the fault is identified at the same time instant at which it occurs, the controller is changed such that the plant is moved to a state satisfying the next requirement of the specification. In [76] the method is revisited for I/O trellis automata, which are defined by unfolding I/O automata.

[62] uses the mega-controller proposed in [43] in order to react to events that have become

unobservable due to the sensor failure. Instead of switching to a predefined controller, the mega-controller adds or removes controller states or changes the output function of the controller.

The reconfiguration method in this thesis also considers the online modification of the nominal controller. However, since the underlying controller structure differs from the ones in the literature, the reconfiguration method will also differ from the proposed ones.

Restart states. In [22–24] the concept of restart states is used for the controller reconfiguration. That is, a set of restart states is computed offline, into which the plant has to be brought manually in case of a fault. Then precomputed restart paths are executed in order to bring the plant back into its last faultless state before the occurrence of the fault. In contrast, the reconfiguration method in this thesis does not require any manual intervention.

1.2.4 Literature on integrated fault-tolerant control

Integrated fault-tolerant methods address the entire work-cycle of the fault-tolerant control loop (cf. Fig. 1.1). However, most of these methods are passive methods, thus not distinguishing between the three operating modes of nominal control, fault diagnosis and controller reconfiguration.

Passive fault-tolerance. In passive fault-tolerance a single controller is used, both, to control the faultless and the faulty plant. Therefore, passive fault-tolerant control can be seen as a special case of robust control [59, 82, 95].

In [42] a method to analyze systems for different kinds of fault-tolerance is presented, where fault-tolerance means that the plant remains in a tolerable state that is sufficiently similar to a faultless state after the occurrence of a fault. In [84] robustness and fault-tolerance is combined by synthesizing a supervisor based on a set of possible plant models which guarantees that the plant can still reach a marked state when a fault occurred. A combination of fault-tolerant robust supervisory control of discrete event systems with time information is discussed in [83].

For the case of sensor failures, [89] proposes to define a (robust) supervisor that still guarantees the fulfillment of the specification after either one of the sensors fails. [119] presents a method to synthesize a supervisor which guarantees that the plant returns to a non-faulty (or equivalent) state within a bounded delay after the occurrence of a fault. This framework is extended in [120] with the notion of weak fault-tolerance. In [45, 121] the synthesized supervisor guarantees the fulfillment of a nominal specification in the faultless case and a possibly degraded specification after a fault occurred. [109] additionally takes the existence of repair-events into account.

[72] focuses on the kind of specification that is useful for the design of a passive fault-tolerant controller. The main contribution is to fix the lower bound specification (e.g., corresponding to the nominal behavior) and to relax the upper bound specification such that it may be violated in case of a fault. Then a controller is synthesized based on the relaxed upper bound specification that is computed based on the fixed lower bound specification.

Usually fault-tolerance with respect to faults in the plant is considered, while the controller is assumed to be fault-free. In [111] however, a distributed controller is considered, which is still able to enforce the specification even if some of the local control units fail.

The fault-tolerant control method in this thesis is an active one. That is, the nominal controller can be defined for the faultless plant and does not have to take into account any a priori information about possible faults.

Active fault-tolerance. In active fault-tolerant control, first, fault diagnosis is performed, before the controller is reconfigured. For the specific case of sensor failures, [86, 87] present a fault-tolerant control framework as an integrated control and diagnosis framework. There is no explicit reconfiguration unit. Instead, the diagnostic result is used to modify the control objectives until they become achievable by the controller.

In [81], a so-called diagnoser-controller consisting of a diagnoser and a set of predefined controllers is introduced. It guarantees, under certain conditions called safe diagnosability and safe controllability, that the fault is identified early enough such that the reconfigured controller can take actions and the plant does not violate any safety constraints.

In [105] an active fault-tolerant control method that is conceptually closely related to the one presented in this thesis is presented. Similar to this thesis, the plant is controlled by a nominal controller in the faultless case and the fault diagnosis is performed by computing a current state estimate for the plant in every time step. If a fault is identified, controllable events that need to be disabled in order to guarantee the safe operation of the system are identified online. The main difference to the method presented in this thesis is the considered system class. Whereas [105] uses finite state machines such that the controller is a supervisor computed with SCT, in this thesis I/O automata are considered.

1.3 Main contributions of the thesis

The main contributions of the thesis with respect to the aim of developing a completely integrated fault-tolerant control method are as follows.

Tracking control. The behavior of many technical systems can be described on an abstracted level as discrete event systems. Different from finite state machines mostly used in literature, I/O automata reflect the natural causality between input signals and output signals of a plant $\mathcal{P}$ in an intuitive way. Therefore, in this thesis deterministic I/O automata are used. When considering the control of real-world systems, it is useful to define a controller to which a desired task for the plant $\mathcal{P}$ can be given at runtime. At the same time there should be a guaranty that the controller indeed enforces the given task in the plant $\mathcal{P}$ but also ensures its safety. In literature there are no controller definitions fulfilling all of these properties (Section 1.2.1).

Therefore, this thesis introduces a new controller structure called tracking controller $\mathcal{C}_T$ for plants $\mathcal{P}$ modeled by deterministic I/O automata (Fig. 3.1). The tracking controller $\mathcal{C}_T$ consists of a trajectory planning unit T and a controller $\mathcal{C}$. The trajectory planning unit T has the task to plan a reference trajectory into a desired final states z_F that avoids transitions which lead to the violation of safety constraints. The controller $\mathcal{C}$ is a deterministic I/O automaton which translates the reference trajectory into an input sequence $V_p(0 \ldots k_e)$ for the plant $\mathcal{P}$.

Controllability conditions based on the reachability of the desired final state z_F are developed. Whenever the plant $\mathcal{P}$ is controllable, it is guaranteed that the tracking controller $\mathcal{C}_T$ steers the plant $\mathcal{P}$ into any desired final state z_F given at runtime, while ensuring that the plant $\mathcal{P}$ does not violate any safety constraints (Theorems 3.3 and 3.4).

That is, the main result on the topic of tracking control is a new method for *defining a tracking controller $\mathcal{C}_T$* based on the model of the faultless plant, which *enforces the fulfillment of the control aim* by the *faultless* plant $\mathcal{P}$.

Fault diagnosis. Active fault diagnosis methods provide a more accurate diagnostic result quicker and more reliably than passive diagnosis methods. Since the accuracy of the diagnostic result $\mathcal{D}^*$ is crucial for the reconfiguration of the tracking controller $\mathcal{C}_T$, active diagnosis is used in this thesis. The few existing active diagnosis methods for discrete-event systems either show a bad computational complexity [40, 116] or require to restrict the nominal controller in order to enforce diagnosability of the plant [80, 96]. The active fault diagnosis method in this thesis overcomes these shortcomings.

Based on testing theory for automata [56], a method for the diagnostic unit $\mathcal{D}$ to select inputs for the nominal controller $\mathcal{C}$ is developed. It is based on the compatibility partition of the state set of the plant under all possible faults. It is shown that, under certain conditions, the application of these inputs and the consistency-based evaluation of the resulting plant outputs lead to an identification of the present fault f and the current state $z_p(k)$ of the faulty plant (Theorems 4.2 and 4.5).

The main result regarding fault diagnosis is a novel *active fault diagnosis* method which allows to *identify* the present fault f and the current state $z_p(k)$ of the faulty plant in a systematic way.

Controller reconfiguration. The controller reconfiguration method developed in this thesis aims at avoiding the complexity of a complete redesign of the controller or of the definition of various controllers for switching control [43, 55, 71, 78]. Instead, the reconfiguration unit $\mathcal{R}$ modifies the nominal tracking controller $\mathcal{C}_T$ based on the diagnostic result $\mathcal{D}^*$ given by the diagnostic unit $\mathcal{D}$. Using basic ideas from [73, 75, 76], a reconfiguration method tailored to the newly introduced tracking controller structure is developed. It effects that only that part of the given nominal tracking controller $\mathcal{C}_T$ is modified, which is responsible for the control of the part of the plant $\mathcal{P}$ affected by the present fault f.

When combining fault diagnosis and reconfiguration, it is not always feasible to assume that the diagnostic unit $\mathcal{D}$ provides an unambiguous diagnostic result $\mathcal{D}^*$ containing only the present fault f. Rather, often a set of fault candidates is provided. Existing methods for robust control of discrete event systems provide ideas on how to deal with uncertainties about the real model of a given plant [59, 95]. Based on these ideas, a common model of the faulty plant is defined, which captures the common part of the behavior of the plant $\mathcal{P}$ under all fault candidates included in an ambiguous diagnostic result $\mathcal{D}^*$. The common model of the faulty plant is used to generalize the reconfiguration method to deal with ambiguous diagnostic results as well.

Reconfigurability conditions relying on the controllability of the faulty plant or the common model of the faulty plant are developed. Whenever the nominal tracking controller $\mathcal{C}_T$ is reconfigurable, it is guaranteed that the reconfigured tracking controller $\mathcal{C}_T^{\mathrm{r}}$ steers the faulty plant $\mathcal{P}$ into any desired final state z_F given at runtime, while ensuring that the plant $\mathcal{P}$ does not violate any safety constraints (Theorems 5.3, 5.4, 5.7 and 5.8).

The main result is a new controller reconfiguration method which *modifies* the nominal tracking controller such that it *enforces the fulfillment of the control aim* by the *faulty* plant $\mathcal{P}$.

Integrated fault-tolerant control method. The main contribution of the thesis is the developed integrated active fault-tolerant control framework for deterministic I/O automata, in which the three operating modes of nominal control, fault diagnosis and controller reconfiguration are executed autonomously. It complements existing approaches from literature which address active fault-tolerant control of finite state machines with similar objectives [81, 105].

The fault-tolerant controller $\mathcal{C}_{\mathrm{FTC}}$ is obtained by combining the developed methods for tracking control, fault diagnosis and controller reconfiguration, and specifying their interfaces. Recoverability conditions for the closed-loop system, which consider diagnosability of the plant $\mathcal{P}$ and reconfigurability of the tracking controller $\mathcal{C}_T$, are developed. The fault-tolerant controller $\mathcal{C}_{\mathrm{FTC}}$ steers the plant $\mathcal{P}$ into any desired final state z_F given at runtime while ensuring that the plant $\mathcal{P}$ does not violate any safety constraints (Theorems 6.3 and 6.5). The fulfillment of this control aim is guaranteed in the faultless case, whenever the faultless plant $\mathcal{P}$ is controllable, but also if a fault f occurs, as long as the closed-loop system is recoverable from this fault f.

Consequently, the main result of the thesis is a systematic and provably effective method for defining an *active fault-tolerant controller* based on *models of the faultless and the faulty plant.*

1.4 Structure of the thesis

Figure 1.2 illustrates the structure of the thesis. **Chapters 2 and 8** frame the main part of the thesis by presenting the fault-tolerant control problem to be solved and summarizing the main results, respectively. In Chapters 3–5 the three elements of the proposed fault-tolerant controller corresponding to the three operating modes of nominal operation, fault diagnosis and controller reconfiguration are introduced.

- **Chapter 3** proposes the tracking controller C_T and analyzes the behavior of the closed-loop system.

- **Chapter 4** considers the diagnostic unit D for which a passive and an active fault diagnosis method are developed.

- **Chapter 5** presents the reconfiguration unit R which modifies the tracking controller C_T from Chapter 3 based on a diagnostic result provided by the diagnostic unit D from Chapter 4.

Chapters 6 and 7 combine the methods from Chapters 3–5 to obtain and evaluate the integrated fault-tolerant control method. In **Chapter 6** it is proved that the proposed fault-tolerant controller solves the fault-tolerant control problem stated in Chapter 2. The effectiveness of the developed integrated fault-tolerant control method is demonstrated in **Chapter 7** by simulations and experiments with a manufacturing cell.

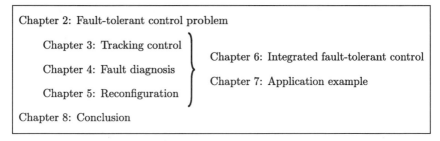

Figure 1.2: Structure of the thesis.

2 Formalization of the fault-tolerant control problem

In this chapter, first, general notations used throughout this thesis are given. Then deterministic I/O automata are introduced, different representations for them are presented and their most important properties are stated. Afterwards the models of the faultless and the faulty plant are introduced and their validity periods are discussed. The control aim is stated and the fault-tolerant control problem to be solved in this thesis is formalized. Finally, a running example is introduced.

2.1 Notation

Scalars are denoted by lower case italics (i.e., z), while sequences of scalars are denoted by upper case italics (e.g., $V(0 \dots k_e)$), where k_e is the time horizon and the sequence is defined by $V(0 \dots k_e) = (v(0), v(1), \dots, v(k_e))$. Vectors are denoted by lower case boldface letters (e.g., z_c), where z_{ci} refers to the i-th element.

Sets are denoted by calligraphic letters (e.g., $\mathcal{Z}$). The notation $2^{\mathcal{X}}$ means the power set of a given set $\mathcal{X}$, that is, the set of all subsets of $\mathcal{X}$. The cardinality of a set $\mathcal{X}$ is denoted by $|\mathcal{X}|$. The symbol $\mathcal{X}^{\infty}$ denotes the set of all sequences $X(0 \dots k_e) = (x(0), \dots, x(k_e))$, $k_e \geq 0$ whose elements $x(k)$ stem from the set $\mathcal{X}$.

The empty symbol is represented by ε. The symbol $\vDash$ means "is modeled by" and the symbol $\overset{!}{=}$ means "shall be equal to".

2.2 Deterministic I/O automata

2.2.1 Definition of deterministic I/O automata

This thesis considers discrete event systems modeled by deterministic I/O automata, which are defined as follows.

Definition 2.1 (Deterministic I/O automaton). *A deterministic I/O automaton $\mathcal{A}$ is defined by the tuple*

$$\mathcal{A} = (\mathcal{Z}, \mathcal{V}, \mathcal{W}, G, H, z_0), \tag{2.1}$$

where the symbols have the following meanings:

- $\mathcal{Z}$ – *state set,*

- $\mathcal{V}$ – *input set,*

- $\mathcal{W}$ – *output set,*

- $G : \mathcal{Z} \times \mathcal{V} \rightarrow \mathcal{Z}$ – *state transition function,*

- $H : \mathcal{Z} \times \mathcal{V} \rightarrow \mathcal{W}$ – *output function,*

- z_0 – *initial state.*

Depending on the element under consideration, all of these variables are labeled with an index, e.g., 'p' for the plant or 'c' for the controller. In the following, for simplicity reasons, deterministic I/O automata will sometimes be called "automata" for short.

The state transition function G maps a state $z(k) \in \mathcal{Z}$ and an input symbol $v(k) \in \mathcal{V}$ to the next state $z(k+1) \in \mathcal{Z}$ also denoted by $z'(k)$, that is,

$$G(z(k), v(k)) = z(k+1) = z'(k). \tag{2.2}$$

Similarly, the output function H maps a state $z(k) \in \mathcal{Z}$ and an input symbol $v(k) \in \mathcal{V}$ to an output symbol $w(k) \in \mathcal{W}$, i.e.,

$$H(z(k), v(k)) = w(k). \tag{2.3}$$

Assume that the empty symbol ε is always contained in the input set $\mathcal{V}$ and the output set $\mathcal{W}$.

Let n denote the cardinality of the state set, p denote the cardinality of the input set and q denote the cardinality of the output set, that is,

$$n = |\mathcal{Z}| \tag{2.4}$$

$$p = |\mathcal{V}| \tag{2.5}$$

$$q = |\mathcal{W}|. \tag{2.6}$$

2.2.2 Complementary descriptions for deterministic I/O automata

Equivalently to using the state transition function G and the output function H, the behavior of an automaton $\mathcal{A}$ can be described by the *behavioral relation*

$$\mathcal{L} = \left\{ (z', w, z, v) \in \mathcal{Z} \times \mathcal{W} \times \mathcal{Z} \times \mathcal{V} : \left[G(z, v) = z' \right] \wedge \left[H(z, v) = w \right] \right\}. \tag{2.7}$$

That is, the behavioral relation $\mathcal{L}$ contains all tuples $(z', w, z, v) \in \mathcal{Z} \times \mathcal{W} \times \mathcal{Z} \times \mathcal{V}$ for which there is a transition from state $z \in \mathcal{Z}$ to state $z' \in \mathcal{Z}$ with the input symbol $v \in \mathcal{V}$ and the output symbol $w \in \mathcal{W}$ in the automaton $\mathcal{A}$.

$\mathcal{V}^\infty$ denotes the set of input sequences $V(0 \ldots k_e) \in \mathcal{V}^\infty$ of any length, where $v(k) \in \mathcal{V}$, $\forall 0 \leq k \leq k_e$. Similarly, $\mathcal{W}^\infty$ is the set of output sequences $W(0 \ldots k_e) \in \mathcal{W}^\infty$, where $w(k) \in \mathcal{W}$, $\forall 0 \leq k \leq k_e$.

The *automaton map* $\Phi : \mathcal{Z} \times \mathcal{V}^\infty \to \mathcal{W}^\infty$ computes the output sequence

$$\Phi(z_0, V(0 \ldots k_e)) = W(0 \ldots k_e) \tag{2.8}$$

that is generated by the automaton $\mathcal{A}$ when it is in state $z_0 \in \mathcal{Z}$ and receives the input sequence $V(0 \ldots k_e) \in \mathcal{V}^\infty$. The state that is thereby reached can be computed by extending the state transition function G in a natural way, such that,

$$G^\infty : \mathcal{Z} \times \mathcal{V}^\infty \to \mathcal{Z} \tag{2.9}$$

$$G^\infty(z_0, V(0 \ldots k_e)) = z(k_e + 1). \tag{2.10}$$

Obviously, for $k_e = 0$, the state returned by the extended state transition function equals the state returned by the usual state transition function:

$$G^\infty(z_0, V(0 \ldots k_e)) = G(z_0, v(0)), \quad \text{if } k_e = 0.$$

The *active input set* $\mathcal{V}_a : \mathcal{Z} \times \mathcal{Z} \to 2^\mathcal{V}$ of an automaton $\mathcal{A}$ contains all input symbols $v \in \mathcal{V}$ leading from the state $z \in \mathcal{Z}$ to the state $z' \in \mathcal{Z}$ and is defined as follows:

$$\mathcal{V}_a(z', z) = \{v \in \mathcal{V} : G(z, v) = z'\}. \tag{2.11}$$

Consequently, the following equivalence between the active input set $\mathcal{V}_a$ and the state transition function G is true:

$$v \in \mathcal{V}_a(z', z) \Leftrightarrow G(z, v) = z'. \tag{2.12}$$

2.2.3 Automata graphs

Automata can be represented by automata graphs consisting of a set $\mathcal{X}$ of vertices and a set $\mathcal{E} \subseteq \mathcal{X} \times \mathcal{X}$ of edges. In the automaton graph, nodes $x \in \mathcal{X}$ correspond to states $z \in \mathcal{Z}$ and edges $e = (z, z') \in \mathcal{E}$ represent transitions from state $z \in \mathcal{Z}$ to state $z' \in \mathcal{Z}$ in an automaton $\mathcal{A}$. That is, the following relation between the edges in an automaton graph and the active input set in (2.11) holds:

$$e = (z, z') \in \mathcal{E} \Leftrightarrow \mathcal{V}_{\mathrm{a}}(z', z) \neq \emptyset. \tag{2.13}$$

In the graphical representation of an automaton graph, edges $e \in \mathcal{E}$ are labeled with their corresponding I/O pair v / w and the initial state z_0 is labeled by an additional arrow. Figure 2.1 shows the automaton graph corresponding to an automaton $\mathcal{A}$ with the following elements:

$$G(z = 1, v = v_1) = 2$$
$$H(z = 1, v = v_1) = w_1$$
$$z_0 = 1.$$

According to (2.7) the behavior of $\mathcal{A}$ can also be expressed by the behavioral relation

$$\mathcal{L} = \{(2, w_1, 1, v_1)\}.$$

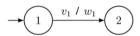

Figure 2.1: Automaton graph of automaton $\mathcal{A}$.

The representation of automata by automata graphs allows for the application of graph theoretical concepts like path search to automata. A path in an automaton graph is defined in the usual way as follows (cf. [66])

Definition 2.2 (Path in automaton graph). *A sequence $((z_1, z_2), (z_2, z_3), \ldots, (z_{l-1}, z_l))$ of edges is called path $P(z, \tilde{z})$ from the state $z \in \mathcal{Z}$ to the state $\tilde{z} \in \mathcal{Z}$ in the graph of an automaton $\mathcal{A}$ if the following conditions hold:*

$$z_1 = z \tag{2.14}$$
$$z_l = \tilde{z} \tag{2.15}$$
$$(z_i, z_{i+1}) \in \mathcal{E}, \qquad \forall i \in \{1, \ldots, l-1\} \tag{2.16}$$
$$z_i \neq z_j, \qquad \forall i, j \in \{1, \ldots, l\}, i \neq j. \tag{2.17}$$

A state $\tilde{z} \in \mathcal{Z}$ is called *reachable* from another state $z \in \mathcal{Z}$ if there exists a path $P(z, \tilde{z})$ from $z \in \mathcal{Z}$ to $\tilde{z} \in \mathcal{Z}$ in the graph of an automaton $\mathcal{A}$.

The automaton corresponding to a subgraph of the automaton graph of $\mathcal{A}$ is called subautomaton of $\mathcal{A}$ and is defined as follows.

Definition 2.3 (Subautomaton). *An automaton $\mathcal{A}_{\mathrm{sub}} = (\mathcal{Z}_{\mathrm{sub}}, \mathcal{V}_{\mathrm{sub}}, \mathcal{W}_{\mathrm{sub}}, G_{\mathrm{sub}}, H_{\mathrm{sub}}, z_{\mathrm{sub}0})$ is called subautomaton of an automaton $\mathcal{A}$ denoted by*

$$\mathcal{A}_{\mathrm{sub}} \subseteq \mathcal{A}, \tag{2.18}$$

if it has the same state set $\mathcal{Z}_{\mathrm{sub}} = \mathcal{Z}$, input set $\mathcal{V}_{\mathrm{sub}} = \mathcal{V}$, output set $\mathcal{W}_{\mathrm{sub}} = \mathcal{W}$ and initial state $z_{\mathrm{sub}0} = z_0$ as $\mathcal{A}$ and the following two implications hold:

$$G_{\mathrm{sub}}(z, v) = z' \Rightarrow G(z, v) = z' \tag{2.19}$$
$$H_{\mathrm{sub}}(z, v) = w \Rightarrow H(z, v) = w. \tag{2.20}$$

An important property of a subautomaton is that all paths in it are also paths in the original automaton as stated by the following proposition.

Proposition 2.1 (Path in subautomaton). *Consider an automaton $\mathcal{A}_{\mathrm{sub}}$ which is a subautomaton of another automaton $\mathcal{A}$, i.e., $\mathcal{A}_{\mathrm{sub}} \subseteq \mathcal{A}$. Then every path $P(z, \tilde{z})$ in the automaton $\mathcal{A}_{\mathrm{sub}}$ is also a path $P(z, \tilde{z})$ in the automaton $\mathcal{A}$.*

Proof. From Definition 2.2 and (2.11), (2.13) and (2.19), it is easy to see that for every path $P(z, \tilde{z}) = ((z_1, z_2), (z_2, z_3), \ldots, (z_{l-1}, z_l))$ in the automaton graph of $\mathcal{A}_{\mathrm{sub}}$ the following holds:

$$
\begin{aligned}
&(z_i, z_{i+1}) \in \mathcal{E}_{\mathrm{sub}}, && \forall\, i \in \{1, \ldots, l-1\} \\
&\Rightarrow \mathcal{V}_{\mathrm{a,sub}}(z_{i+1}, z_i) \neq \emptyset \\
&\Rightarrow (\exists v \in \mathcal{V})\, G_{\mathrm{sub}}(z_i, v) = z_{i+1} \\
&\Rightarrow (\exists v \in \mathcal{V})\, G(z_i, v) = z_{i+1} \\
&\Rightarrow \mathcal{V}_{\mathrm{a}}(z_{i+1}, z_i) \neq \emptyset, && \forall\, i \in \{1, \ldots, l-1\} \\
&\Rightarrow (z_i, z_{i+1}) \in \mathcal{E}, && \forall\, i \in \{1, \ldots, l-1\}.
\end{aligned}
$$

Hence, since the other requirements for a path according to Definition 2.2 are not impacted, the proof is completed. $\square$

2.2.4 Properties of deterministic I/O automata

An automaton $\mathcal{A}$ is called *completely defined* if its state transition function G and its output function H are defined for all state-input pairs $(z, v) \in \mathcal{Z} \times \mathcal{V}$.

Two states $z_1, z_2 \in \mathcal{Z}$ in an automaton $\mathcal{A}$ are called *equivalent* and are denoted by $z_1 \sim z_2$ if the following equation holds:

$$\Phi(z_1, V(0 \ldots k_e)) = \Phi(z_2, V(0 \ldots k_e)), \quad \forall V(0 \ldots k_e) \in \mathcal{V}^\infty, \tag{2.21}$$

that is, they will always produce the same output sequence. Otherwise z_1 and z_2 are called *distinguishable*. Since equivalence is a transitive relation, that is,

$$(z_1 \sim z_2 \wedge z_2 \sim z_3) \Rightarrow z_1 \sim z_3,$$

it is also possible to define sets of equivalent states.

An input sequence $V_S(0 \ldots k_e)$ is called *separating sequence* for a distinguishable state pair (z_1, z_2) if

$$\Phi(z_1, V_S(0 \ldots k_e)) \neq \Phi(z_2, V_S(0 \ldots k_e)). \tag{2.22}$$

An I/O-pair $(V(0 \ldots k_e), W(0 \ldots k_e))$ is called *consistent* with an automaton $\mathcal{A}$ if

$$(\exists z_0 \in \mathcal{Z}) \quad \Phi(z_0, V(0 \ldots k_e)) = W(0 \ldots k_e). \tag{2.23}$$

Similarly, an I/O-pair $(V(0 \ldots k_e), W(0 \ldots k_e))$ is called consistent with a state $z \in \mathcal{Z}$ if (2.23) holds and

$$(\exists z_0 \in \mathcal{Z}) \quad G^\infty(z_0, V(0 \ldots k_e)) = z. \tag{2.24}$$

For automata whose state transition function G and output function H are not completely defined some of the above definitions have to be modified to become meaningful. An input sequence $V(0 \ldots k_e)$ is called *acceptable* to a state $z_0 \in \mathcal{Z}$ if $G^\infty(z_0, V(0 \ldots k_e))$ is defined, i.e., if

$$G(z(k), v(k))!, \quad \forall 0 \leq k \leq k_e,$$

where $z(k) = G^\infty(z_0, V(0 \ldots k-1))$.

The notion of equivalent states has to be adapted as follows. Two states z_1 and z_2 are called *compatible* if for all input sequences $V(0 \ldots k_e) \in \mathcal{V}^\infty$ that are acceptable to both z_1 and z_2

$$\Phi(z_1, V(0 \ldots k_e)) = \Phi(z_2, V(0 \ldots k_e)) \tag{2.25}$$

holds. If there is at least one input sequence that is acceptable to both z_1 and z_2 which yields different output sequences, z_1 and z_2 are called *incompatible*. Note that, in contrast to equivalence, compatibility is not a transitive relation, because the input sequences that are acceptable to both z_1 and z_2 are not necessarily acceptable to both z_1 and z_3. Therefore, it is only possible to consider the compatibility of state pairs, but not the compatibility of sets of states, unless all considered states accept the same input sequences.

A *separating sequence* $V_s(0 \ldots k_e) \in \mathcal{V}^\infty$ for an incompatible state pair (z_1, z_2) has to be acceptable to both z_1 and z_2 and has to fulfill

$$\Phi(z_1, V_s(0 \ldots k_e)) \neq \Phi(z_2, V_s(0 \ldots k_e)).$$

2.3 Plant models

The fault-tolerant control method presented in this thesis is model-based. Therefore, the model of the faultless plant and a model of the plant under every possible fault have to be known in advance. The construction and / or identification of such models goes beyond the scope of this thesis. Instead, the model of the faultless plant and the effects of all possible faults on the plant are assumed to be given.

2.3.1 General assumptions

All faults f under consideration belong to a known *fault set* $\mathcal{F} = \{1, \ldots, F\}$, where F denotes the total number of faults. Three different kinds of faults are considered, namely actuator faults, sensor faults and plant faults. Actuator faults can either lead to a corruption of an input symbol or they can manifest themselves in form of a failure, such that no input at all is generated. The same differentiation holds true for sensor faults, while plant faults change the internal dynamics of the plant.

The class of faults under consideration is restricted by the following assumption:

Assumption 1 (Class of considered faults). *All considered faults $f \in \mathcal{F}$ are persistent and only a single fault occurs.*

Furthermore, it is assumed that faults affect the plant $\mathcal{P}$, but not the fault-tolerant controller.

2.3.2 Relation between the faultless and the faulty plant

The faultless plant is modeled by the completely defined deterministic I/O automaton

$$\mathcal{A}_0 = (\mathcal{Z}_0, \mathcal{V}_0, \mathcal{W}_0, G_0, H_0, z_{00}). \tag{2.26}$$

For each fault $f \in \mathcal{F}$ one deterministic I/O automaton

$$\mathcal{A}_f = (\mathcal{Z}_f, \mathcal{V}_f, \mathcal{W}_f, G_f, H_f, z_{f0}), \quad (f \in \mathcal{F}) \tag{2.27}$$

describing the behavior of the plant $\mathcal{P}$ in the presence of fault f is defined as well. These automata $\mathcal{A}_f$, $(f \in \mathcal{F})$ can be obtained by applying so-called *error relations* E_{zf}, E_{vf} and E_{wf} to the model $\mathcal{A}_0$ of the faultless plant in (2.26) according to

$$\mathcal{A}_f = E(\mathcal{A}_0, E_{zf}, E_{vf}, E_{wf}), \quad (f \in \mathcal{F}). \tag{2.28}$$

The error relations describe the physical impact that a given fault has on the plant and are described in more detail in Section 2.3.3. The mapping E in (2.28) is detailed in Section 2.3.4.

2.3.3 Error relations

Error relations map the nominal inputs, outputs or states to the faulty ones (cf. [74]). For example, if an actuator whose impact on the plant is represented by the input symbol $v \in \mathcal{V}_0$ fails, the corresponding error relation replaces the input symbol $v \in \mathcal{V}_0$ by the empty symbol ε.

Therefore, for actuator faults, the error relation $E_{vf} : \mathcal{V}_0 \to \mathcal{V}_f$ is defined as follows:

$$E_{vf}(v) = \begin{cases} v & \text{faultless input,} \\ v_\# \neq v, \ v_\# \in \mathcal{V}_f & \text{faulty input,} \\ \varepsilon & \text{input failure.} \end{cases} \tag{2.29}$$

Similarly, sensor faults are described by the error relation $E_{wf} : \mathcal{W}_0 \to \mathcal{W}_f$ defined as follows:

$$E_{wf}(w) = \begin{cases} w & \text{faultless output,} \\ w_\# \neq w, \ w_\# \in \mathcal{W}_f & \text{faulty output,} \\ \varepsilon & \text{output failure.} \end{cases} \tag{2.30}$$

Note that it is assumed that even in case of an output failure, i.e., $E_{wf}(w) = \varepsilon$, it can be noticed that some transition occurred.

Plant faults can be described by the error relation $E_{zf} : \mathcal{Z}_0 \times \mathcal{V}_0 \rightarrow \mathcal{Z}_f$ with

$$E_{zf}(z,v) = \begin{cases} G_0(z,v) & \text{faultless transition,} \\ z_\# \neq G_0(z,v), \ z_\# \in \mathcal{Z}_f & \text{faulty transition.} \end{cases} \quad (2.31)$$

2.3.4 Models of the faulty plant

In this section the function E, that is, the way that the elements $(\mathcal{Z}_0, \mathcal{V}_0, \mathcal{W}_0, G_0, H_0, z_{00})$ of the automaton $\mathcal{A}_0$ and the error relations E_{zf}, E_{vf} and E_{wf} in (2.29)–(2.31) are mapped to the elements $(\mathcal{Z}_f, \mathcal{V}_f, \mathcal{W}_f, G_f, H_f, z_{f0})$ of the automata $\mathcal{A}_f$, $(f \in \mathcal{F})$, is described (cf. (2.28)).

The state set $\mathcal{Z}_0$, input set $\mathcal{V}_0$ and output set $\mathcal{W}_0$ of the model $\mathcal{A}_0$ have to be chosen such that they already contain all possible symbols and states, even those which can only be reached or generated in case of a fault. Hence, the state set, input set and output set of the faultless plant are kept as follows:

$$\mathcal{Z}_f = \mathcal{Z}_0, \quad \forall f \in \mathcal{F} \quad (2.32)$$

$$\mathcal{V}_f = \mathcal{V}_0, \quad \forall f \in \mathcal{F} \quad (2.33)$$

$$\mathcal{W}_f = \mathcal{W}_0, \quad \forall f \in \mathcal{F}. \quad (2.34)$$

The state transition function G_f for the models $\mathcal{A}_f$, $(f \in \mathcal{F})$ is given by

$$G_f(z,v) = E_{zf}(z, E_{vf}(v)). \quad (2.35)$$

Depending on the nature of the plant, the output function H_f of the faulty plant can be defined in different ways. In the most general case, the output w of the plant depends on its current state z, its input v and its next state z'. In order to account for the possibility of actuator and plant faults, the output function of the model $\mathcal{A}_f$, $(f \in \mathcal{F})$ of the faulty plant then results from the output function of the model $\mathcal{A}_0$ of the faultless plant by

$$H_f(z,v) = E_{wf}(H_0(\tilde{z},\tilde{v})), \text{ if } [z = \tilde{z}] \wedge [E_{vf}(v) = \tilde{v}] \wedge [G_f(z,v) = G_0(\tilde{z},\tilde{v})]. \quad (2.36)$$

Substituting the terms from the right hand side of the equation, the following output function results:

$$H_f(z,v) = E_{wf}(H_0(z, E_{vf}(v))), \text{ if } G_f(z,v) = G_0(z, E_{vf}(v)). \quad (2.37)$$

Note that this definition is very restricting in the sense that the output function $H_f(z,v)$ of the faulty plant will remain undefined for many pairs $(z,v) \in \mathcal{Z}_f \times \mathcal{V}_f$. For all of these pairs, the output function $H_f(z,v)$ has to be specified manually.

However, if the output of the plant does not depend on its current state z, its input v *and* its next state z', the respective conditions can be removed from (2.36). Consequently the number of pairs $(z,v) \in \mathcal{Z}_f \times \mathcal{V}_f$ for which $H_f(z,v)$ needs to be defined manually, is reduced. For example, if the output depends on the input and the next state of the plant but not on its current state, i.e.,

$$G_0(z,v) = G_0(\tilde{z},v) \Rightarrow H_0(z,v) = H_0(\tilde{z},v), \quad \forall z,\tilde{z} \in \mathcal{Z}_0, v \in \mathcal{V}_0,$$

the output function of the faulty plant derived from (2.36) without the condition $[z = \tilde{z}]$ is given by

$$H_f(z,v) = E_{wf}(H_0(\tilde{z}, E_{vf}(v))) \text{ if } (\exists \tilde{z} \in \mathcal{Z}_0) \, G_f(z,v) = G_0(\tilde{z}, E_{vf}(v)). \tag{2.38}$$

If the output depends on the current state and the next state of the plant but not on its input, i.e.,

$$G_0(z,v) = G_0(z,\tilde{v}) \Rightarrow H_0(z,v) = H_0(z,\tilde{v}), \quad \forall z \in \mathcal{Z}_0, v,\tilde{v} \in \mathcal{V}_0,$$

the output function of the faulty plant derived from (2.36) without the condition $[E_{vf}(v) = \tilde{v}]$ is given by

$$H_f(z,v) = E_{wf}(H_0(z,\tilde{v})), \text{ if } (\exists \tilde{v} \in \mathcal{V}_0) \, G_f(z,v) = G_0(z,\tilde{v}). \tag{2.39}$$

Similar equations can be found for the other possible dependencies of the output of the plant.

Since the model $\mathcal{A}_0$ of the faultless plant is assumed to be completely defined, the models $\mathcal{A}_f$, $(f \in \mathcal{F})$ of the faulty plant are also completely defined, which is summarized in the following assumption.

Assumption 2 (Completeness of plant models). *The model $\mathcal{A}_0$ of the faultless plant and all models $\mathcal{A}_f$, $(f \in \mathcal{F})$ of the faulty plant are completely defined.*

For a simplified notation, the *overall model $\mathcal{A}_\Delta$ of the faulty plant* is introduced, which contains all models $\mathcal{A}_f$, $(f \in \mathcal{F})$ as isolated parts. The state of $\mathcal{A}_\Delta$ is augmented by the information about its corresponding fault as follows:

$$\boldsymbol{z}_\Delta := \begin{pmatrix} z & f \end{pmatrix}^\top. \tag{2.40}$$

Hence, the automaton $\mathcal{A}_\Delta$ contains

$$n_\Delta = n \cdot F \tag{2.41}$$

states. Recall that n is the number of states in the model $\mathcal{A}_0$ of the faultless plant and F is the number of faults. Consequently, the overall model $\mathcal{A}_\Delta$ of the faulty plant is a deterministic I/O automaton defined by

$$
\mathcal{A}_\Delta : \begin{cases}
\mathcal{Z}_\Delta = \bigcup_{f \in \mathcal{F}} \mathcal{Z}_f \times \{f\} & \text{(2.42a)} \\[2mm]
\mathcal{V}_\Delta = \mathcal{V}_f & \text{(2.42b)} \\[2mm]
\mathcal{W}_\Delta = \mathcal{W}_f & \text{(2.42c)} \\[2mm]
G_\Delta(\boldsymbol{z}_\Delta, v) = G_\Delta\left(\begin{pmatrix} z \\ f \end{pmatrix}, v\right) = \begin{pmatrix} G_f(z, v) \\ f \end{pmatrix} & \text{(2.42d)} \\[3mm]
H_\Delta(\boldsymbol{z}_\Delta, v) = H_\Delta\left(\begin{pmatrix} z \\ f \end{pmatrix}, v\right) = H_f(z, v). & \text{(2.42e)}
\end{cases}
$$

From Assumption 2 it easily follows that the overall model $\mathcal{A}_\Delta$ of the faulty plant is also completely defined.

The use of error relations accounts for the fact that a fault $f \in \mathcal{F}$ usually only affects a part of the plant, but does not change its entire behavior. Therefore, it is reasonable to describe the behavior of the faulty plant based on the model $\mathcal{A}_0$ of the faultless plant. Hence, on the one hand, the use of error relations is a simplification, for the modeling as well as for the representation of the behavior of the plant at runtime. On the other hand, the usage of error relations restricts the behavior that can be represented. For example, it is not possible to model a fault that changes the nominal output w of a sensor in some states to a symbol $w_1 \neq w$ and in other states to another symbol $w_2 \neq w_1$.

The following fundamental assumption summarizes that all plant models are assumed to describe behavior of the plant without any deviations, including the initial state z_{00} of the faultless plant, and all possible faults are contained in the fault set $\mathcal{F}$.

Assumption 3 (Correctness of the plant models). *The model $\mathcal{A}_0 = (\mathcal{Z}_0, \mathcal{V}_0, \mathcal{W}_0, G_0, H_0, z_{00})$ describes the behavior of the faultless plant correctly, while the behavior of the faulty plant under all possible faults is correctly described by the models $\mathcal{A}_f = (\mathcal{Z}_f, \mathcal{V}_f, \mathcal{W}_f, G_f, H_f)$, $(f \in \mathcal{F})$.*

2.3.5 Plant model before and after the occurrence of a fault

Depending on the presence of a fault, the plant is either modeled by the model $\mathcal{A}_0$ (i.e., $\mathcal{P} \vDash \mathcal{A}_0$) or, if the fault $\bar{f} \in \mathcal{F}$ is present, by the model $\mathcal{A}_{\bar{f}}$ (i.e., $\mathcal{P} \vDash \mathcal{A}_{\bar{f}}$). In any case, the state of the plant at time k is denoted by $z_p(k)$, while its input and output symbol are given by $v_p(k)$ and $w_p(k)$, respectively.

The time at which a fault $\bar{f} \in \mathcal{F}$ manifests itself at the plant for the first time is called *fault occurrence time* k_f. That is, at the fault occurrence time k_f, the model of the plant changes from $\mathcal{A}_0$ to $\mathcal{A}_{\bar{f}}$. This fact is illustrated by the time bar in Fig. 2.2. The current state $z_p(k_f)$ of the plant $\mathcal{P}$ is assumed to be unaffected by the fault, such that it remains the same as in the faultless case until a transition is performed. After the fault occurrence time k_f the plant can also be modeled by the overall model $\mathcal{A}_\Delta$ of the faulty plant in (2.42), whose current state is given by

$$z_\Delta(k) = \begin{pmatrix} z_p(k) \\ \bar{f} \end{pmatrix}, \quad \forall k \geq k_f. \tag{2.43}$$

Since all faults are assumed to be persistent (Assumption 1), the model $\mathcal{A}_{\bar{f}}$, $(\bar{f} \in \mathcal{F})$ or, equivalently, $\mathcal{A}_\Delta$ is valid for all times $k \geq k_f$.

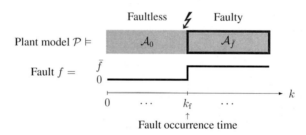

Figure 2.2: Plant model before and after occurrence of a fault.

2.4 Problem statement

2.4.1 Control aim: Reaching a desired final state

In this thesis the specification to be fulfilled by the controlled plant is given in form of a desired final state $z_F \in \mathcal{Z}_0$ (recall that $\mathcal{Z}_f = \mathcal{Z}_0, \forall f \in \mathcal{F}$). The desired final state z_F corresponds, for example, to the completion of a task by the plant. In order to achieve fault-tolerance, the desired final state z_F should not only be reached in the faultless case (i.e., if $\mathcal{P} \vDash \mathcal{A}_0$), but also when a fault $f \in \mathcal{F}$ occurs. Therefore, the control aim is given by:

The controlled plant $\mathcal{P}$ shall reach the desired final state $z_F \in \mathcal{Z}_0$ at some finite time $k_F \geq 0$, that is,

$$(\exists k_F \geq 0) \quad z_p(k_F) = z_F, \tag{2.44}$$

even when a fault $f \in \mathcal{F}$ occurs.

Additionally, it is possible to consider safety constraints in form of a set

$$\mathcal{E}_{ill} \subset \mathcal{Z}_0 \times \mathcal{Z}_0 \tag{2.45}$$

of illegal transitions for the plant $\mathcal{P}$. Illegal transitions $(z, z') = (z, G_0(z, v)) \in \mathcal{E}_{ill}$ are present if the input $v \in \mathcal{V}_0$ must not be applied to the plant $\mathcal{P}$ if the plant is in state $z \in \mathcal{Z}_0$. In this case, the control aim (2.44) is modified as follows:

The controlled plant $\mathcal{P}$ shall reach the desired final state $z_F \in \mathcal{Z}_0$ at some finite time $k_F \geq 0$, that is,

$$(\exists k_F \geq 0) \quad z_p(k_F) = z_F, \tag{2.46}$$

while not executing any illegal transitions

$$(z_p(k), z_p(k+1)) \in \mathcal{E}_{ill}, \ (0 \leq k < k_F) \tag{2.47}$$

even when a fault $f \in \mathcal{F}$ occurs.

For the avoidance of illegal transitions by the plant $\mathcal{P}$ it is useful to define the legal part $\mathcal{A}_{leg}$ of a plant model $\mathcal{A}$, which results from removing all illegal transitions $(z, z') \in \mathcal{E}_{ill}$ from $\mathcal{A}$.

Definition 2.4 (Legal part of plant model). *Given a plant $\mathcal{P} \models \mathcal{A} = (\mathcal{Z}, \mathcal{V}, \mathcal{W}, G, H, z_0)$ and a set $\mathcal{E}_{ill} \subset \mathcal{Z} \times \mathcal{Z}$ of illegal transitions, the subautomaton $\mathcal{A}_{leg} = (\mathcal{Z}, \mathcal{V}, \mathcal{W}, G_{leg}, H_{leg}, z_0) \subset \mathcal{A}$ of $\mathcal{A}$ with*

$$G_{leg}(z, v) = \begin{cases} G(z, v) & \text{if } (z, G(z, v)) \notin \mathcal{E}_{ill}, \\ \text{undefined} & \text{otherwise,} \end{cases} \tag{2.48}$$

$$H_{leg}(z, v) = \begin{cases} H(z, v) & \text{if } (z, G(z, v)) \notin \mathcal{E}_{ill}, \\ \text{undefined} & \text{otherwise,} \end{cases} \tag{2.49}$$

is called legal part of the plant model $\mathcal{A}$.

2.4.2 Fault-tolerant control problem

The control aim (2.44) shall be achieved within the fault-tolerant control loop in Fig. 2.3. Therefore, the problem to be solved in this thesis can be summarized as follows.

Problem 2.1 (Fault-tolerant control problem). *Given the model $\mathcal{A}_0$ of the faultless plant and the set $\{E_{zf}, E_{vf}, E_{wf}, (f \in \mathcal{F})\}$ of error relations, find a fault-tolerant controller as shown in Fig. 2.3 such that the controlled plant, which might be subject to a fault $f \in \mathcal{F}$, reaches the desired final state $z_F \in \mathcal{Z}_0$ at some finite time $k_F \geq 0$.*

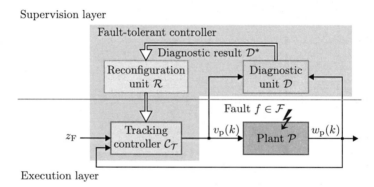

Figure 2.3: Fault-tolerant control loop.

Including a set $\mathcal{E}_{ill}$ of illegal transitions to be avoided by the plant $\mathcal{P}$ at all times, the control aim is given by (2.46) and (2.47) and the safe fault-tolerant control problem is defined as follows.

Problem 2.2 (Safe fault-tolerant control problem). *Given the model $\mathcal{A}_0$ of the faultless plant and the set $\{E_{zf}, E_{vf}, E_{wf}, (f \in \mathcal{F})\}$ of error relations, find a fault-tolerant controller as shown in Fig. 2.3 such that the controlled plant, which might be subject to a fault $f \in \mathcal{F}$, reaches the desired final state $z_F \in \mathcal{Z}_0$ at some finite time $k_F \geq 0$ while not executing any illegal transitions $(z_p(k), z_p(k+1)) \in \mathcal{E}_{ill}, (0 \leq k < k_F)$.*

The following chapters solve the fault-tolerant control problems stated in Problem 2.1 and Problem 2.2 by first considering each component of the FTC loop individually (Chapters 3–5) and then combining them to obtain an integrated FTC method (Chapter 6).

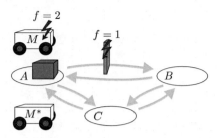

Figure 2.4: Transport of parcels in automated warehouse.

2.5 Running example: Automated warehouse

In this section a small running example is introduced, which will be used to demonstrate the main ideas of the thesis in the following chapters. Consider some area in an automated warehouse, where parcels can be transported between positions A, B and C by using the mobile robots M and M^* (cf. Fig. 2.4).

Model of the faultless plant. The model $\mathcal{A}_0$ of the faultless plant is shown in Fig. 2.5. Its states

$$z \in \mathcal{Z}_0 = \{A, B, C\} \tag{2.50}$$

reflect the position of a parcel, where the initial state is

$$z_{00} = A. \tag{2.51}$$

The inputs

$$v \in \mathcal{V}_0 = \{AB, AC, BA, BC, CA, CB, AB^*, AC^*, BA^*, BC^*, CA^*, CB^*\} \tag{2.52}$$

request a transport from one position to another position by a specific robot. For example, the input $v = AB^*$ requests a transport from position A to position B using robot M^*. The outputs

$$w \in \mathcal{W}_0 = \{\text{ok}, \text{np}, \text{BkA}, \text{BkB}, \text{BkC}\} \tag{2.53}$$

provide a feedback, whether the transport was successful ($w = \text{ok}$) or the transport was not possible ($w = \text{np}$). The outputs $w = \text{BkA}$, $w = \text{BkB}$ and $w = \text{BkC}$ are only generated in case of a fault and represent a blocking at position A, B or C, respectively.

The I/O-pairs at the self-loops in Fig. 2.5 are omitted for readability. The self-loops occur for all input symbols $v \in \mathcal{V}_0$ for which no labeled transition starts at the respective state in Fig. 2.5. They always lead to the output $w = \text{np}$. Hence, in conformance with Assumption 2, the model $\mathcal{A}_0$ of the faultless plant is completely defined.

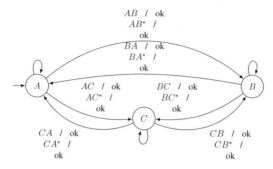

Figure 2.5: Model $\mathcal{A}_0$ for faultless transport of parcels in the automated warehouse.

Legal part of the plant model.   Assume that sometimes the direct route between position A and position B becomes hazardous, because some construction work is done there. In this case, the transitions from state $z = A$ to state $z' = B$ and vice versa become illegal transitions, that is,

$$\mathcal{E}_{\text{ill}} = \{(A, B), (B, A)\}. \tag{2.54}$$

The legal part of the plant model $\mathcal{A}_0$ from which, according to Definition 2.4, illegal transitions have been removed is shown in Fig. 2.6.

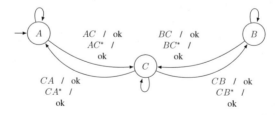

Figure 2.6: Legal part of model $\mathcal{A}_0$ for faultless transport of parcels with respect to $\mathcal{E}_{\text{ill}}$ in (2.54).

Models of the faulty plant.   Two faults are considered in the fault set $\mathcal{F} = \{1, 2\}$ illustrated by the lightnings in Fig. 2.4:

- **Fault** $f = 1$: The direct route between position A and B is blocked such that the transporting robot has to stop immediately after picking up the parcel.

- **Fault** $f = 2$: The robot M breaks down and can not conduct any transport.

Of course there are multiple other faults that could be considered, like the breakdown of robot M^* or the failure of one of the sensors.

Fault $f = 1$ is a plant fault. Therefore, according to Section 2.3.3, the error relations modeling the impact of this fault on the plant are given as follows:

$$E_{v1}(v) = v, \qquad \forall v \in \mathcal{V}_0, \tag{2.55a}$$

$$E_{w1}(w) = w, \qquad \forall w \in \mathcal{W}_0, \tag{2.55b}$$

$$E_{z1}(z,v) = \begin{cases} A & \text{if } (z,v) \in \{(A, AB), (A, AB^*)\}, \\ G_0(z,v) & \text{otherwise.} \end{cases} \tag{2.55c}$$

In contrast, fault $f = 2$ is modeled as an actuator fault as follows:

$$E_{v2}(v) = \begin{cases} \varepsilon & \text{if } v \in \{AB, AC, BA, BC, CA, CB\}, \\ v & \text{otherwise,} \end{cases} \tag{2.56a}$$

$$E_{w2}(w) = w, \qquad \forall w \in \mathcal{W}_0, \tag{2.56b}$$

$$E_{z2}(z,v) = G_0(z,v), \qquad \forall (z,v) \in \mathcal{Z}_0 \times \mathcal{V}_0. \tag{2.56c}$$

Based on these error relations, the state transition functions G_1 and G_2 are computed using (2.35). In this example, the output is fixed for a given input and next state, therefore, the definition for the output function H_f of the faulty plant in (2.38) applies. To deal with the newly occurring transitions, the output function H_f resulting from (2.38) is completed as follows:

$$H_f(z,v) = \begin{cases} H_f(z,v) & \text{if } H_f(z,v)! \text{ according to (2.38),} \\ \text{BkA} & \text{if } \left[G_f(z,v) = A\right] \wedge \left[\neg H_f(z,v)! \text{ according to (2.38)}\right], \\ \text{BkB} & \text{if } \left[G_f(z,v) = B\right] \wedge \left[\neg H_f(z,v)! \text{ according to (2.38)}\right], \\ \text{BkC} & \text{if } \left[G_f(z,v) = C\right] \wedge \left[\neg H_f(z,v)! \text{ according to (2.38)}\right]. \end{cases}$$

Figure 2.7 shows the overall model $\mathcal{A}_\Delta$ of the faulty plant, where the upper part of the automaton graph corresponds to the model $\mathcal{A}_1$ of the faulty plant for the fault $f = 1$ and the lower part to the model $\mathcal{A}_2$ for the fault $f = 2$. According to (2.42), the state set $\mathcal{Z}_\Delta$ of the overall model $\mathcal{A}_\Delta$ of the faulty plant is given by

$$\mathcal{Z}_\Delta = \bigcup_{f\in\{1,2\}} \{A,B,C\} \times \{f\}$$

$$= \left\{ \begin{pmatrix} A \\ 1 \end{pmatrix}, \begin{pmatrix} B \\ 1 \end{pmatrix}, \begin{pmatrix} C \\ 1 \end{pmatrix}, \begin{pmatrix} A \\ 2 \end{pmatrix}, \begin{pmatrix} B \\ 2 \end{pmatrix}, \begin{pmatrix} C \\ 2 \end{pmatrix} \right\}. \tag{2.57}$$

Again, missing inputs lead to self-loops with output $w = \text{np}$. Recall that the initial state of the models of the faulty plant is unknown.

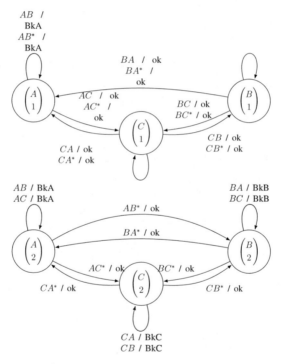

Figure 2.7: Overall model $\mathcal{A}_\Delta$ of faulty automated warehouse, where upper part corresponds to fault $f = 1$ and lower part corresponds to fault $f = 2$.

3 Tracking control of deterministic I/O automata

In this chapter a novel tracking controller for deterministic I/O automata is proposed which controls the plant in the faultless case. First, a controllability definition is given and a controllability condition that is based on the existence of a path in the automaton graph of the plant is derived. Then the structure of the tracking controller and the definitions of its two elements, the trajectory planning unit and the controller, are presented. As the main result of the chapter it is proved that in the closed-loop system consisting of the proposed tracking controller and the plant, the control aim is fulfilled if the plant is controllable and the tracking controller is defined based on the correct model of the plant.

3.1 Controllability analysis

The controllability of a deterministic I/O automaton $\mathcal{A}$ corresponding to the control aim (2.44), that is, the controllability with respect to a desired final state z_F, is defined as follows. It is based on the more general controllability definition for deterministic I/O automata in [65].

Definition 3.1 (Controllability). *A deterministic I/O automaton $\mathcal{A} = (\mathcal{Z}, \mathcal{V}, \mathcal{W}, G, H, z_0)$ is said to be controllable with respect to the final state $z_\mathrm{F} \in \mathcal{Z}$ if it can be steered from the initial state $z(0) = z_0$ into the final state $z(k_\mathrm{e} + 1) = z_\mathrm{F}$ within a finite time k_e by applying an appropriate input sequence $V(0 \ldots k_\mathrm{e})$.*

In order to develop a controllability condition, it is useful to introduce the notion of an admissible state sequence first. An admissible state sequence is a state sequence $Z(0 \ldots k_\mathrm{e}+1) \in \mathcal{Z}^\infty$ corresponding to a path in the automaton graph of an automaton $\mathcal{A}$, as given by the following definition.

Definition 3.2 (Admissible state sequence). *A state sequence $Z(0 \ldots k_e + 1)$ is called admissible state sequence for an automaton $\mathcal{A} = (\mathcal{Z}, \mathcal{V}, \mathcal{W}, G, H, z_0)$ if there exists a path $P(z, \tilde{z}) = ((z_1, z_2), (z_2, z_3), \ldots, (z_{l-1}, z_l))$ with $l = k_e + 2$ in the automaton graph of $\mathcal{A}$ such that*

$$z_{k+1} = z(k), \quad \forall\, 0 \leq k \leq k_e + 1 \tag{3.1}$$

holds.

Consequently, for an admissible state sequence $Z(0 \ldots k_e + 1)$ the following relation holds:

$$\mathcal{V}_a(z(k+1), z(k)) \neq \emptyset, \quad \forall\, 0 \leq k \leq k_e. \tag{3.2}$$

Based on this definition the following controllability condition can be stated.

Theorem 3.1 (Controllability). *An automaton $\mathcal{A} = (\mathcal{Z}, \mathcal{V}, \mathcal{W}, G, H, z_0)$ is controllable with respect to a final state z_F if and only if there exists an admissible state sequence $Z(0 \ldots k_e + 1) \in \mathcal{Z}^\infty$ with $z(0) = z_0$ and $z(k_e + 1) = z_F$ for it.*

Proof. If an automaton $\mathcal{A}$ is controllable with respect to a final state z_F, there exists an input sequence

$$V(0 \ldots k_e) = (v(0), \ldots, v(k_e)) \tag{3.3}$$

such that if $z(0) = z_0$ is given, then $z(k_e + 1) = z_F$ holds (cf. Definition 3.1). That is, $G^\infty(z_0, V(0 \ldots k_e)) = z_F$. With (2.10), it follows that there exists an admissible state sequence $Z(0 \ldots k_e + 1)$ with $z(0) = z_0$ and $z(k_e + 1) = z_F$, where the input symbols $v(k)$ required in (3.2) are given in (3.3).

Reciprocally, if there exists an admissible state sequence

$$Z(0 \ldots k_e + 1) \in \mathcal{Z}^\infty$$

with $z(0) = z_0$ and $z(k_e + 1) = z_F$, there exists an input symbol $v(k) \in \mathcal{V}$ for all $0 \leq k \leq k_e$ such that

$$G(z(k), v(k)) = z(k+1)$$

(cf. (3.2)). That is, the input sequence $V(0 \ldots k_e) = (v(0), \ldots v(k_e))$ steers the automaton $\mathcal{A}$ from state $z(0) = z_0$ into state $z(k_e + 1) = z_F$, which concludes the proof. $\qquad \square$

Avoiding illegal transitions. If a set $\mathcal{E}_{\text{ill}}$ of illegal transitions as defined in (2.45) needs to be respected, the controllability definition is modified as follows.

Definition 3.3 (Safe controllability). *An automaton $\mathcal{A} = (\mathcal{Z}, \mathcal{V}, \mathcal{W}, G, H, z_0)$ is said to be safely controllable with respect to the final state $z_F \in \mathcal{Z}$ and a set $\mathcal{E}_{\text{ill}} \subset \mathcal{Z} \times \mathcal{Z}$ of illegal transitions if it can be steered from the initial state $z(0) = z_0$ into the final state $z(k_e + 1) = z_F$ within a finite time k_e by applying an appropriate input sequence $V(0 \ldots k_e)$ while not executing any illegal transitions $(z(k), z(k + 1)) \in \mathcal{E}_{\text{ill}}, \ (0 \leq k < k_F)$*

It is desirable to derive a condition for the safe controllability similar to the one in Theorem 3.1 corresponding to the notion of safe controllability in Definition 3.3. Based on the definition of the legal part of the plant model in Definition 2.4 and the result of Proposition 2.1 on the existence of a path in a subautomaton, the following condition for safe controllability can be stated.

Theorem 3.2 (Safe controllability). *An automaton $\mathcal{A}$ is safely controllable with respect to a final state $z_F \in \mathcal{Z}$ and a set $\mathcal{E}_{\text{ill}} \subset \mathcal{Z} \times \mathcal{Z}$ of illegal transitions if and only if the legal part $\mathcal{A}_{\text{leg}}$ of the plant model $\mathcal{A}$ is controllable with respect to the final state z_F.*

Proof. If the automaton $\mathcal{A}$ is safely controllable with respect to the final state $z_F \in \mathcal{Z}$ and a set $\mathcal{E}_{\text{ill}} \subset \mathcal{Z} \times \mathcal{Z}$ of illegal transitions, there exists an input sequence

$$V(0 \ldots k_e) = (v(0), \ldots, v(k_e)) \tag{3.4}$$

such that if $z(0) = z_0$ is given, then $z(k_e + 1) = z_F$ holds and the automaton $\mathcal{A}$ passes no illegal transitions $(z(k), z(k+1)) \in \mathcal{E}_{\text{ill}}, \ (0 \leq k \leq k_e)$ (cf. Definition 3.3). That is, there exists an admissible state sequence $Z(0 \ldots k_e + 1)$ with $z(0) = z_0$ and $z(k_e + 1) = z_F$ according to Definition 3.2, where the input symbols $v(k)$ in (3.4) guarantee the fulfillment of (3.2) and

$$\Big(z(k), G(z(k), v(k))\Big) \notin \mathcal{E}_{\text{ill}}, \quad \forall \, 0 \leq k \leq k_e. \tag{3.5}$$

Consequently, the state sequence $Z(0 \ldots k_e + 1)$ is also an admissible state sequence for the legal part $\mathcal{A}_{\text{leg}}$ of the plant model in Definition 2.4 (compare (2.48)). Hence, the legal part $\mathcal{A}_{\text{leg}}$ of the plant model is controllable with respect to the state z_F (Definition 3.1).

Reciprocally, if the legal part $\mathcal{P}_{\text{leg}}$ of the plant is controllable with respect to state z_F, there exists an admissible state sequence $Z_{\text{leg}}(0 \ldots k_e + 1)$ with $z_{\text{leg}}(0) = z_0$ and $z_{\text{leg}}(k_e + 1) = z_F$ for it. This state sequence $Z_{\text{leg}}(0 \ldots k_e + 1)$ is also an admissible state sequence for the automaton $\mathcal{A}$, because each admissible state sequences corresponds to a path in an automaton graph (Definition 3.2) and a path in a subautomaton is also a path in the original automaton (Proposition 2.1). Hence, there exists an input symbol $v(k) \in \mathcal{V}$ for all $0 \leq k \leq k_e$ such that

$$G_{\mathrm{leg}}(z_{\mathrm{leg}}(k), v(k)) = z_{\mathrm{leg}}(k+1) \text{ and } G(z_{\mathrm{leg}}(k), v(k)) = z_{\mathrm{leg}}(k+1)$$

(cf. Definition 3.2). With (2.48) it follows that

$$(z_{\mathrm{leg}}(k), z_{\mathrm{leg}}(k+1)) \neq \mathcal{E}_{\mathrm{ill}}, \quad \forall 0 \leq k \leq k_e.$$

Consequently, the input sequence $V(0 \ldots k_e)$ steers the plant from its initial state $z_0 = z_{\mathrm{leg}}(0)$ into the desired final state $z_F = z_{\mathrm{leg}}(k_e + 1)$ while avoiding illegal transitions. $\qquad\square$

Example 3.1 *Tracking control for automated warehouse*

Consider the automated warehouse example introduced in Section 2.5. The controllability of the faultless plant whose model $\mathcal{A}_0$ is shown in Fig. 2.5 shall be determined. It can be seen that there exists a path from the initial state $z_0 = A$ to the states A, B and $C \in \mathcal{Z}_0$, hence, there also exist admissible state sequences $Z(0 \ldots k_e + 1) \in \mathcal{Z}_0^\infty$ with $z(0) = A$ and $z(k_e + 1) = A, B, C$ between these states. Therefore, according to Theorem 3.1, the plant is controllable with respect to the initial state $z_0 = A$ and any final state $z_F \in \{A, B, C\}$.

Avoiding illegal transitions. Now, also the set $\mathcal{E}_{\mathrm{ill}}$ of illegal transitions in (2.54) is taken into account. In the legal part of the plant model in Fig. 2.6 it can be seen that the states A, B and $C \in \mathcal{Z}_0$ remain reachable from the initial state $z_0 = A$, hence according to Theorem 3.2 the automaton $\mathcal{A}$ is controllable with respect to the initial state $z_0 = A$, the set $\mathcal{E}_{\mathrm{ill}}$ of illegal transitions in (2.54) and any final state $z_F \in \{A, B, C\}$. $\qquad\square$

3.2 Structure of the tracking controller

In order to fulfill the control aim (2.44) in the faultless case (i.e., $\mathcal{P} \models \mathcal{A}_0$), in the execution layer of the fault-tolerant control loop in Fig. 2.3 the tracking controller structure shown in Fig. 3.1 is used.

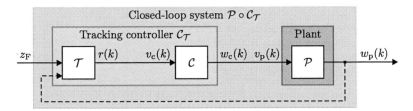

Figure 3.1: Nominal tracking control setting.

The tracking controller C_T consists of two elements:

- the trajectory planning unit T and

- the controller C.

The idea of tracking control is to find a reference trajectory $R(0 \ldots k_e)$ leading into the desired final state z_F and steer the plant P along this trajectory. The computation of the reference trajectory $R(0 \ldots k_e)$ is performed by the trajectory planning unit T. At every time step k the controller C receives the reference signal $r(k) = v_c(k)$ as an input. The controller C has to be designed such that it selects the input $v_p(k) = w_c(k)$ for the plant P in order to ensure that the plant follows the given reference trajectory $R(0 \ldots k_e) = V_c(0 \ldots k_e)$. That is, it is aimed to guarantee that

$$z_p(k+1) = r(k), \quad \forall 0 \le k < k_e. \tag{3.6}$$

Since deterministic systems are considered, the tracking controller C_T only needs to know *that* the plant generated an output symbol and hence completed one transition, but not *which* output $w_p(k)$ was generated. This fact is illustrated by the dashed line in Fig. 3.1. To account for the resulting structure of the control loop, the model of the *closed-loop system* consisting of the tracking controller C_T and the plant P is denoted by $P \circ C_T$.

In the following, the tracking controller C_T designed for a plant $P \vDash A$ will be denoted by $C_T(A)$, wherein the trajectory planning unit is denoted by $T(A)$ and the controller by $C(A)$. When additionally considering a set $\mathcal{E}_{ill}$ of illegal transitions, the tracking controller is denoted by $C_T(A, \mathcal{E}_{ill})$.

The control problem to be solved in the absence of illegal transitions can be formally stated as follows.

Problem 3.1 (Tracking control problem). *Given the model A of a plant P, find a tracking controller $C_T(A)$ consisting of a trajectory planning unit $T(A)$ and a controller $C(A)$ such that the controlled plant $P \vDash A$ reaches the desired final state $z_F \in Z$ at some finite time $k_F \ge 0$.*

If additionally the avoidance of a set $\mathcal{E}_{ill}$ of illegal transitions by the plant shall be guaranteed, the following safe tracking control problem has to be solved.

Problem 3.2 (Safe tracking control problem). *Given the model A of a plant P and a set $\mathcal{E}_{ill}$ of illegal transitions, find a tracking controller $C_T(A, \mathcal{E}_{ill})$ consisting of a trajectory planning unit $T(A, \mathcal{E}_{ill})$ and a controller $C(A, \mathcal{E}_{ill})$ such that the controlled plant $P \vDash A$ reaches the desired final state $z_F \in Z$ at some finite time $k_F \ge 0$, while not executing any illegal transitions $(z_p(k), z_p(k+1)) \in \mathcal{E}_{ill}, (0 \le k < k_F)$.*

3.3 Controller for tracking a reference trajectory

The aim of the controller $\mathcal{C}(\mathcal{A})$ in the tracking controller $\mathcal{C}_T(\mathcal{A})$ is to generate an input sequence $V(0\ldots k_e)$ for the plant $\mathcal{P} \vDash \mathcal{A} = (\mathcal{Z}, \mathcal{V}, \mathcal{W}, G, H, z_0)$ such that the state $z_p(k)$ of the plant follows the reference trajectory $R(0\ldots k_e)$.

3.3.1 Definition of the controller

The main idea is to define the controller $\mathcal{C}(\mathcal{A})$ in form of a deterministic I/O automaton $\mathcal{A}_c$ (i.e., $\mathcal{C}(\mathcal{A}) \vDash \mathcal{A}_c$), whose state $z_c(k)$ corresponds to the current state $z_p(k)$ of the plant $\mathcal{P} \vDash \mathcal{A}$ at all times $k \geq 0$. Based on its current state $z_c(k)$ and input $v_c(k) = r(k)$, the controller $\mathcal{C}(\mathcal{A})$ shall generate an output symbol $w_c(k) = v_p(k)$, which steers the plant $\mathcal{P} \vDash \mathcal{A}$ from its current state $z_p(k)$ into the next desired state given by the current value $r(k)$ of the reference trajectory $R(0\ldots k_e)$. Therefore, the following definition for the controller $\mathcal{C}(\mathcal{A})$ is proposed:

$$
\mathcal{C}(\mathcal{A}) \vDash \mathcal{A}_c : \begin{cases}
\mathcal{Z}_c = \mathcal{Z}, & \text{(3.7a)} \\[4pt]
\mathcal{V}_c = \mathcal{Z}, & \text{(3.7b)} \\[4pt]
\mathcal{W}_c = \mathcal{V}, & \text{(3.7c)} \\[4pt]
G_c(z_c, v_c) = \begin{cases} v_c & \text{if } \mathcal{V}_a(v_c, z_c) \neq \emptyset, \\ \text{undefined} & \text{otherwise,} \end{cases} & \text{(3.7d)} \\[12pt]
H_c(z_c, v_c) = \begin{cases} w_c \in \mathcal{V}_a(v_c, z_c) & \text{if } \mathcal{V}_a(v_c, z_c) \neq \emptyset, \\ \text{undefined} & \text{otherwise,} \end{cases} & \text{(3.7e)} \\[12pt]
z_{c0} = z_0. & \text{(3.7f)}
\end{cases}
$$

Corresponding to the desired functionality of the controller $\mathcal{C}(\mathcal{A})$, the state set $\mathcal{Z}_c$ of the controller in (3.7a) equals the state set $\mathcal{Z}$ of the plant $\mathcal{P}$ and their respective initial states are identical, i.e., $z_{c0} = z_0$ (3.7f). Since the inputs $v_c(k) \in \mathcal{V}_c$ for the controller $\mathcal{C}(\mathcal{A})$ are desired next states for the plant $\mathcal{P}$, the input set $\mathcal{V}_c$ of the controller in (3.7b) also equals the state set $\mathcal{Z}$ of the plant, while its output set $\mathcal{W}_c$ in (3.7c) is the same as the input set $\mathcal{V}$ of the plant (cf. Fig. 3.1).

A transition from a state $z_c(k)$ with input $v_c(k)$ in the controller $\mathcal{C}(\mathcal{A})$ is defined (that is, $G_c(z_c(k), v_c(k))!$ and $H_c(z_c(k), v(k))!$) if and only if the active input set $\mathcal{V}_a(v_c(k), z_c(k))$ of the plant is not empty. This is the case if there is an input symbol $v_p(k) \in \mathcal{V}$ leading to a transition from the state $z_p(k) = z_c(k)$ to the state $z_p(k+1) = v_c(k)$ in the model $\mathcal{A}$ of the plant $\mathcal{P}$.

If a transition from a state $z_c(k)$ with input $v_c(k)$ in the controller $\mathcal{C}(\mathcal{A})$ is defined, the next state $z_c(k+1)$ of the controller $\mathcal{C}(\mathcal{A})$ is determined by its current input symbol $v_c(k)$ in (3.7d), while its output symbol $w_c(k)$ has to be chosen from the active input set $\mathcal{V}_a(v_c(k), z_c(k))$ of the plant as given in (3.7e). The choice of the output symbol $w_c(k)$ of the controller $\mathcal{C}(\mathcal{A})$ is the only degree of freedom that occurs during the definition of $\mathcal{C}(\mathcal{A})$. It becomes necessary when there are multiple possible input symbols $v_p(k) \in \mathcal{V}$ leading to a transition from the state $z_p(k) \in \mathcal{Z}$ to the next state $z_p(k + 1) \in \mathcal{Z}$ in the model $\mathcal{A}$ of the plant $\mathcal{P}$. In order to make the controller $\mathcal{C}(\mathcal{A})$ deterministic, a decision for one of the possible input symbols has to be made when the controller is defined.

The definition of the controller $\mathcal{C}(\mathcal{A})$ in (3.7) supports the assumption in Section 3.2 that the controller $\mathcal{C}(\mathcal{A})$ is independent from the output $w_p(k) = H(z_p(k), v_p(k))$ of the plant $\mathcal{P}$.

3.3.2 Controller construction algorithm

The controller $\mathcal{C}(\mathcal{A})$ defined in (3.7) can be systematically deduced from a given model $\mathcal{A}$ of the plant $\mathcal{P}$ by means of Algorithm 3.1.

The following proposition proves that Algorithm 3.1 indeed yields a controller $\mathcal{C}(\mathcal{A})$ as defined in (3.7).

Proposition 3.1 (Controller construction algorithm)**.** *If Algorithm 3.1 is applied to an automaton $\mathcal{A} = (\mathcal{Z}, \mathcal{V}, \mathcal{W}, G, H, z_0)$, the controller $\mathcal{C}(\mathcal{A})$ as defined in (3.7) results.*

Proof. The initialization step of the algorithm guarantees the fulfillment of (3.7a) –(3.7c) and (3.7f). The first part of (3.7d) (without the "if") is fulfilled by the statement in Line 5. The first part of (3.7e) (without the "if") is equivalent to stating that

$$G(z_c(k), H_c(z_c(k), v_c(k))) = v_c(k) \tag{3.8}$$

Algorithm 3.1: Construction of controller $\mathcal{C}(\mathcal{A})$ defined in (3.7).

Given: Model $\mathcal{A} = (\mathcal{Z}, \mathcal{V}, \mathcal{W}, G, H, z_0)$ of the plant $\mathcal{P}$
Initialize: $\mathcal{Z}_c = \mathcal{Z}$, $\mathcal{V}_c = \mathcal{Z}$, $\mathcal{W}_c = \mathcal{V}$, $z_{c0} = z_0$

1 **forall** $z \in \mathcal{Z}$
2 │   **forall** $v \in \mathcal{V}$
3 │   │ $z' = G(z, v)$
4 │   │   **if** $G_c(z, z')$ *is not defined* **then**
5 │   │   │ $G_c(z, z') = z'$
6 │   │   │ $H_c(z, z') = v$
7 │   │   **end**
8 │   **end**
9 **end**

Result: Controller $\mathcal{C}(\mathcal{A}) = (\mathcal{Z}_c, \mathcal{V}_c, \mathcal{W}_c, G_c, H_c, z_{c0})$

(cf. (2.12)). Line 6 of the algorithm yields $H_c(z, z') = v$ with $z' = G(z, v)$, that is,

$$H_c(z, G(z, v)) = v. \tag{3.9}$$

By substituting $z_c := z$ and $v_c := G(z, v)$ in (3.8), the following results:

$$G(z, H_c(z, G(z, v))) = G(z, v).$$

With (3.9), the equivalence becomes visible immediately.

Finally, the second part of (3.7d) and (3.7e) has to be considered. That is, the controller $C(\mathcal{A})$ has to be defined exactly for all pairs (z_c, v_c) for which $\mathcal{V}_a(v_c, z_c) \neq \emptyset$ holds. This is true, because in Line 3 of the algorithm a state z' for which $\mathcal{V}_a(z', z) = \{v\} \neq \emptyset$ holds is found and the pair (z, z') is used for the definition of G_c and H_c in Lines 5 and 6. Furthermore, all combinations of states z and inputs v of the plant are considered (Lines 1 and 2).

Therefore, all equations (3.7a)–(3.7f) are fulfilled by the controller $C(\mathcal{A})$ resulting from Algorithm 3.1. $\qquad \square$

3.3.3 Properties of the controller

In this section some interesting properties of the controller $C(\mathcal{A})$ are stated, which will become useful at a later point of the thesis.

It is possible to show that the controller $C(\mathcal{A})$ has the same structure as the automaton $\mathcal{A}$ based on which it is defined. That is, the automata graphs of $C(\mathcal{A})$ and $\mathcal{A}$ are identical when the I/O labels v/w at their respective edges are neglected. This fact is proved in the following proposition.

Proposition 3.2 (Structure of controller). *There is a transition from state z to state z' in the controller $C(\mathcal{A})$ if and only if there is a transition from state z to state z' in the automaton $\mathcal{A}$.*

Proof. A transition from state z to state z' is represented by an edge $e = (z, z')$ in the automaton graph of the corresponding automaton. In (2.13) the relation between an edge $e = (z, z') \in \mathcal{E}$ and the active input set $\mathcal{V}_a(z', z)$ of an automaton $\mathcal{A}$ has been stated. Therefore, it is possible to perform the following steps:

$$e = (z, z') \in \mathcal{E} \Leftrightarrow \mathcal{V}_a(z', z) \neq \emptyset \qquad \qquad |\text{By (3.7d)}$$
$$\Leftrightarrow G_c(z, z') = z' \qquad \qquad |\text{By (2.13)}$$
$$\Leftrightarrow e(z, z') \in \mathcal{E}_c. \qquad \qquad \square$$

The following proposition proves that the output function H_c of the controller $\mathcal{C}(\mathcal{A})$ is an injective function for any fixed state $z_c \in \mathcal{Z}_c$. Hence, it is possible to reconstruct the input $v_c(k)$ to the controller from its output $w_c(k)$ and its current state $z_c(k)$.

Proposition 3.3 (Injectivity of output function of controller)**.** *The output function H_c of the controller $\mathcal{C}(\mathcal{A})$ is injective for any fixed state $z_c \in \mathcal{Z}_c$, that is,*

$$\forall z_c \in \mathcal{Z}_c : H_c(z_c, v_{c1}) = H_c(z_c, v_{c2}) \Rightarrow v_{c1} = v_{c2}. \qquad (3.10)$$

Proof. For any controller $\mathcal{C}(\mathcal{A})$ defined as in (3.7) the following implications hold:

$$H_c(z_c, v_{c1}) = H_c(z_c, v_{c2})$$
$$\Rightarrow \mathcal{V}_a(v_{c1}, z_c) \cap \mathcal{V}_a(v_{c2}, z_c) \neq \emptyset$$
$$\Rightarrow \{v \in \mathcal{V} : G(z_c, v) = v_{c1}\} \cap \{v \in \mathcal{V} : G(z_c, v) = v_{c2}\} \neq \emptyset$$
$$\Rightarrow (\exists v \in \mathcal{V})\left(\left[G(z_c, v) = v_{c1}\right] \wedge \left[G(z_c, v) = v_{c2}\right]\right)$$

from which due to the determinism of the plant $v_{c1} = v_{c2}$ follows. $\qquad \square$

Due to the injectivity of the output function of the controller it is possible to define the *inverted automaton map* $\Phi_c^{-1} : \mathcal{Z}_c \times \mathcal{W}_c^\infty \rightarrow \mathcal{V}_c^\infty$ of the controller by

$$\Phi_c^{-1}(z_c, W_c(0 \ldots k_e)) = V_c(0 \ldots k_e) \Leftrightarrow \Phi_c(z_c, V_c(0 \ldots k_e)) = W_c(0 \ldots k_e). \qquad (3.11)$$

Based on this inverted automaton map, the current state $z_c(k)$ of the controller $\mathcal{C}(\mathcal{A})$ can always be reconstructed based on its output sequence $W_c(0 \ldots k - 1)$ as shown by the following proposition.

Proposition 3.4 (Current state of controller)**.** *The current state $z_c(k)$ of the controller $\mathcal{C}(\mathcal{A})$ is given by*

$$z_c(k) = G_c^\infty(z_{c0}, \Phi_c^{-1}(z_{c0}, W_c(0 \ldots k - 1))). \qquad (3.12)$$

Proof. By definition, the current state of the controller $\mathcal{C}(\mathcal{A})$ is given by

$$z_c(k) = G_c^\infty(z_{c0}, V_c(0 \ldots k - 1)),$$

where the input sequence $V_c(0 \ldots k-1)$ is unknown here. The corresponding output sequence is given by

$$W_c(0 \ldots k-1) = \Phi_c(z_{c0}, V_c(0 \ldots k-1)).$$

Therefore, from the definition of the inverted automaton map in (3.11) it follows that the input sequence can be computed by

$$V_c(0 \ldots k-1) = \Phi_c^{-1}(z_{c0}, W_c(0 \ldots k_e))$$

such that it is proved that (3.12) is correct.   □

The controller $C(\mathcal{A})$ is usually not completely defined, because it only accepts inputs corresponding to states in the plant $\mathcal{P} \vDash \mathcal{A}$ that are reachable within one step. However, the following proposition can be stated.

Proposition 3.5 (Existence of an acceptable input symbol). *For every state $z_c \in \mathcal{Z}_c$ in the controller $C(\mathcal{A})$ there is at least one acceptable input symbol $v_c \in \mathcal{V}_c$, i.e.,*

$$\forall z_c \in \mathcal{Z}_c : (\exists v_c \in \mathcal{V}_c) \text{ such that } G_c(z_c, v_c)!$$

Proof. The above equation can be reformulated as follows:

$$\forall z_c \in \mathcal{Z}_c : (\exists v_c \in \mathcal{V}_c) \text{ such that } G_c(z_c, v_c)!$$
$$\Leftrightarrow \forall z_c \in \mathcal{Z}_c : (\exists v_c \in \mathcal{V}_c) \text{ such that } \mathcal{V}_a(v_c, z_c) \neq \emptyset$$
$$\Leftrightarrow \forall z_c \in \mathcal{Z}_c : (\exists v_p \in \mathcal{V}_p) \text{ such that } G(z_c, v_p)!$$

The last statement is always true, because $\mathcal{Z}_c = \mathcal{Z}$ (cf. (3.7a)) and according to Assumption 2 any model $\mathcal{A}$ of the plant $\mathcal{P}$ is assumed to be completely defined.   □

The above proposition shows that the controller $C(\mathcal{A})$ contains no absorbing states, which would lead to a deadlock of the closed-loop system.

3.3.4 Behavior of the controlled plant

In the following proposition, the behavior of a plant $\mathcal{P} \vDash \mathcal{A}$ that is controlled by a controller $C(\mathcal{A})$ as defined in (3.7) which receives an admissible state sequence for the plant as an input is analyzed.

Proposition 3.6 (Tracking an admissible state sequence). *Consider a plant* $\mathcal{P} \models \mathcal{A}$ *and an admissible state sequence* $Z(0 \ldots k_e + 1)$ *with* $z(0) = z_0$ *for it. Then the controller* $\mathcal{C}(\mathcal{A})$ *defined in (3.7) with input sequence* $V_c(0 \ldots k_e) = Z(1 \ldots k_e + 1)$ *generates an input sequence* $V_p(0 \ldots k_e) = W_c(0 \ldots k_e)$ *for the plant* $\mathcal{P}$ *such that the state* $z_p(k)$ *of the plant* $\mathcal{P}$ *follows the admissible state sequence* $Z(0 \ldots k_e + 1)$ *exactly:*

$$z_p(k) = z(k), \quad \forall 0 \le k \le k_e + 1. \tag{3.13}$$

Proof. (By induction) Starting from $z_c(0) = z_0 = z(0)$ in (3.7f) it can be easily shown that

$$z_c(k) = z(k), \quad \forall 0 \le k \le k_e + 1 \tag{3.14}$$

as follows:

$$
\begin{aligned}
z_c(k + 1) = G_c(z_c(k), v_c(k)) &= v_c(k) & \text{if } \mathcal{V}_a(v_c(k), z_c(k)) \ne \emptyset \\
&= z(k + 1) & \text{if } \mathcal{V}_a(z(k + 1)), z_c(k)) \ne \emptyset.
\end{aligned}
$$

Since $Z(0 \ldots k_e)$ is an admissible state sequence for $\mathcal{A}$, $\mathcal{V}_a(z(k + 1), z(k)) \ne \emptyset$ holds for all $0 \le k \le k_e$. Therefore, from $z_c(k) = z(k)$ it follows directly that $z_c(k + 1) = z(k + 1)$, $(0 \le k \le k_e)$.

The correctness of (3.13) can then also be proved by induction based on $z_p(0) = z_0 = z(0)$. For $z_p(k) = z(k)$,

$$z_p(k + 1) = G(z_p(k), v_p(k)) = G(z(k), w_c(k))$$

is obtained, where (3.7e) and (3.14) specify that

$$w_c(k) \in \mathcal{V}_a(v_c(k), z_c(k)) = \mathcal{V}_a(z(k + 1), z(k)),$$

such that $z_p(k + 1) = z(k + 1)$, $(0 \le k \le k_e)$ follows (cf. (2.12)). $\qquad \square$

Proposition 3.6 shows that a plant $\mathcal{P} \models \mathcal{A}$ controlled by the controller $\mathcal{C}(\mathcal{A})$ defined in (3.7) follows any admissible state sequence $Z(0 \ldots k_e + 1)$ that is used as an input sequence $V_c(0 \ldots k_e)$ for the controller exactly, as long as the states z_0 and $z(0)$ coincide. This fact plays an important role for the specification of a reference trajectory $R(0 \ldots k_e)$ by the trajectory planning unit $\mathcal{T}$ as described in the next section.

From Proposition 3.6 it also follows that the series connection $\mathcal{P} \circ \mathcal{C}(\mathcal{A})$ of the plant and the controller realizes the identity map $\mathcal{I} : \mathcal{V}_c \rightarrow \mathcal{W}_p$ with $\mathcal{V}_c = \mathcal{Z}_p, \mathcal{W}_p = \mathcal{Z}_p$ and

$$\mathcal{I}(z) = z, \tag{3.15}$$

if the state of the plant $\mathcal{P} \vDash \mathcal{A}$ is measurable, that is, $w_p(k) = z_p(k+1)$ holds for all $k > 0$.

Note that the controller $\mathcal{C}$ is completely independent of the specification, that is, of the desired final state z_F and the set $\mathcal{E}_{ill}$ of illegal transitions. Therefore, it is universally applicable and does not have to be changed when the specification changes. Consequently, the solution to the safe tracking control problem in Problem 3.2 is a tracking controller $\mathcal{C}_T(\mathcal{A}, \mathcal{E}_{ill})$ whose controller is given by $\mathcal{C}(\mathcal{A})$ instead of $\mathcal{C}(\mathcal{A}, \mathcal{E}_{ill})$.

Example 3.1 (cont.)   *Tracking control for automated warehouse*

Figure 3.2 shows the controller $\mathcal{C}(\mathcal{A}_0)$ for the faultless automated warehouse, which results from applying the definition in (3.7) to the model $\mathcal{A}_0$ in Fig. 2.5. Therefore, its state set

$$\mathcal{Z}_c = \{A, B, C\}$$

equals the state set $\mathcal{Z}_0$ of $\mathcal{A}_0$, its input set is given by

$$\mathcal{V}_c = \mathcal{Z}_0 = \{A, B, C\}$$

and its output set is given by

$$\mathcal{W}_c = \mathcal{V}_0 = \{AB, AC, BA, BC, CA, CB, AB^*, AC^*, BA^*, BC^*, CA^*, CB^*\}.$$

For the state transition function G_c and the output function H_c exemplarily the transition from the state $z_c = A$ with the input $v_c = B$ is presented here. According to (3.7d) the next state is given by

$$G_c(z_c = A, v_c = B) = v_c = B,$$

because from the definition of the active input set in (2.11) it follows that

$$\mathcal{V}_{a0}(v_c = B, z_c = A)$$
$$= \{v \in \mathcal{V}_0 : G_0(z_c = A, v) = v_c = B\}$$
$$= \{AB, AB^*\} \neq \emptyset.$$

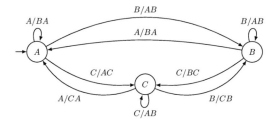

Figure 3.2: Controller $\mathcal{C}(\mathcal{A}_0)$ for transport of parcels in automated warehouse.

Consequently, by (3.7e) the output is given either by

$$H_c(z_c = A, v_c = B) = AB \in V_{a0}(v_c = B, z_c = A) = \{AB, AB^*\}$$

or by

$$H_c(z_c = A, v_c = B) = AB^* \in V_{a0}(v_c = B, z_c = A) = \{AB, AB^*\}.$$

Here the output AB is chosen arbitrarily, which means that robot M instead of robot M^* is used for the transport. For this example, the controller $C(\mathcal{A}_0)$ is completely defined, because in the model $\mathcal{A}_0$ of the faultless plant in Fig. 2.5 every state is reachable from every other state in one step.

According to (3.7f), the initial state of the controller $C(\mathcal{A}_0)$ is set to

$$z_{c0} = z_{00} = A. \tag{3.16}$$

□

3.4 Planning a reference trajectory

The aim of the trajectory planning unit $T(\mathcal{A})$ in the tracking controller $C_T(\mathcal{A})$ is to find and output a reference trajectory $R(0 \ldots k_e)$ which ends in the desired final state z_F. This reference trajectory shall be followed by the plant $\mathcal{P} \vDash \mathcal{A}$. Proposition 3.6 suggests to use an admissible state sequence $Z(0 \ldots k_e + 1)$ for the plant $\mathcal{P} \vDash \mathcal{A}$ with $z(0) = z_0$ and $z(k_e + 1) = z_F$ to define the reference trajectory as

$$R(0 \ldots k_e) = Z(1 \ldots k_e + 1). \tag{3.17}$$

Consequently, the trajectory planning unit $T(\mathcal{A})$ needs to find a state sequence $Z(0 \ldots k_e + 1)$ which

1. is an admissible state sequence for the plant $\mathcal{P} \vDash \mathcal{A}$,

2. starts with the initial state of the plant, i.e., $z(0) = z_0$ and

3. ends with the desired final state z_F, i.e., $z(k_e + 1) = z_F$.

A state sequence $Z(0 \ldots k_e + 1)$ that fulfills the above requirements corresponds to a path in the automaton graph of $\mathcal{A}$ as stated by the following proposition.

Proposition 3.7 (Properties of paths in plant model). *Any path from the initial state z_0 of the plant to the desired final state z_F in the automaton graph of the model $\mathcal{A}$ of the plant corresponds to a state sequence $Z(0 \ldots k_e + 1)$ fulfilling requirements 1–3 stated above.*

Proof. By Definition 3.2, a path in the automaton graph always corresponds to an admissible state sequence, such that the first requirement is fulfilled. If the path starts at state z_0 and ends at state z_F, the other two requirements are fulfilled as well. □

Because of the result in this proposition, the trajectory planning unit $\mathcal{T}(\mathcal{A})$ performs a search for a path from the state z_0 to the state z_F in the automaton graph of the model $\mathcal{A}$ of the plant. The employed search algorithm has to be *complete* and *correct*, that is, if such a path exists, the algorithm has to find it and the algorithm may not yield any results that do not correspond to such a path. To solve this task, there are various well-established graph search methods, which originate in the field of graph theory and are, for example, described in [52].

In summary, the trajectory planning unit $\mathcal{T}(\mathcal{A})$ is represented by

$$\mathcal{T}(\mathcal{A}) = \begin{cases} \mathcal{A} = (\mathcal{Z}, \mathcal{V}, \mathcal{W}, G, H, z_0) \\ \text{Complete and correct graph search algorithm to be applied to } \mathcal{A}. \end{cases} \tag{3.18}$$

In order to steer the plant $\mathcal{P}$ into the desired final state z_F in the minimal number of steps, the reference trajectory $R(0 \dots k_e)$ should be as short as possible. Therefore, for example, Breadth-first-search can be used to find a shortest path from the state z_0 to the state z_F. Breadth-first-search is a complete and correct search method.

However, if desired, it is also possible to incorporate additional information about the plant or additional requirements on the state sequence $Z(0 \dots k_e + 1)$ into the search. For example, if the time that is required for the completion of every transition in the model $\mathcal{A}$ of the plant is known and it is aimed to reach the desired final state z_F as soon as possible, Dijkstra's algorithm can be used for the search. In this case each transition in the automaton graph of $\mathcal{A}$ gets assigned a cost corresponding to its time consumption and Dijkstra's algorithm finds an optimal path from the state z_0 to the state z_F that minimizes the total time. That is, Dijkstra's algorithm is not only complete and correct, but also optimal with respect to the path's costs.

Avoiding illegal transitions.   If a set $\mathcal{E}_{ill}$ of illegal transitions as defined in (2.45) needs to be respected, the state sequence $Z(0 \dots k_e + 1)$ to be found has to fulfill the following additional requirement:

4. The state sequence $Z(0 \dots k_e + 1)$ must not contain any illegal transitions, that is,

$$(z(k), z(k+1)) \notin \mathcal{E}_{ill}, \quad \forall 0 \leq k \leq k_e.$$

In order to find such a state sequence, the trajectory planning unit $\mathcal{T}(\mathcal{A}, \mathcal{E}_{ill})$, which now also depends on the set $\mathcal{E}_{ill}$ of illegal transitions, performs its search not on the model $\mathcal{A}$ of the plant itself, but on its legal part $\mathcal{A}_{leg}$ given in Definition 2.4. If the search for a path from the initial state z_0 of the plant to the desired final state z_F is performed in the automaton graph of $\mathcal{A}_{leg}$ in exactly the same way as described before, a state sequence $Z(0 \dots k_e + 1)$ fulfilling all of the above requirements is found as stated by the following proposition.

Proposition 3.8 (Properties of paths in legal part of plant model)**.** *Any path from the initial state z_0 of the plant to the desired final state z_F in the automaton graph of the legal part $\mathcal{A}_{leg}$ of the plant model $\mathcal{A}$ corresponds to a state sequence $Z(0 \ldots k_e + 1)$ fulfilling requirements 1–4 stated above.*

Proof. By definition, a path in the legal part $\mathcal{A}_{leg}$ of the plant model does not contain any illegal transitions $(z, z') \in \mathcal{E}_{ill}$ (cf. (2.48)), hence requirement 4 is fulfilled. Since $\mathcal{A}_{leg}$ is a subautomaton of $\mathcal{A}$ (Definition 2.4), it follows from (2.19) in Definition 2.3 that any path in the automaton graph of $\mathcal{A}_{leg}$ is also a path in the automaton graph of $\mathcal{A}$. Therefore, together with Proposition 3.7, the fulfillment of requirements 1–3 is also guaranteed. $\square$

Based on these considerations, the trajectory planning unit $\mathcal{T}(\mathcal{A}, \mathcal{E}_{ill})$ results to

$$\mathcal{T}(\mathcal{A}, \mathcal{E}_{ill}) = \begin{cases} \mathcal{A}_{leg} = (\mathcal{Z}_{leg}, \mathcal{V}_{leg}, \mathcal{W}_{leg}, G_{leg}, H_{leg}, z_{leg0}) \\ \text{Complete and correct graph search algorithm to be applied to } \mathcal{A}_{leg}. \end{cases} \tag{3.19}$$

Example 3.1 (cont.) *Tracking control for automated warehouse*

Now the trajectory planning for the automated warehouse example is considered. When the control aim for the automated warehouse is to transport a parcel from the initial position A to position B, the desired final state is

$$z_F = B.$$

The shortest path from the initial state $z_{00} = A$ to the desired final state $z_F = B$ in the automaton graph of $\mathcal{A}_0$ in Fig. 2.5 is given by $P(A, B) = ((A, B))$. According to Proposition 3.7, this path corresponds to the state sequence

$$Z(0 \ldots 1) = (A, B),$$

which fulfills the requirements 1–3 such that the reference trajectory

$$R(0 \ldots k_e) = Z(1 \ldots k_e + 1) = B \tag{3.20}$$

results from (3.17). Since it contains only one element, its time horizon k_e is equal to zero.

Avoiding illegal transitions. If additionally the set $\mathcal{E}_{ill}$ of illegal transitions in (2.54) is considered, as suggested by Proposition 3.8, the trajectory planning unit $\mathcal{T}(\mathcal{A}_0, \mathcal{E}_{ill})$ searches for a path from the initial state $z_{00} = A$ to the desired final state $z_F = B$ in the automaton graph of the legal part of the plant shown in Fig. 2.6. Now the shortest path is given by $P(A, B) = ((A, C), (C, B))$, such that the reference trajectory

$$R(0 \ldots k_e) = Z(1 \ldots k_e + 1) = (C, B) \tag{3.21}$$

results from (3.17). That is, because of the presence of illegal transitions, a detour over position C is proposed in order to avoid the direct route between position A and position B. $\square$

3.5 Fulfillment of the control aim in the closed-loop system

In the previous sections the two elements of the tracking controller $\mathcal{C}_T(\mathcal{A})$, namely the trajectory planning unit $\mathcal{T}(\mathcal{A})$ and the controller $\mathcal{C}(\mathcal{A})$, have been introduced. Now these two elements are combined to obtain the tracking controller $\mathcal{C}_T(\mathcal{A})$. If the resulting tracking controller $\mathcal{C}_T(\mathcal{A})$ is used to control the plant $\mathcal{P} \vDash \mathcal{A}$, the closed-loop system $\mathcal{P} \circ \mathcal{C}_T(\mathcal{A})$ shown in Fig. 3.1 occurs. The following theorem states that the proposed tracking controller $\mathcal{C}_T(\mathcal{A})$ solves the tracking control problem in Problem 3.1 if the model $\mathcal{A}$ of the plant $\mathcal{P}$ is controllable according to Definition 3.1.

Theorem 3.3 (Fulfillment of control aim). *Consider a plant $\mathcal{P}$ modeled by the automaton $\mathcal{A}$ that is controllable with respect to the state $z_F \in \mathcal{Z}$. Then the plant $\mathcal{P} \vDash \mathcal{A}$ in the closed-loop system $\mathcal{P} \circ \mathcal{C}_T(\mathcal{A})$ reaches the desired final state z_F at time $k_F = k_e + 1$, where k_e is the time horizon of the reference trajectory in (3.17).*

Proof. The trajectory planning unit $\mathcal{T}(\mathcal{A})$ in the tracking controller $\mathcal{C}_T(\mathcal{A})$ uses a complete and correct search algorithm for the search of a path from the state z_0 to the state z_F in the automaton graph of $\mathcal{A}$. Hence, if such a path exists, it is found. According to Proposition 3.7, such a path corresponds to an admissible state sequence $Z(0 \ldots k_e + 1) \in \mathcal{Z}^\infty$ with $z(0) = z_0$ and $z(k_e + 1) = z_F$. Since the plant $\mathcal{P} \vDash \mathcal{A}$ is controllable with respect to the state $z_F \in \mathcal{Z}$, it is guaranteed that such an admissible state sequence $Z(0 \ldots k_e + 1) \in \mathcal{Z}^\infty$ with $z(0) = z_0$ and $z(k_e + 1) = z_F$ exists (Theorem 3.1), hence also the path exists and is found by the trajectory planning unit $\mathcal{T}(\mathcal{A})$.

By (3.17), the reference trajectory that serves as an input for the controller $\mathcal{C}(\mathcal{A})$ in the tracking controller $\mathcal{C}_T(\mathcal{A})$ is given by

$$V_c(0 \ldots k_e) = R(0 \ldots k_e) = Z(1 \ldots k_e + 1).$$

According to Proposition 3.6, the plant $\mathcal{P} \vDash \mathcal{A}$ in the closed-loop system $\mathcal{P} \circ \mathcal{C}_T(\mathcal{A})$ follows the state sequence $Z(0 \ldots k_e + 1)$ exactly, because $Z(0 \ldots k_e + 1)$ is an admissible state sequence for $\mathcal{A}$ and $z(0) = z_0$ holds. Hence, because $z(k_e + 1) = z_F$, the plant $\mathcal{P}$ reaches the desired final state z_F at time $k_F = k_e + 1$. □

Theorem 3.3 shows that the proposed structure of the tracking controller $\mathcal{C}_T$, combining a trajectory planning unit $\mathcal{T}$ and a controller $\mathcal{C}$, does not restrict the controllability of the plant $\mathcal{P}$. Whenever the plant is controllable, hence there exists a controller with arbitrary structure that can steer the plant into the desired final state, the proposed tracking controller $\mathcal{C}_T$ will also fulfill this task.

Avoiding illegal transitions. If additionally a set $\mathcal{E}_{\text{ill}}$ of illegal transitions as defined in (2.45) needs to be respected, the following theorem results from Theorem 3.3, which states that the proposed tracking controller $\mathcal{C}_T(\mathcal{A}, \mathcal{E}_{\text{ill}})$ solves the safe tracking control problem in Problem 3.2.

> **Theorem 3.4** (Safe fulfillment of control aim). *Consider a plant $\mathcal{P}$ modeled by the automaton $\mathcal{A}$ that is controllable with respect to the state $z_F \in \mathcal{Z}$ and the set $\mathcal{E}_{\text{ill}} \subset \mathcal{Z} \times \mathcal{Z}$ of illegal transitions. Then the plant $\mathcal{P} \vDash \mathcal{A}$ in the closed-loop system $\mathcal{P} \circ \mathcal{C}_T(\mathcal{A}, \mathcal{E}_{\text{ill}})$ reaches the desired final state z_F at time $k_F = k_e + 1$, where k_e is the time horizon of the reference trajectory in (3.17), while not executing any illegal transitions $(z_p(k), z_p(k+1)) \in \mathcal{E}_{\text{ill}}$, $(0 \le k < k_F)$.*

Proof. The proof is analogous to the one of Theorem 3.3, where the additional avoidance of illegal transitions is guaranteed by Proposition 3.8. □

Theorem 3.3 or, if additionally illegal transitions are considered, Theorem 3.4, state that the control aim in Section 2.4.1 is fulfilled by the plant $\mathcal{P} \vDash \mathcal{A}$ in the closed-loop system $\mathcal{P} \circ \mathcal{C}_T$ if the plant is controllable and the model $\mathcal{A}$ of the plant is used for the definition of the tracking controller $\mathcal{C}_T(\mathcal{A})$ or $\mathcal{C}_T(\mathcal{A}, \mathcal{E}_{\text{ill}})$, respectively. That is, in order to fulfill the control aim, the correct model of the plant, including its initial state $z_p(0) = z_0$ has to be known.

Example 3.1 (cont.) *Tracking control for automated warehouse*

If the automated warehouse modeled by the automaton $\mathcal{A}_0$ in Fig. 2.5 is controlled by the tracking controller $\mathcal{C}_T(\mathcal{A}_0)$ whose elements are defined as described in the previous sections, the for the closed-loop system the behavior shown in Fig. 3.3 results. The first subplot displays the reference trajectory $R(0 \ldots k_e)$. In the second to fourth subplot the input sequence $V_p(0 \ldots k_e)$, the state sequence $Z_p(0 \ldots k_e)$ and the output sequence $W_p(0 \ldots k_e)$ of the plant $\mathcal{P}$ are shown. The fifth subplot shows the distance to the desired final state z_F, that is, the length of the shortest path from the current state $z_p(k)$ of the plant $\mathcal{P}$ to the desired final state z_F.

For the desired final state $z_F = B$, the trajectory planning unit $\mathcal{T}(\mathcal{A}_0)$ generates the reference trajectory $R(0) = B$, which consists only of one element such that $k_e = 0$ (cf. (3.20)). Therefore, the controller $\mathcal{C}(\mathcal{A}_0)$ in Fig. 3.2 passes through the state sequence

$$Z_c(0 \ldots 1) = (A, B)$$

and generates the input

$$v_p(0) = w_c(0) = AB \tag{3.22}$$

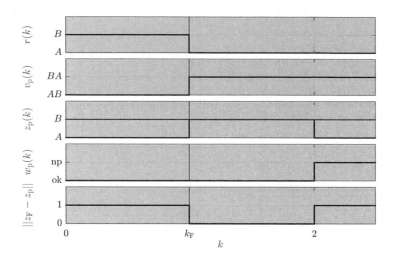

Figure 3.3: Tracking control of automated warehouse.

for the plant $\mathcal{P} \models \mathcal{A}_0$. As a result the plant passes from its initial state $z(0) = z_{00} = A$ to state

$$z_p(1) = B$$

and generates the output $w_p(0) = \text{ok}$. That means that a parcel is successfully transported from position A to position B using robot M. Hence, the automated warehouse reaches the desired final state $z_F = B$ at time $k_F = 1 = k_e + 1$. This is exactly the result that has been stated in Theorem 3.3.

Avoiding illegal transitions. If additionally the set $\mathcal{E}_{\text{ill}}$ of illegal transitions in (2.54) is considered, the trajectory planning unit $\mathcal{T}(\mathcal{A}_0, \mathcal{E}_{\text{ill}})$ generates the reference trajectory in (3.21) such that the controller $\mathcal{C}(\mathcal{A}_0)$ in Fig. 3.2 passes through the state sequence

$$Z_c(0\ldots 2) = (A, C, B)$$

and generates the input sequence

$$V_p(0\ldots 1) = W_c(0\ldots 1) = (AC, CB) \tag{3.23}$$

for the plant $\mathcal{P} \models \mathcal{A}_0$. As a result the plant passes through the states

$$Z_p(0\ldots 2) = (A, C, B)$$

and generates the output sequence $W_p(0\ldots 1) = (\text{ok}, \text{ok})$. That is, the parcel is transported to position B via position C in order to avoid the hazardous direct route and the desired final state $z_F = B$ is reached at time $k_F = k_e + 1 = 2$. These results are summarized in Fig. 3.4. The illegal transitions between state A and state B are visualized by the red arrows right of the third subplot showing the state sequence $Z_p(0\ldots k_e)$ of the plant $\mathcal{P}$.   □

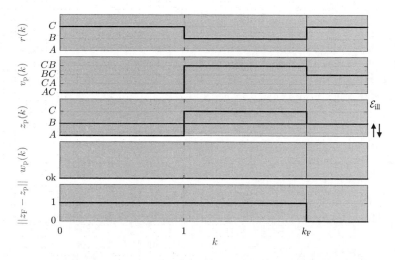

Figure 3.4: Tracking control of automated warehouse avoiding illegal transitions.

3.6 Complexity of the tracking control method

In this section it is shown that the complexity of the proposed tracking control method, both, for the controllability analysis, as well as for the construction of the tracking controller and its online execution, is quite low such that the method is applicable to large systems.

3.6.1 Complexity of the controllability analysis

According to Theorem 3.1 and Theorem 3.2 a necessary and sufficient condition for the controllability of a plant $\mathcal{P}$ modeled by an automaton $\mathcal{A}$ is the existence of a path from the initial state z_0 of the plant into the desired final state z_F in $\mathcal{A}$ or the legal part $\mathcal{A}_{\text{leg}}$ of $\mathcal{A}$, respectively. The construction of $\mathcal{A}_{\text{leg}}$ from $\mathcal{A}$ according to Definition 2.4 requires the consideration of every transition in $\mathcal{A}$. Therefore, its complexity is given by $\mathcal{O}(n \cdot p)$.

Beyond, the complexity of the controllability analysis only depends on the complexity of the search algorithm that is used to find such a path. For example, Breadth-first-search has a time complexity of $\mathcal{O}(|\mathcal{N}| + |\mathcal{E}|)$ and a space complexity of $\mathcal{O}(|\mathcal{N}|)$ if applied to a graph composed of the nodes $\mathcal{N}$ and the edges $\mathcal{E}$ (see [52]). Hence, considering the controllability analysis, the time complexity is $\mathcal{O}(n + n \cdot p) = \mathcal{O}(n \cdot p)$ and the space complexity $\mathcal{O}(n)$, because for each state $z \in \mathcal{Z}$ there is one transition for every input symbol $v \in \mathcal{V}$. That is, the complexity of the controllability analysis is linear in the number of states and input symbols of the automaton $\mathcal{A}$.

3.6.2 Complexity of the construction of the tracking controller

For the definition of the trajectory planning unit $\mathcal{T}$ in the tracking controller $\mathcal{C}_\mathcal{T}(\mathcal{A})$, only the model $\mathcal{A}$ of the plant or the legal part $\mathcal{A}_{\text{leg}}$ of the plant model $\mathcal{A}$ need to be stored. The construction of $\mathcal{A}_{\text{leg}}$ is again of time complexity $\mathcal{O}(n \cdot p)$, but the complexity for the definition of the trajectory planning unit $\mathcal{T}$ itself is negligible.

As proved in Proposition 3.1, the controller $\mathcal{C}(\mathcal{A})$ defined in (3.7) can be constructed by means of Algorithm 3.1. In Algorithm 3.1 two for-loops over the state set $\mathcal{Z}$ and the input set $\mathcal{V}$ of the plant model $\mathcal{A}$ are nested. Therefore, its time complexity is $\mathcal{O}(n \cdot p)$. Consequently, the complexity of constructing the tracking controller $\mathcal{C}_\mathcal{T}(\mathcal{A})$ is again linear in the number of states and input symbols of the automaton $\mathcal{A}$.

3.6.3 Complexity of the online execution of the control method

In the trajectory planning unit $\mathcal{T}$, the model $\mathcal{A}$ of the plant or the legal part $\mathcal{A}_{\text{leg}}$ of the plant model $\mathcal{A}$ needs to be stored. According to [52], the space requirement for the storage of any graph is $\mathcal{O}(|\mathcal{N}| + |\mathcal{E}|)$. Therefore, the space requirement of the trajectory planning unit $\mathcal{T}(\mathcal{A})$ is $\mathcal{O}(n + n \cdot p) = \mathcal{O}(n \cdot p)$. Since the controller $\mathcal{C}(\mathcal{A})$ has the same structure as the plant model $\mathcal{A}$ (Proposition 3.2), the space requirement of the controller $\mathcal{C}(\mathcal{A})$ is also $\mathcal{O}(n \cdot p)$.

The time complexity for planning a reference trajectory $R(0 \ldots k_\text{e})$ in the trajectory planning unit $\mathcal{T}$ depends on the employed search method. The search has to be executed once whenever a new desired final state z_F is given. Again, using Breadth-first-search has a time complexity of $\mathcal{O}(n \cdot p)$. Of course, this time complexity might be considerably reduced if heuristic search methods dedicated to the given application are used. Since the controller $\mathcal{C}(\mathcal{A})$ is a deterministic I/O automaton, the time for evaluating the state transition function G_c and the output function H_c in order to determine its next state $z_\text{c}(k + 1)$ and output symbol $w_\text{c}(k)$ in every time step k is negligible.

4 Active fault diagnosis

This chapter considers the fault diagnosis problem for deterministic I/O automata. Both, the case that the nominal tracking controller generates inputs for the faulty plant (passive diagnosis) and the case that the diagnostic unit generates a diagnostic trajectory for the controller (active diagnosis) are considered. First, the respective diagnosability definitions are given and diagnosability criteria are developed. Afterwards the diagnostic procedure is described, which in the case of active diagnosis relies on the generation of separating sequences. It is proved that an unambiguous, complete and correct diagnostic result is obtained by the described methods if the plant is diagnosable. Finally, an alternative approach that avoids the use of illegal transitions during the fault diagnosis is proposed.

4.1 Consistency-based fault diagnosis settings

The aim of the diagnostic unit $\mathcal{D}$ is to detect whether a fault occurred, and, if this is the case, identify which fault $\bar{f} \in \mathcal{F}$ is actually present in the plant $\mathcal{P}$. The *diagnostic result*

$$\mathcal{D}^* \subset \mathcal{Z}_\Delta = \bigcup_{f \in \mathcal{F}} \mathcal{Z}_f \times \{f\} \tag{4.1}$$

reflects the knowledge about the faulty plant $\mathcal{P} \models \mathcal{A}_{\bar{f}}$ after the end of the fault diagnosis. The diagnostic result $\mathcal{D}^*$ should be *complete* and *correct*. That is, the diagnostic result needs to contain the present fault $\bar{f} \in \mathcal{F}$, but must not contain any faults $f \in \mathcal{F}$ that do not match the previous measurements. Of course it is desirable to obtain an *unambiguous* diagnostic result $\mathcal{D}^*$ that *only* contains the present fault $\bar{f} \in \mathcal{F}$.

Consistency-based fault diagnosis relies on the comparison between the actual behavior of a plant $\mathcal{P}$ reflected by the I/O-pairs $(v_{\mathrm{p}}(k), w_{\mathrm{p}}(k))$ on the one hand and the behavior described by some models of the plant on the other hand. For the fault detection the model $\mathcal{A}_0$ of the faultless plant can be used, while for the identification of the fault the models $\mathcal{A}_f$, $(f \in \mathcal{F})$ of the faulty plant need to be available.

In order to allow for the fault diagnosis, it is crucial to excite the plant $\mathcal{P}$ by some input sequence $V_\mathrm{p}(0 \ldots k_\mathrm{e})$. During the fault detection the input sequence $V_\mathrm{p}(0 \ldots k_\mathrm{e})$ for the plant $\mathcal{P}$ is provided by the nominal tracking controller $\mathcal{C}_\mathcal{T}(\mathcal{A}_0)$ during the nominal operation of the closed-loop system $\mathcal{P} \circ \mathcal{C}_\mathcal{T}(\mathcal{A}_0)$ (Fig. 4.1a). If it is detected that some fault occurred, two possible strategies for the identification of the present fault $\bar{f} \in \mathcal{F}$ are thinkable:

- **Passive fault diagnosis** keeps the original control structure, in which the nominal tracking controller $\mathcal{C}_\mathcal{T}(\mathcal{A}_0)$ generates inputs $v_\mathrm{p}(k)$ for the faulty plant $\mathcal{P} \vDash \mathcal{A}_{\bar{f}}$ (Fig. 4.1a).

- **Active fault diagnosis** changes the control structure such that the diagnostic unit $\mathcal{D}$ generates a *diagnostic trajectory* for the controller $\mathcal{C}(\mathcal{A}_0)$ in the nominal tracking controller $\mathcal{C}_\mathcal{T}(\mathcal{A}_0)$ which then generates inputs $v_\mathrm{p}(k)$ for the faulty plant $\mathcal{P} \vDash \mathcal{A}_{\bar{f}}$ (Fig. 4.1b).

While passive diagnosis does not require any additional intervention into the closed-loop

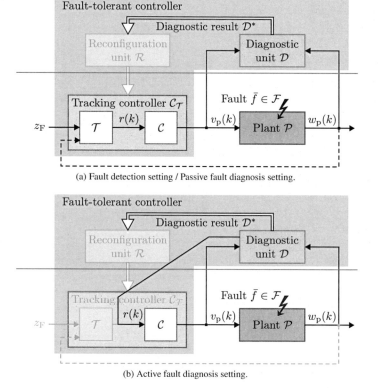

(a) Fault detection setting / Passive fault diagnosis setting.

(b) Active fault diagnosis setting.

Figure 4.1: Fault detection, passive and active fault diagnosis setting.

system, active diagnosis has the benefit of allowing for a precise influence on the faulty plant $\mathcal{P}$ such that the identification of the fault is accelerated or even becomes possible at all.

The fault diagnosis problems to be solved with passive or active fault diagnosis, respectively, can be formalized as follows. If the fault diagnosis is considered as a closed task, usually the detection and identification of a fault $\bar{f} \in \mathcal{F}$ possibly present in the plant $\mathcal{P}$ is its only aim. If, however, the fault diagnosis is performed by a diagnostic unit $\mathcal{D}$ within a fault-tolerant controller, additionally the current state $z_\text{p}(k)$ of the faulty plant $\mathcal{P} \vDash \mathcal{A}_{\bar{f}}$ might be of interest.

Problem 4.1 (Passive fault diagnosis problem). *Given the model $\mathcal{A}_0$ of the faultless plant, the set $\{\mathcal{A}_f, \, (f \in \mathcal{F})\}$ of models of the faulty plant and the I/O-pair $(V_\text{p}(0 \ldots k), W_\text{p}(0 \ldots k))$ of the plant $\mathcal{P}$ in the closed-loop system $\mathcal{P} \circ \mathcal{C}_\mathcal{T}(\mathcal{A}_0)$, detect whether some fault occurred and, if this is the case, identify the present fault $\bar{f} \in \mathcal{F}$ and the current state $z_\text{p}(k)$ of the faulty plant $\mathcal{P} \vDash \mathcal{A}_{\bar{f}}$.*

Problem 4.2 (Active fault diagnosis problem). *Given the model $\mathcal{A}_0$ of the faultless plant, the set $\{\mathcal{A}_f, \, (f \in \mathcal{F})\}$ of models of the faulty plant and the I/O-pair $(V_\text{p}(0 \ldots k), W_\text{p}(0 \ldots k))$ of the plant $\mathcal{P}$ in the closed-loop system $\mathcal{P} \circ \mathcal{C}_\mathcal{T}(\mathcal{A}_0)$, detect whether some fault occurred and, if this is the case, apply an input sequence $V_\text{c}(k_\text{f}^* \ldots k_\text{e})$ to the controller $\mathcal{C}(\mathcal{A}_0)$ and identify the present fault $\bar{f} \in \mathcal{F}$ and the current state $z_\text{p}(k)$ of the faulty plant $\mathcal{P} \vDash \mathcal{A}_{\bar{f}}$.*

Avoiding illegal transitions. When additionally a set $\mathcal{E}_\text{ill}$ of illegal transitions needs to be respected during the fault diagnosis, only the use of active diagnosis gives the possibility to guarantee the avoidance of these transitions. Therefore, the following problem statement is applicable.

Problem 4.3 (Active safe fault diagnosis problem). *Given the model $\mathcal{A}_0$ of the faultless plant, the set $\{\mathcal{A}_f, \, (f \in \mathcal{F})\}$ of models of the faulty plant, a set $\mathcal{E}_\text{ill} \subset \mathcal{Z}_0 \times \mathcal{Z}_0$ of illegal transitions and the I/O-pair $(V_\text{p}(0 \ldots k), W_\text{p}(0 \ldots k))$ of the plant $\mathcal{P}$ in the closed-loop system $\mathcal{P} \circ \mathcal{C}_\mathcal{T}(\mathcal{A}_0, \mathcal{E}_\text{ill})$, detect whether some fault occurred and, if this is the case, apply an input sequence to the controller $\mathcal{C}(\mathcal{A}_0)$ and identify the present fault $\bar{f} \in \mathcal{F}$ (and the current state $z_\text{p}(k)$ of the faulty plant $\mathcal{P} \vDash \mathcal{A}_{\bar{f}}$), while guaranteeing that the plant $\mathcal{P}$ does not execute any illegal transitions $(z_\text{p}(k), z_\text{p}(k+1)) \in \mathcal{E}_\text{ill}, \, (k \geq 0)$.*

4.2 Testing deterministic I/O automata

The passive fault diagnosis problem in Problem 4.1, the active fault diagnosis problem in Problem 4.2 and the active safe fault diagnosis problem in Problem 4.3 are closely related to a problem from the theory of testing deterministic I/O automata. Therefore, some basic notions from this area are introduced in this section.

4.2.1 Homing sequences

After the detection of a fault, the internal condition of the faulty plant $\mathcal{P} \vDash \mathcal{A}_{\bar{f}}$ shall be reconstructed from the I/O-pair $(V_{\mathrm{p}}(0 \ldots k), W_{\mathrm{p}}(0 \ldots k))$ of the plant $\mathcal{P}$ (see Problems 4.1–4.3). The problems are related to the so called *homing experiments* in the theory of testing deterministic I/O automata (cf. [56]). The aim of a homing experiment is to steer an automaton with known model $\mathcal{A}$ from an unknown initial state within the *initial uncertainty* $\mathcal{Z}^* \subseteq \mathcal{Z}$ into an arbitrary but known state. The *current uncertainty* during a homing experiment is a state set $\mathcal{Z}^*(k) \subseteq \mathcal{Z}$ containing all possible current states of the automaton $\mathcal{A}$ when considering the previous inputs $V(0 \ldots k-1)$ and outputs $W(0 \ldots k-1)$ of $\mathcal{A}$.

The input sequence used in a homing experiment is called homing sequence. It can be distinguished between *preset* and *adaptive* homing sequences. A preset homing sequence is completely specified before the homing experiment and is hence independent from the real initial state $z_0 \in \mathcal{Z}^*$ of $\mathcal{A}$. In contrast, an adaptive homing sequence can be adapted during the homing experiment based on the previous outputs $W(0 \ldots k)$ of the automaton $\mathcal{A}$. Therefore, for different initial states $z_0 \in \mathcal{Z}^*$ of $\mathcal{A}$, different input sequences may result. Preset homing sequences are formally defined as follows.

Definition 4.1 (Preset homing sequence). *An input sequence* $V_{\mathrm{H}}(0 \ldots k_{\mathrm{e}}) \in \mathcal{V}^\infty$ *is called preset homing sequence for a completely defined automaton* $\mathcal{A}$ *with respect to an initial uncertainty* $\mathcal{Z}^* \subseteq \mathcal{Z}$ *if for all states* $z_1, z_2 \in \mathcal{Z}^*$ *the following implication holds:*

$$G^\infty(z_1, V_{\mathrm{H}}(0 \ldots k_{\mathrm{e}})) \neq G^\infty(z_2, V_{\mathrm{H}}(0 \ldots k_{\mathrm{e}})) \Rightarrow \Phi(z_1, V_{\mathrm{H}}(0 \ldots k_{\mathrm{e}})) \neq \Phi(z_2, V_{\mathrm{H}}(0 \ldots k_{\mathrm{e}})).$$
$$(4.2)$$

That is, for any state pair (z_1, z_2) within the initial uncertainty $\mathcal{Z}^* \subseteq \mathcal{Z}$ the homing sequence $V_{\mathrm{H}}(0 \ldots k_{\mathrm{e}})$ leads to different output sequences

$$W_i(0 \ldots k_{\mathrm{e}}) = \Phi(z_i, V_{\mathrm{H}}(0 \ldots k_{\mathrm{e}})), \ (i \in \{1, 2\})$$

whenever the reached states

$$z_i(k_{\mathrm{e}} + 1) = G^\infty(z_i, V_{\mathrm{H}}(0 \ldots k_{\mathrm{e}})), \ (i \in \{1, 2\})$$

differ. Consequently it is possible to deduce the current state $z(k_{\mathrm{e}} + 1)$ of the automaton $\mathcal{A}$ after the application of the preset homing sequence $V_{\mathrm{H}}(0 \ldots k_{\mathrm{e}})$ from the generated output sequence $W(0 \ldots k_{\mathrm{e}})$. If the automaton $\mathcal{A}$ is not completely defined, the preset homing sequence $V_{\mathrm{H}}(0 \ldots k_{\mathrm{e}})$ has to be acceptable to all states within the initial uncertainty $\mathcal{Z}^*$.

It is known that a preset homing sequence exists for every completely defined automaton $\mathcal{A}$

without equivalent states (cf. [56]). However, for automata that are not completely defined no similar result can be found because of the restriction to input sequences acceptable to all states within the initial uncertainty $\mathcal{Z}^*$.

In this case, the use of an adaptive homing sequence that is more flexible might be necessary. Furthermore, even if a preset homing sequence exists, the corresponding adaptive homing sequence is usually shorter. Since adaptive homing sequences depend on the previous outputs of the automaton $\mathcal{A}$, they are given in form of a decision tree instead of a real sequence and are defined as follows.

Definition 4.2 (Adaptive homing sequence). *An adaptive homing sequence for a completely defined automaton $\mathcal{A}$ with respect to an initial uncertainty $\mathcal{Z}^* \subseteq \mathcal{Z}$ is a tree*

- *whose internal vertices are labeled with input symbols $v \in \mathcal{V}$ and edges are labeled with output symbols $w \in \mathcal{W}$,*

- *whose leaves are labeled with sets $\mathcal{Z}_\mathsf{l} \subseteq \mathcal{Z}^*$ of states and a single state $z_\mathsf{l} \in \mathcal{Z}$, such that every state $z \in \mathcal{Z}^*$ is contained in the state set $\mathcal{Z}_\mathsf{l}$ of exactly one leaf,*

- *in which edges starting in a common vertex are labeled with distinct output symbols and*

- *in which for every leaf labeled with a set $\mathcal{Z}_\mathsf{l}$ and a state z_l, the following equations hold for all states $z \in \mathcal{Z}_\mathsf{l}$:*

$$G^\infty(z, V(0 \ldots k_\mathrm{e})) = z_\mathsf{l}, \tag{4.3}$$

$$\Phi(z, V(0 \ldots k_\mathrm{e})) = W(0 \ldots k_\mathrm{e}), \tag{4.4}$$

where $V(0 \ldots k_\mathrm{e})$ and $W(0 \ldots k_\mathrm{e})$ are the input and output sequence formed by the vertex and edge labels, respectively, on the path from the root to the leaf.

Figure 4.2 illustrates the structure of an adaptive homing sequence. Roughly speaking, an adaptive homing sequence indicates by the labels in its vertices the next input symbol $v(k)$ to be applied to the automaton $\mathcal{A}$ based on its previous output symbol $w(k-1)$. Furthermore, it reveals the current state $z_\mathsf{l} = z(k)$ of the automaton $\mathcal{A}$ when a leaf is reached. For automata $\mathcal{A}$ that are not completely defined the additional property that the input symbol v in every internal vertex x is acceptable to all states z_l for which (4.3) and (4.4) hold, where $V(0 \ldots k_\mathrm{e})$ and $W(0 \ldots k_\mathrm{e})$ are the input and output sequence formed by the vertex and edge labels, respectively, on the path from the root to the vertex x, occurs. That is, the input symbol $v \in \mathcal{V}$ in the root vertex has to be acceptable for all states within the initial uncertainty $\mathcal{Z}^*$, while the input symbols in all other vertices only have to be acceptable to a subset of states depending on the previous outputs.

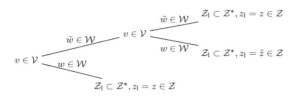

Figure 4.2: Structure of an adaptive homing sequence.

The construction of an adaptive homing sequence will be described in Sections 4.2.2 and 4.2.3. Based on a given adaptive homing sequence for an automaton $\mathcal{A}$ with respect to the initial uncertainty $\mathcal{Z}^*$, a homing experiment can be conducted as described in Algorithm 4.1.

The following proposition shows that Algorithm 4.1 indeed reveals the current state $z(k)$ of the automaton $\mathcal{A}$.

Proposition 4.1 (Adaptive homing experiment). *Let $\mathcal{A} = (\mathcal{Z}, \mathcal{V}, \mathcal{W}, G, H, z_0)$ be a deterministic I/O automaton with unknown initial state $z(0) = z_0$. If an adaptive homing sequence with respect to an initial uncertainty $\mathcal{Z}^* \subseteq \mathcal{Z}$ for $\mathcal{A}$ is given, the result of Algorithm 4.1 is the current state $z(k)$ of the automaton $\mathcal{A}$ if $z_0 \in \mathcal{Z}^*$.*

Proof. Since an adaptive homing sequence is a tree such that it has exactly one root vertex and by definition its input symbol $v \in \mathcal{V}$ is acceptable to all states within the initial uncertainty $\mathcal{Z}^*$, Line 1 can always be executed.

Algorithm 4.1: Adaptive homing experiment.

Given: Automaton $\mathcal{A}$ with unknown initial state $z(0) = z_0 \in \mathcal{Z}^*$.
Adaptive homing sequence for $\mathcal{A}$ with respect to the initial uncertainty $\mathcal{Z}^*$.

1 Apply the input $v(0) = v$ from the root vertex to the automaton $\mathcal{A}$.
2 **for** *level $l = 0, 1, \ldots$ in the adaptive homing sequence*
3   Observe the output $w(l)$ of the automaton $\mathcal{A}$.
4   Follow the edge labeled with the observed output $w(l)$ of the automaton $\mathcal{A}$ to the next vertex x in the adaptive homing sequence.
5   **if** *x is a leaf* **then**
6   Current state of the automaton $\mathcal{A}$ is given by the state z_1 in the label of the vertex x.
7   **else**
8   Use the input symbol $v \in \mathcal{V}$ in the vertex x as the next input $v(l+1)$ for the automaton $\mathcal{A}$.
9   **end**
10 **end**

Result: Current state $z(l) = z_1$ of the automaton $\mathcal{A}$.

As a result the automaton $\mathcal{A}$ passes into a state

$$z(1) = G(z(0), v(0))$$

and generates the output symbol

$$w(0) = H(z(0), v(0)),$$

where $z(0) = z_0$ and $v(0) = v$.

Then the for-loop in Lines 2–9 is executed for the time steps $l = 0, 1, \ldots$. By the third property in Definition 4.2 it is guaranteed that Line 3 always leads to a unique next vertex x. If x is not a leaf, by definition, it is labeled with an input symbol $v \in \mathcal{V}$, which is acceptable to all possible current states $z(l) \in \mathcal{Z}$ of the automaton $\mathcal{A}$. That is, when Line 7 is executed, the input $v(l) = v$ is applied to the automaton $\mathcal{A}$ which passes to state

$$z(l+1) = G(z(l), v(l))$$

and generates the output symbol

$$w(l) = H(z(l), v(l)).$$

Since the adaptive homing sequence is a tree with a finite number of vertices, it is guaranteed that eventually a vertex x that is a leaf is reached after an input sequence $V(0 \ldots k_e)$ has been applied to the automaton $\mathcal{A}$ and the output sequence $W(0 \ldots k_e)$ has been generated. Then Line 5 becomes applicable, which yields the state z_1 in the label of the vertex x as a result. According to (4.3), this state equals the result of the extended state transition function $G^\infty(z, V(0 \ldots k_e))$ of the automaton $\mathcal{A}$ for any state z within the set $\mathcal{Z}_1$ in the vertex x. Additionally, (4.4) states that *only* states from which the output sequence $W(0 \ldots k_e)$ is generated upon the application of $V(0 \ldots k_e)$ are included in the set $\mathcal{Z}_1$. At the same time the third property in Definition 4.2 ensures that *all* states from which the output sequence $W(0 \ldots k_e)$ is generated upon the application of $V(0 \ldots k_e)$ are included in the set $\mathcal{Z}_1$.

Consequently, if the initial state $z(0) = z_0$ of the automaton $\mathcal{A}$ is included in the initial uncertainty $\mathcal{Z}^*$, and $\mathcal{A}$ received the input sequence $V(0 \ldots k_e)$ and generated the output sequence $W(0 \ldots k_e)$, it is known that the current state of the automaton $\mathcal{A}$ is given by $z(k_e + 1) = z_1$. Therefore, the result of Algorithm 4.1 is correct. $\qquad\square$

The input sequence $V(0 \ldots k_e)$ that is applied to the automaton $\mathcal{A}$ during the adaptive homing experiment performed based on an adaptive homing sequence is called *realization* of the adaptive homing sequence with respect to the initial state $z_0 \in \mathcal{Z}^*$ of $\mathcal{A}$. Each realization of the adaptive homing sequence corresponds to a path from the root to one leaf in the tree describing the adaptive homing sequence. Hence, roughly speaking, the adaptive homing sequence is the union of its realizations for all possible initial states $z_0 \in \mathcal{Z}^*$ of $\mathcal{A}$.

Example 4.1   *Construction of homing sequences*

The concept of preset and adaptive homing sequences as well of adaptive homing experiments are illustrated by an academic example. Consider the automaton $\mathcal{A}$ in Fig. 4.3 with unknown initial state $z_0 \in \{1, 2, 3, 4\} = \mathcal{Z}^*$.

A shortest preset homing sequence for this automaton $\mathcal{A}$ is, for example, the sequence

$$V(0 \dots 2) = (v_1, v_1, v_2). \tag{4.5}$$

The output sequences $W(0 \dots 2) = \Phi(z_0, V(0 \dots 2))$ that the automaton $\mathcal{A}$ generates when this input sequence $V(0 \dots 2)$ is applied for different initial states $z_0 \in \mathcal{Z}^*$ are shown in Table 4.1. It can be seen that from the output of the automaton $\mathcal{A}$ its current state $z(3)$ can be reconstructed. For example, if the output sequence $W(0 \dots 2) = (w_1, w_1, w_1)$ is generated, the current state is known to be $z(3) = 1$. In this example also the initial state z_0 can be deduced from the observed output sequence, however, this is not a general property of homing sequences.

Initial state z_0	Output sequence $W(0 \dots 2)$	Current state $z(3)$
1	(w_1, w_1, w_1)	1
2	(w_2, w_1, w_1)	1
3	(w_1, w_1, w_2)	1
4	(w_2, w_2, w_2)	3

Table 4.1: Output sequences and current states corresponding to different initial states.

Figure 4.4 shows an *adaptive* homing sequence for the same automaton $\mathcal{A}$ with the same initial uncertainty $\mathcal{Z}^* = \{1, 2, 3, 4\}$. It shows that initially the input symbol $v(0) = v_1$ has to be applied, whereas the choice of the second input symbol $v(1) \in \{v_1, v_2\}$ depends on the previous output $w(0) \in \{w_2, w_1\}$ of the automaton $\mathcal{A}$.

Based on the adaptive homing sequence in Fig. 4.4, an adaptive homing experiment as described in

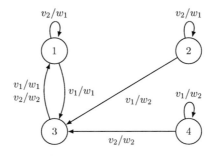

Figure 4.3: Automaton $\mathcal{A}$ with unknown initial state $z_0 \in \{1, 2, 3, 4\}$.

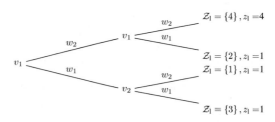

Figure 4.4: Adaptive homing sequence for automaton $\mathcal{A}$ in Fig. 4.3.

Algorithm 4.1 can be conducted. Suppose that the automaton $\mathcal{A}$ is in the unknown initial state

$$z(0) = z_0 = 1.$$

First, the input $v(0) = v_1$ from the root vertex of the adaptive homing sequence is applied to $\mathcal{A}$ (Line 1). Thereby, automaton $\mathcal{A}$ generates the output $w(0) = w_1$ and goes to state $z(1) = 3$. Consequently the next vertex x to be considered is the lower one in the first level of the adaptive homing sequence (Lines 3 and 4). Since the vertex x is no leaf, its input symbol v_2 is used as the next input $v(1)$ for the automaton $\mathcal{A}$ (Lines 5 and 8). As a result, automaton $\mathcal{A}$ generates the output $w(1) = w_2$ and goes back to state $z(2) = 1$. Based on this output, the next vertex in the adaptive homing sequence is the one labeled $\mathcal{Z}_1 = \{1\}, z_1 = 1$ (Lines 3 and 4). Since this vertex is a leaf, it is now known that the current state of the automaton $\mathcal{A}$ is given by $z_1 = 1$, which equals its true current state $z(2) = 1$.   □

4.2.2 Homing tree

In [53] it is described how to obtain a preset homing sequence from the so-called *homing tree* of the automaton $\mathcal{A} = (\mathcal{Z}, \mathcal{V}, \mathcal{W}, G, H, z_0)$. The homing tree can be constructed by analyzing the I/O behavior of the automaton $\mathcal{A}$. The vertices of the homing tree correspond to all possible input sequences $V(0 \ldots k_e) \in \mathcal{V}^\infty$ and contain all sets of possible current state of the automaton $\mathcal{A}$ for a given input sequence $V(0 \ldots k_e) \in \mathcal{V}^\infty$, where states are included in the same set if and only if they are reached with the same output sequence $W(0 \ldots k_e) \in \mathcal{W}^\infty$.

However, in this thesis a method for the construction of an *adaptive* homing sequence is required for which an extended version of the homing tree has to be defined. In the proposed extended homing tree branches are not only labeled by input symbols $v \in \mathcal{V}$, but by I/O-pairs $(v, w) \in \mathcal{V} \times \mathcal{W}$. That is, from the root vertex x in level $l = 0$ containing the initial uncertainty $\mathcal{Z}_x = \mathcal{Z}^*$, at most $p \cdot q$ branches emerge, which correspond to all possible I/O-pairs $(v, w) \in \mathcal{V} \times \mathcal{W}$. A branch labeled (v, w) leads to a new vertex x' containing a state set

$$\mathcal{Z}_{x'} = \left\{ z' \in \mathcal{Z} : (\exists z \in \mathcal{Z}_x) \left[z' = G(z, v) \right] \wedge \left[w = H(z, v) \right] \right\}. \tag{4.6}$$

From all resulting vertices again all branches labeled $(v, w) \in \mathcal{V} \times \mathcal{W}$ leading to new vertices x' with state sets according to (4.6) are added. Without reducing the information content, a new vertex x' containing the empty set $\mathcal{Z}_{x'} = \emptyset$ is not added to the extended homing tree.

There are two reasons for a vertex x in level l of the extended homing tree to become a leaf vertex:

1. the state set $\mathcal{Z}_x$ in the vertex x is already present in some vertex in a preceding level $l^- < l$

2. the vertex x is labeled by a singleton state set $\mathcal{Z}_x = \{z'\}$.

It follows that the state set $\mathcal{Z}_x$ of any vertex x in the extended homing tree contains all states that result from the states within the initial uncertainty $\mathcal{Z}^*$ while generating the output sequence $W_x(0 \ldots l-1)$ when the input sequence $V_x(0 \ldots l-1)$ is applied:

$$\mathcal{Z}_x = \left\{ z' \in \mathcal{Z} : (\exists z \in \mathcal{Z}^*) \left[z' = G^\infty(z, V_x(0 \ldots l-1)) \right] \right.$$
$$\left. \wedge \left[W_x(0 \ldots l-1) = \Phi(z, V_x(0 \ldots l-1)) \right] \right\}. \qquad (4.7)$$

The I/O-pair $(V_x(0 \ldots l-1), W_x(0 \ldots l-1))$ is the sequence of the I/O-pairs (v, w) at the branches of the path leading from the root vertex to the vertex x in the l-th level of the extended homing tree.

Example 4.1 (cont.)   *Construction of homing sequences*

Figure 4.5 shows the extended homing tree for automaton $\mathcal{A}$ in Fig. 4.3. From the root vertex labeled with the initial uncertainty $\mathcal{Z}^* = \{1, 2, 3, 4\}$ four branches emerge, which are labeled with all possible I/O-pairs $(v, w) \in \{v_1, v_2\} \times \{w_1, w_2\}$. For example, the edge labeled v_1/w_1 leads to a vertex labeled $\{1, 3\}$. As described by (4.6), this state set $\mathcal{Z}_{x'}$ is obtained, because states $z = 1$ and $z = 3$ generate the output $w = w_1$ when receiving the input $v = v_1$ and reach the states $z' = 1$ or $z' = 3$, respectively. The construction of the extended homing tree ends after level $l = 2$, because the state sets in all vertices in this level are either present in some preceding level (e.g., $\mathcal{Z}_x = \{1, 3\}$) or are singletons (e.g., $\mathcal{Z}_x = \{1\}$).

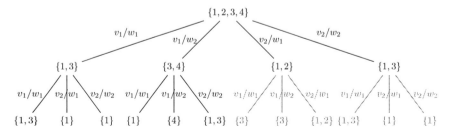

Figure 4.5: Extended homing tree for automaton $\mathcal{A}$ in Fig. 4.3.

□

4.2.3 Construction of homing sequences from a homing tree

An adaptive homing sequence as defined in Definition 4.2 can be constructed simultaneously with the extended homing tree. Whenever a new vertex x is added to the homing tree, it is examined for its contribution to the adaptive homing sequence. The idea for the construction of an adaptive homing sequence is to start with a linear graph which contains exactly one leaf and corresponds to the realization of an adaptive homing sequence for one state within the initial uncertainty $\mathcal{Z}^*$. Then, when realizations for other states within the initial uncertainty are found, new branches ending in new leaves are added to this tree in accordance with the properties of an adaptive homing sequence in Definition 4.2.

Whenever a vertex x which contains a singleton state set $\mathcal{Z}_x$ is added to the homing tree, a state set for which (4.3) and (4.4) in Definition 4.2 are fulfilled can be constructed. The following proposition shows how this state set results from the I/O-pair $(V_x(0 \ldots l-1), W_x(0 \ldots l-1))$ formed by the edge labels on the path from the root vertex to the vertex x in the homing tree.

Proposition 4.2 (Set corresponding to leaf in adaptive homing sequence). *Given a vertex x in an extended homing tree, whose corresponding state set $\mathcal{Z}_x = \{z'\}$ is a singleton, and an I/O-pair $(V_x(0 \ldots l-1), W_x(0 \ldots l-1))$ formed by the edge labels on the path from the root vertex to the vertex x in the extended homing tree, all states z from the set*

$$
\begin{aligned}
\mathcal{Z}_x^{\mathrm{pre}} = \Big\{ z \in \mathcal{Z}^* : \Big[& G\left(z, V_x(0 \ldots l-1)\right) \in \mathcal{Z}_x \Big] \\
& \wedge \Big[\Phi\left(z, V_x(0 \ldots l-1)\right) = W_x(0 \ldots l-1) \Big] \Big\}
\end{aligned}
\tag{4.8}
$$

of predecessors of states in $\mathcal{Z}_x$ fulfill (4.3) and (4.4) in Definition 4.2 with respect to the state $z_1 = z' \in \mathcal{Z}_x$ and the sequences $V(0 \ldots k_e) = V_x(0 \ldots l-1)$ and $W(0 \ldots k_e) = W_x(0 \ldots l-1)$.

Proof. If $\mathcal{Z}_x = \{z'\}$ in (4.7) is a singleton and the sequences $V_x(0 \ldots l-1)$ and $W_x(0 \ldots l-1)$ are used as the vertex and edge labels, respectively, to form a path in a tree, it follows directly from the definition of $\mathcal{Z}_x^{\mathrm{pre}}$ in (4.8) that (4.3) and (4.4) are fulfilled for all states $z \in \mathcal{Z}_x^{\mathrm{pre}}$, where $z_1 = z'$, $V(0 \ldots k_e) = V_x(0 \ldots l-1)$ and $W(0 \ldots k_e) = W_x(0 \ldots l-1)$. $\qquad \square$

Consequently, it is possible to associate the state set $\mathcal{Z}_1 = \mathcal{Z}_x^{\mathrm{pre}}$ in (4.8) and the state $z_1 \in \mathcal{Z}_x$ with a leaf in the tree corresponding to the adaptive homing sequence in Definition 4.2. By this procedure, realizations of an adaptive homing sequence corresponding any state $z \in \mathcal{Z}^*$ within the initial uncertainty $\mathcal{Z}^*$ can be found. However, in order to combine all of these realizations to an adaptive homing sequence as given in Definition 4.2, two input sequences $V_x(0 \ldots l-1)$ and $V_{\tilde{x}}(0 \ldots \tilde{l}-1)$ containing different symbols at some position k have to correspond to two output sequences $W_x(0 \ldots l-1)$ and $W_{\tilde{x}}(0 \ldots \tilde{l}-1)$ that differ before time k such that it is possible to select a unique next input for the automaton $\mathcal{A}$ at runtime.

Therefore, a set S of trees which are candidates for the adaptive safe homing sequence is constructed. Whenever a vertex x whose state set Z_x is a singleton is defined in the extended homing tree, a new tree is added to the set S. This new tree is a linear graph whose internal vertices correspond to the input symbols in the sequence $V_x(0 \ldots l-1)$ and whose leaf is labeled with the set $Z_1 = Z_x^{\text{pre}}$ in (4.8) and the state $z_1 = z' \in Z_x$. Its edges are labeled with the output symbols from the sequence $W_x(0 \ldots l-1)$.

Additionally, where possible, the new tree is united with every other tree in S and the resulting tree is again added to S. Two trees can be combined whenever their root vertices contain the same input symbol and for all coinciding edges the input symbols in the labels of the reached vertices are again the same. Thereby it is guaranteed that the resulting tree fulfills the third property in Definition 4.2. The resulting tree contains all vertices and edges of the two original trees, where identical vertices and edges are only present once.

An adaptive homing sequence has been found, when the sets of state pairs in the leaves in one of the trees contain all states $z \in \tilde{Z}^*$ within the initial uncertainty as their first element, which is proved in the following proposition.

Proposition 4.3 (Adaptive homing sequence). *A tree, which is constructed by the method described above and whose leafs contain state sets Z_1 for which*

$$Z^* = \bigcup_{\text{all leaves}} Z_1 \tag{4.9}$$

holds, is an adaptive homing sequence for the automaton A with respect to the initial uncertainty $Z^* \subseteq Z$ according to Definition 4.2.

Proof. The fulfillment of the first condition for an adaptive homing sequence follows directly from the construction of the trees. The fulfillment of the second condition follows from the fact that leaves contain the state sets $Z_x^{\text{pre}} \subseteq Z^*$ and the states $z_1 \in Z_x \subset Z$ and it is claimed that the sets Z_1 in the leaves of the given tree contain all states $z \in Z^*$. The third condition for an adaptive homing sequence is obviously always fulfilled for new trees given in form of linear graphs, but is also preserved when two trees are united according to the procedure described above. The fourth condition for an adaptive homing sequence is fulfilled according to Proposition 4.2. $\square$

The construction of an adaptive homing sequence based on an extended homing tree is summarized in Algorithm 4.2.

It can be shown that $\frac{1}{2}n(n-1)$ is an upper bound for the length of a homing sequence and hence for the number of levels in a homing tree corresponding to an automaton A with $|Z| = n$ states. Each level $l = 0, \ldots, \frac{1}{2}n(n-1)$ of a homing tree contains p^l vertices. Therefore, the complexity of the construction of the homing tree is of order $\mathcal{O}(p^{n^2})$. Consequently, for plants with a large number of states the construction of the homing tree is not practical.

Algorithm 4.2: Construction of an adaptive homing sequence.

Given: Extended homing tree for automaton $\mathcal{A}$ with respect to initial uncertainty $\mathcal{Z}^* \subseteq \mathcal{Z}$.
Initialize: Set $\mathcal{S} = \emptyset$ of candidates for the adaptive homing sequence

1 **for** *levels* $l = 0, 1, \ldots$ *in the extended homing tree*
2   **forall** *vertices* x *in level* l *of the extended homing tree*
3   **if** $\mathcal{Z}_x$ *corresponding to* x *is a singleton* **then**
4   Read I/O-pair $(V_x(0 \ldots l - 1), W_x(0 \ldots l - 1))$ from edge labels on path from root vertex to vertex x in homing tree.
5   Add new tree T_{new} in form of linear graph with internal vertices labeled by input symbols in $V_x(0 \ldots l - 1)$, edges labeled by output symbols in $W_x(0 \ldots l - 1)$ and leaf labeled by set $\mathcal{Z}_1 = \mathcal{Z}_x^{\text{pre}}$ of states in (4.8) and state $z_1 \in \mathcal{Z}_x$ to set $\mathcal{S}$.
6   **forall** *trees* T *in* $\mathcal{S} \setminus \{T_{\text{new}}\}$
7   **if** *root vertices of* T_{new} *and* T *contain same input symbol and input symbols in vertices reached by coinciding edges are the same* **then**
8   Combine T_{new} and T to a single tree and add it to set $\mathcal{S}$.
9   **end**
10   **if** *in the state sets* $\mathcal{Z}_1$ *of the leaves of one tree* $T \in \mathcal{S}$ *all states* $z \in \mathcal{Z}^*$ *are contained* **then**
11   Adaptive homing sequence is given by T.
12   Goto Line 17.
13   **end**
14   **end**
15   **end**
16   **end**
17 **end**

Result: Adaptive homing sequence for $\mathcal{A}$ with respect to $\mathcal{Z}^*$.

Example 4.1 (cont.) *Construction of homing sequences*

Consider the extended homing tree in Fig. 4.5. One vertex x containing a singleton state set is the second leftmost one in the level $l = 2$ with $\mathcal{Z}_x = \{1\}$. The corresponding I/O-pair is given by

$$(V_x(0 \ldots l - 1), W_x(0 \ldots l - 1)) = ((v_1, v_2), (w_1, w_1)).$$

According to (4.8), the set of predecessors for this vertex results to

$$\mathcal{Z}_x^{\text{pre}} = \{3\}.$$

That is, only from state $z = 3$ the output sequence $W_x(0 \ldots 1) = (w_2, w_2)$ is generated when the input sequence $V_x(0 \ldots 1) = (v_2, v_2)$ is received.

When this vertex x is added to the extended homing tree, the topmost linear graph in Fig. 4.6 is included into the set $\mathcal{S}$. The second vertex with a singleton state set in the extended homing tree in Fig. 4.5 corresponds to the second linear graph in Fig. 4.6. The lowermost graph shows the combination of the two trees. The procedure of defining new vertices in the extended homing tree and adding trees to

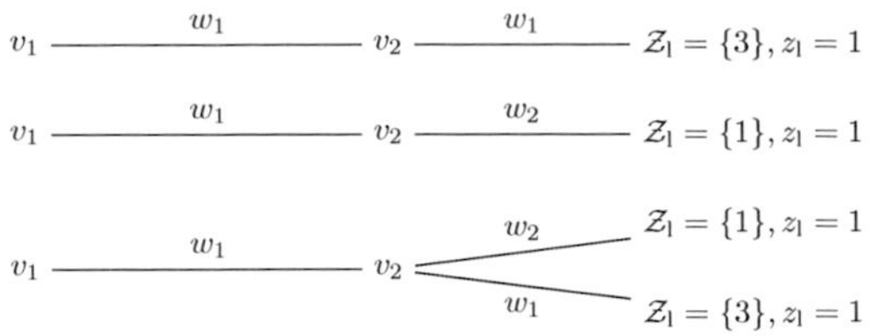

$$v_1 \xrightarrow{\quad w_1 \quad} v_2 \xrightarrow{\quad w_1 \quad} \mathcal{Z}_1 = \{3\}, z_1 = 1$$

$$v_1 \xrightarrow{\quad w_1 \quad} v_2 \xrightarrow{\quad w_2 \quad} \mathcal{Z}_1 = \{1\}, z_1 = 1$$

$$v_1 \xrightarrow{\quad w_1 \quad} v_2 \begin{array}{l} \xrightarrow{\quad w_2 \quad} \mathcal{Z}_1 = \{1\}, z_1 = 1 \\ \xrightarrow{\quad w_1 \quad} \mathcal{Z}_1 = \{3\}, z_1 = 1 \end{array}$$

Figure 4.6: Trees within set $\mathcal{S}$ after two vertices with singleton sets have been found.

the set $\mathcal{S}$ is continued until the adaptive homing sequence in Fig. 4.4 is found. At that time the extended homing tree consists of the black part in Fig. 4.5.   □

4.3 Diagnosability analysis

Diagnosability with respect to a fault set $\mathcal{F}$ is a property of the plant $\mathcal{P}$ that combines the ability to detect and to identify the present fault from the I/O-pair $(V_p(0 \ldots k), W_p(0 \ldots k))$ of the plant $\mathcal{P}$ in the closed-loop system $\mathcal{P} \circ \mathcal{C}_T(\mathcal{A}_0)$.

4.3.1 Detectability of a fault

Detectability is the ability to decide from the I/O-pair $(V_p(0 \ldots k), W_p(0 \ldots k))$ of the plant $\mathcal{P}$ whether or not a fault occurred. Of course the detectability depends on the input sequence $V_p(0 \ldots k_e)$ to the plant $\mathcal{P}$, because a fault that has no impact on the behavior of the plant in its current operating point can not be detected. Therefore, detectability is defined as follows.

Definition 4.3 (Detectability). *A fault $f \in \mathcal{F}$ is said to be detectable in the closed-loop system $\mathcal{P} \circ \mathcal{C}_T(\mathcal{A}_0)$ with respect to the reference trajectory $R(0 \ldots k_e)$ generated by the trajectory planning unit $T(\mathcal{A}_0)$ if it is possible to decide from the I/O-pair $(V_p(0 \ldots k), W_p(0 \ldots k))$ of the plant $\mathcal{P}$ whether a fault has occurred or not.*

Detectability is given whenever the fault $f \in \mathcal{F}$ changes the I/O-behavior of the plant $\mathcal{P}$ in its current operating point compared to the faultless case, which leads to the following condition.

Proposition 4.4 (Detectability). *Let $V_p(0 \ldots k_e) = \Phi_c(z_{c0}, R(0 \ldots k_e))$ be the input sequence for the plant $\mathcal{P}$ generated by the nominal tracking controller $\mathcal{C}_T(\mathcal{A}_0)$. Let $z_p(k_f)$ be the state of the plant $\mathcal{P}$ at the fault occurrence time $k_f \geq 0$. Then a fault $f \in \mathcal{F}$ is detectable in the closed-loop system $\mathcal{P} \circ \mathcal{C}_T(\mathcal{A}_0)$ with respect to the reference trajectory $R(0 \ldots k_e)$ generated by the trajectory planning unit $T(\mathcal{A}_0)$ if and only if there exists a time $k_f^* \leq k_e$ such that*

$$\Phi_f(z_p(k_f), V_p(k_f \ldots k_f^*)) \neq \Phi_0(z_p(k_f), V_p(k_f \ldots k_f^*)). \tag{4.10}$$

Proof. If, as stated by (4.10), the fault $f \in \mathcal{F}$ changes the output sequence generated by the plant $\mathcal{P}$ compared to the faultless case, it is obvious that a fault occurred whenever a changed output sequence is observed. At the same time, it is known that whenever the output sequence expected from the faultless plant is observed, no fault occurred. Conversely, if the fault $f \in \mathcal{F}$ is detectable, it has to change the output sequence of the plant $\mathcal{P}$ such that a decision whether a fault occurred or not is possible. □

There are two possible reasons for a fault $f \in \mathcal{F}$ to be not detectable. First, the fault might have no impact on the behavior of the plant $\mathcal{P}$ in its current operating point. In this case it is acceptable for the fault to remain undiscovered. Second, however, the fault might change the behavior of the plant $\mathcal{P}$, but not its output sequence $W_\mathrm{p}(k_\mathrm{f} \ldots k_\mathrm{e})$ compared to the faultless case. In this case the missing detectability of the fault will lead to a non-fulfillment of the control aim (2.44).

Example 4.2 *Fault diagnosis of automated warehouse*

Consider the automated warehouse described in Section 2.5 together with its nominal tracking controller $\mathcal{C}_T(\mathcal{A}_0)$ from Example 3.1. As shown in (3.22), the plant $\mathcal{P}$ receives the input $v_\mathrm{p}(0) = AB$ from the nominal tracking controller $\mathcal{C}_T(\mathcal{A}_0)$ in order to reach the desired final state $z_\mathrm{F} = B$. For the fault occurrence time $k_\mathrm{f} = 0$, both faults $f \in \{1, 2\}$ are detectable within the closed-loop system $\mathcal{P} \circ \mathcal{C}_T(\mathcal{A}_0)$ according to Definition 4.4, because

$$\Phi_1(z_\mathrm{p}(0) = A, AB) = \mathrm{BkA} \neq \Phi_0(z_\mathrm{p}(0) = A, AB) = \mathrm{ok} \tag{4.11}$$

and

$$\Phi_2(z_\mathrm{p}(0) = A, AB) = \mathrm{BkA} \neq \Phi_0(z_\mathrm{p}(0) = A, AB) = \mathrm{ok}. \tag{4.12}$$

That is, if fault $f = 1$ or fault $f = 2$ occurs at time $k_\mathrm{f} = 0$, the plant $\mathcal{P}$ does not generate the expected output $w_\mathrm{p}(0) = \mathrm{ok}$ that the transport of the parcel was successful, but states that a blocking at position A occurs ($w_\mathrm{p}(0) = \mathrm{BkA}$) such that both faults are detectable with $k_\mathrm{f}^* = 0 \leq k_\mathrm{e} = 0$.

For an arbitrary fault occurrence time k_f the fault $f = 1$ is only detectable if a transport from position A to position B is requested afterwards, while fault $f = 2$ is detectable whenever any transport is requested, because the controller $\mathcal{C}(\mathcal{A}_0)$ in Fig. 3.2 always tries to use the faulty robot M. □

4.3.2 Identifiability of a fault

Identifiability is the ability to decide from the inputs $V_\mathrm{p}(k_\mathrm{f} \ldots k)$ and outputs $W_\mathrm{p}(k_\mathrm{f} \ldots k)$ of the faulty plant $\mathcal{P}$ which fault $f \in \mathcal{F}$ occurred. Like detectability, identifiability obviously depends on the input sequence $V_\mathrm{p}(k_\mathrm{f} \ldots k_\mathrm{e})$ that the plant receives after the fault occurrence time k_f as well. Since the fault occurrence time k_f is usually unknown, the fault detection time k_f^* is used in the following identifiability definition. Compared to usual identifiability definitions it is extended, because it also requires the identification of the current state $z_\mathrm{p}(k)$ of the faulty plant $\mathcal{P} \vDash \mathcal{A}_f$ (cf. Section 4.1).

Definition 4.4 (Identifiability). *A fault $f \in \mathcal{F}$ is said to be identifiable in the closed-loop system $\mathcal{P} \circ \mathcal{C}_\mathcal{T}(\mathcal{A}_0)$ with respect to the reference trajectory $R(0 \ldots k_e)$ generated by the trajectory planning unit $\mathcal{T}(\mathcal{A}_0)$ if it is possible to decide from the I/O-pair $(V_p(k_f^* \ldots k), W_p(k_f^* \ldots k))$ of the plant $\mathcal{P}$ whether the fault f has occurred or not and which is the current state $z_p(k)$ of the faulty plant.*

Identifiability of a fault $f \in \mathcal{F}$ is given, when it can be distinguished from all other faults $\bar{f} \in \mathcal{F} \setminus \{f\}$. That is, the output sequences that the faulty plant generates when receiving the input sequence $V_p(k_f^* \ldots k_e)$ after the fault detection time k_f^* have to differ from each other. In passive diagnosis the input sequence $V_p(k_f^* + 1 \ldots k_e)$ is generated by the nominal tracking controller $\mathcal{C}_\mathcal{T}(\mathcal{A}_0)$. In contrast, in active diagnosis this input sequence is generated by the controller $\mathcal{C}(\mathcal{A}_0)$, which receives a diagnostic trajectory $R(k_f^* + 1 \ldots k_e)$ from the diagnostic unit $\mathcal{D}$.

First, a criterion for identifiability with passive diagnosis is stated. It relies on the fact that if the input sequence for the plant $\mathcal{P}$ generated by the nominal tracking controller $\mathcal{C}_\mathcal{T}(\mathcal{A}_0)$ is a preset homing sequence (cf. Definition 4.1) for the overall model $\mathcal{A}_\Delta$ of the faulty plant in (2.42), its state $z_\Delta(k)$ will eventually be known such that the current state $z_p(k)$ of the faulty plant $\mathcal{P}$ and the present fault $\bar{f}$ are identified (cf. (2.43)).

Proposition 4.5 (Identifiability with passive diagnosis). *Let $z_c(k_f^*) = G_c^\infty(z_{c0}, R(0 \ldots k_f^* - 1))$ be the state of the controller $\mathcal{C}(\mathcal{A}_0)$ at the fault detection time k_f^*. Then all faults $f \in \mathcal{F}$ are identifiable with passive diagnosis in the closed-loop system $\mathcal{P} \circ \mathcal{C}_\mathcal{T}(\mathcal{A}_0)$ with respect to the reference trajectory $R(0 \ldots k_e)$ generated by the trajectory planning unit $\mathcal{T}(\mathcal{A}_0)$ if and only if the input sequence*

$$V_p(k_f^* \ldots k_e) = \Phi_c(z_c(k_f^*), R(k_f^* \ldots k_e)) \tag{4.13}$$

is a preset homing sequence for the overall model $\mathcal{A}_\Delta$ of the faulty plant in (2.42) with respect to the initial uncertainty $\mathcal{Z}^* = \mathcal{Z}_\Delta$.

Proof. First, the "if"-direction is proved. If the input sequence $V_p(k_f^* \ldots k_e)$ in (4.13) that the controller $\mathcal{C}(\mathcal{A}_0)$ generates for the plant $\mathcal{P}$ after the fault detection time k_f^* is a preset homing sequence for the overall model $\mathcal{A}_\Delta$ of the faulty plant with respect to the initial uncertainty $\mathcal{Z}^* = \mathcal{Z}_\Delta$, the state z_Δ of the overall model $\mathcal{A}_\Delta$ of the faulty plant will be known unambiguously at time $k_e + 1$. Since the state of the overall model $\mathcal{A}_\Delta$ contains an information about the current state of the plant and about the present fault (cf. (2.43)), according to Definition 4.4 all faults $f \in \mathcal{F}$ are identifiable using passive diagnosis in the closed-loop system $\mathcal{P} \circ \mathcal{C}_\mathcal{T}(\mathcal{A}_0)$.

Now, the "only if"-direction is proved. According to Definition 4.4, if all faults $f \in \mathcal{F}$ are identifiable in the closed-loop system $\mathcal{P} \circ \mathcal{C}_\mathcal{T}(\mathcal{A}_0)$, from the models $\{\mathcal{A}_f, (f \in \mathcal{F})\}$ of the faulty plant and the inputs $V_p(k_f^* \ldots k_e)$ and outputs $W_p(k_f^* \ldots k_e)$ of the controlled plant the present fault $\bar{f}$ and the current state $z_p(k)$ of the plant can be identified. Recall from (2.42) that the overall model $\mathcal{A}_\Delta$ of the faulty plant is a combination of all models $\{\mathcal{A}_f, (f \in \mathcal{F})\}$ of the faulty plant. That is, the current state $z_\Delta(k)$ of the overall model $\mathcal{A}_\Delta$ of the faulty plant can be identified (cf. (2.43)), which is only possible if for all other possible current states

$$G_\Delta(z_\Delta, V_p(k_f^* \ldots k_e)) \neq z_\Delta(k), \ (z_\Delta \in \mathcal{Z}_\Delta)$$

different output sequences

$$W_p(k_f^* \ldots k_e) = H_\Delta(z_\Delta(k_f^*), V_p(k_f^* \ldots k_e)) \neq H_\Delta(z_\Delta, V_p(k_f^* \ldots k_e))$$

are generated. The current state $z_\Delta(k_f^*)$ of the overall model $\mathcal{A}_\Delta$ of the faulty plant at the fault detection time k_f^* can be any state within the state set $\mathcal{Z}_\Delta$. Therefore, the input sequence $V_p(k_f^* \ldots k_e)$ is a preset homing sequence for the overall model $\mathcal{A}_\Delta$ of the faulty plant with respect to the initial uncertainty $\mathcal{Z}^* = \mathcal{Z}_\Delta$ (see Definition 4.1). $\qquad \square$

When the model of the controller $\mathcal{C}(\mathcal{A}_0) \models \mathcal{A}_c$ in (3.7) and the overall model $\mathcal{A}_\Delta$ of the faulty plant in (2.42) are combined in a series connection, the *overall model $\bar{\mathcal{A}}_\Delta$ of the closed-loop system* is obtained as follows:

$$\bar{\mathcal{A}}_\Delta : \begin{cases} \bar{\mathcal{Z}}_\Delta = \mathcal{Z}_c \times \mathcal{Z}_\Delta, & \text{(4.14a)} \\[2mm] \bar{\mathcal{V}}_\Delta = \mathcal{V}_c, & \text{(4.14b)} \\[2mm] \bar{\mathcal{W}}_\Delta = \mathcal{W}_\Delta, & \text{(4.14c)} \\[2mm] \bar{G}(\bar{z}_\Delta, \bar{v}_\Delta) = \begin{cases} \begin{pmatrix} G_c(\bar{z}_{\Delta 1}, \bar{v}_\Delta) \\ G_\Delta(\bar{z}_{\Delta 2}, H_c(\bar{z}_{\Delta 1}, \bar{v}_\Delta)) \end{pmatrix} & \text{if } G_c(\bar{z}_{\Delta 1}, \bar{v}_\Delta)!, \\ \text{undefined} & \text{otherwise,} \end{cases} & \text{(4.14d)} \\[6mm] \bar{H}(\bar{z}_\Delta, \bar{v}_\Delta) = \begin{cases} H_\Delta(\bar{z}_{\Delta 2}, H_c(\bar{z}_{\Delta 1}, \bar{v}_\Delta)) & \text{if } H_c(\bar{z}_{\Delta 1}, \bar{v}_\Delta)!, \\ \text{undefined} & \text{otherwise.} \end{cases} & \text{(4.14e)} \end{cases}$$

All states

$$\bar{z}_\Delta = \begin{pmatrix} z_c & z_\Delta \end{pmatrix}^\mathsf{T} \in \bar{\mathcal{Z}}_\Delta \qquad (4.15)$$

of $\bar{\mathcal{A}}_\Delta$ are vectors whose first element is a state $z_c \in \mathcal{Z}_c$ of the controller $\mathcal{C}(\mathcal{A}_0)$ and whose second element is a state $z_\Delta = \begin{pmatrix} z & f \end{pmatrix}^\mathsf{T} \in \mathcal{Z}_\Delta$ of the overall model $\mathcal{A}_\Delta$ of the faulty plant. Since the overall model $\mathcal{A}_\Delta$ of the faulty plant is completely defined (Assumption 2), an input sequence $\bar{V}(0 \ldots k_e) \in \bar{\mathcal{V}}_\Delta^\infty$ is acceptable to a state $\bar{z}_\Delta \in \bar{\mathcal{Z}}_\Delta$ in (4.15) if and only if it is acceptable to the state z_c in the nominal controller $\mathcal{C}(\mathcal{A}_0)$.

The overall model $\bar{A}_\Delta$ of the closed-loop system contains

$$\bar{n}_\Delta = |\mathcal{Z}_c| \cdot |\mathcal{Z}_\Delta| = n \cdot n_\Delta = n^2 \cdot F \tag{4.16}$$

states (cf. (2.41)), where $n = |\mathcal{Z}_0|$ is the number of states in the model $\mathcal{A}_0$ of the faultless plant and $F = |\mathcal{F}|$ is the number of faults.

Based on the overall model $\bar{A}_\Delta$ of the closed-loop system in (4.14), the identifiability criterion in Proposition 4.5 can be restated as follows. Note that the identifiability criteria differ with respect to the used models, but always yield the same result.

Proposition 4.6 (Identifiability with passive diagnosis based on model of closed-loop system).
Let $z_c(k_f^*) = G_c^\infty(z_{c0}, R(0 \ldots k_f^* - 1))$ be the state of the controller $\mathcal{C}(\mathcal{A}_0)$ at the fault detection time k_f^*. Then all faults $f \in \mathcal{F}$ are identifiable with passive diagnosis in the closed-loop system $\mathcal{P} \circ \mathcal{C}_\mathcal{T}(\mathcal{A}_0)$ with respect to the reference trajectory $R(0 \ldots k_e)$ generated by the trajectory planning unit $\mathcal{T}(\mathcal{A}_0)$ if and only if the reference trajectory $R(k_f^* \ldots k_e)$ after the fault detection time k_f^* is a preset homing sequence for the overall model $\bar{A}_\Delta$ of the closed-loop system in (4.14) with respect to the initial uncertainty $\mathcal{Z}^* = \{z_c(k_f^*)\} \times \mathcal{Z}_\Delta \subset \bar{\mathcal{Z}}_\Delta$.

Proof. First it has to be shown that $R(k_f^* \ldots k_e)$ is an acceptable input sequence for all states $\bar{z}_\Delta$ within the initial uncertainty $\mathcal{Z}^*$. By construction, the reference trajectory $R(k_f^* \ldots k_e)$ is acceptable to the state $z_c(k_f^*)$ of the controller $\mathcal{C}(\mathcal{A}_0)$. Since the considered initial uncertainty $\mathcal{Z}^* = \{z_c(k_f^*)\} \times \mathcal{Z}_\Delta$ contains only states $\bar{z}_\Delta$ with identical first element $z_c(k_f^*)$, it follows that $R(k_f^* \ldots k_e)$ is acceptable to all states $\bar{z}_\Delta \in \mathcal{Z}^*$.

From Proposition 4.5 and the definition of the overall model $\bar{A}_\Delta$ of the closed-loop system in (4.14) the proof therefore follows directly. $\qquad\square$

Now, identifiability with active diagnosis is considered. Therefore, the identifiability criterion with passive diagnosis in Proposition 4.6 has to be modified such that the reference trajectory $R(k_f^* + 1 \ldots k_e)$ after the detection of a fault is not fixed any longer, but becomes a free parameter. Consequently, identifiability with passive diagnosis implies identifiability with active diagnosis but not the other way around.

The value $r(k_f^*)$ of the reference trajectory at the fault detection time k_f^* has already been applied to the faulty plant such that it is fixed but unknown during the identifiability analysis. Therefore, the criterion for identifiability with active diagnosis is given as follows.

Proposition 4.7 (Identifiability with active diagnosis). *Let $\tilde{\mathcal{V}}_c(z_c) = \{v_c \in \mathcal{V}_c : G_c(z_c, v_c)!\}$ be the set of all possible values of the reference trajectory at the fault detection time. Then all faults $f \in \mathcal{F}$ are identifiable with active diagnosis in the closed-loop system $\mathcal{P} \circ \mathcal{C}_T(\mathcal{A}_0)$ if and only if for the overall model $\bar{A}_\Delta$ of the closed-loop system in (4.14) there exists an adaptive homing sequence with first input symbol v_c with respect to any initial uncertainty $\mathcal{Z}^* = \{z_c\} \times \mathcal{Z}_\Delta, (z_c \in \mathcal{Z}_c)$ and every $v_c \in \tilde{\mathcal{V}}_c(z_c)$.*

Proof. If there exists an adaptive homing sequence for $\bar{A}_\Delta$ with respect to every initial uncertainty $\mathcal{Z}^* = \{z_c\} \times \mathcal{Z}_\Delta, (z_c \in \mathcal{Z}_c)$ and every input symbol $v_c \in \mathcal{V}_c$ for which $G_c(z_c, v_c)$ is defined, an adaptive homing experiment as described in Algorithm 4.1 can be performed once a fault is detected, where the first input symbol of the adaptive homing sequence has already been applied. By definition, during the adaptive homing experiment only acceptable input symbols are used.

According to Proposition 4.1 the adaptive homing experiment reveals the current state $\bar{z}_\Delta(k)$ of the overall model $\bar{A}_\Delta$ of the closed-loop system whenever its state

$$\bar{z}_\Delta(k_f^*) = \left(z_c(k_f^*) \quad \boldsymbol{z}_\Delta(k_f^*) \right)^\top$$

at the begin of the experiment lies in the initial uncertainty $\mathcal{Z}^* = \{z_c\} \times \mathcal{Z}_\Delta, (z_c \in \mathcal{Z}_c)$. That is, since the current state $z_c(k)$ of the controller $\mathcal{C}$ can always be reconstructed from its output sequence $W_c(0 \ldots k - 1)$, which is the measured input sequence $V_p(0 \ldots k - 1)$ to the plant $\mathcal{P}$ (Proposition 3.4), the present fault $\bar{f}$ and the current state $z_p(k)$ of the plant $\mathcal{P}$ can be identified. Therefore, all faults $f \in \mathcal{F}$ are identifiable with active diagnosis in the closed-loop system $\mathcal{P} \circ \mathcal{C}_T(\mathcal{A}_0)$ according to Definition 4.4.

If with respect to one state $z_c \in \mathcal{Z}_c$ of the controller $\mathcal{C}(\mathcal{A}_0)$ no adaptive homing sequence exists, the state $\bar{z}_\Delta(k)$ of the overall model $\bar{A}_\Delta$ of the closed-loop system can not be identified such that either the present fault $\bar{f}$ or the current state $z_p(k)$ of the plant $\mathcal{P}$ (or both) remain unknown. Therefore, some faults $f \in \mathcal{F}$ are not identifiable with active diagnosis in this case, which concludes the proof. $\qquad\square$

It is possible to state the following sufficient condition for identifiability of all faults $f \in \mathcal{F}$ with active diagnosis, which gives an easily provable sufficient condition for the existence of a homing sequence, namely the absence of compatible states.

Proposition 4.8 (Sufficient condition for identifiability with active diagnosis). *All faults $f \in \mathcal{F}$ are identifiable with active diagnosis in the closed-loop system $\mathcal{P} \circ \mathcal{C}_T(\mathcal{A}_0)$ with respect to any reference trajectory $R(0 \ldots k_e)$ generated by the trajectory planning unit $T(\mathcal{A}_0)$ if there are no compatible states in $\{z_c\} \times \mathcal{Z}_\Delta \subset \bar{\mathcal{Z}}_\Delta$ for any $z_c \in \mathcal{Z}_c$.*

Proof. It is well-known that every completely defined automaton without equivalent states has an adaptive homing sequence, because it is always possible to find a separating sequence that distinguishes two states in the current uncertainty and hence removes at least one state from the current uncertainty in every step until it becomes a singleton set (cf. [53]).

The argumentation for incompletely defined automata like the overall model $\bar{A}_\Delta$ of the closed-loop system in (4.14) is similar with the difference that it has to be guaranteed that the separating sequences are acceptable to all states in the current uncertainty. This condition is always fulfilled, because the state $z_c(k)$ of the controller $C(A_0)$ can always be reconstructed (Proposition 3.4) such that all states in any current uncertainty have an identical first entry. $\square$

In order to test the identifiability criterion in Proposition 4.8, it has to be checked, whether there are compatible states within the state sets $\mathcal{Z}^* = \{z_c\} \times \mathcal{Z}_\Delta \subset \bar{\mathcal{Z}}_\Delta$ for any $z_c \in \mathcal{Z}_c$. The test is based on a method in [64], which tests a completely defined automaton $\mathcal{A}$ for equivalent states. This method is slightly adapted here such that the state sets $\{z_c\} \times \mathcal{Z}_\Delta \subset \bar{\mathcal{Z}}_\Delta$, $(z_c \in \mathcal{Z}_c)$ in the overall model $\bar{A}_\Delta$ of the closed-loop system can be tested for compatible states. In [64] the state set $\mathcal{Z}$ of the automaton $\mathcal{A}$ is partitioned into sets of equivalent states based on a recursive equivalence definition. That is, the state set $\mathcal{Z}$ is successively partitioned into sets $\mathcal{Z}_i^k$, $(i = 1, \ldots, m_k)$ of k-equivalent states for $k = 0, \ldots, n - 2$, where n is the cardinality of the state set $\mathcal{Z}$. Two states z_1 and z_2 are called k-equivalent if they generate the same output sequence for all input sequences $V(0 \ldots l) \in \mathcal{V}^\infty$ with $l \leq k$.

Since every state in the set $\mathcal{Z}^*$ accepts the same input sequences $V(0 \ldots k_e) \in \bar{\mathcal{V}}^\infty$, it is possible to extend the compatibility notion introduced in Section 2.2.4 from pairs of states to sets of states here. Therefore, the procedure described above can be used in order to partition each state set $\{z_c\} \times \mathcal{Z}_\Delta, (z_c \in \mathcal{Z}_c)$ into sets $\mathcal{Z}_i^k$, $(k = 0, \ldots, \bar{n}_\Delta - 2, i = 1, \ldots, m_k)$ of k-compatible states, with the only difference that only acceptable input sequences are considered for the partitioning.

By Proposition 3.5 it is guaranteed that at least one acceptable input symbol exists in every partitioning step. As a result for each state $z_c \in \mathcal{Z}_c$ a set of state sets is obtained, where each state set contains only compatible states. Hence, if all of these state sets are singletons, there are no compatible states in any state set $\{z_c\} \times \mathcal{Z}_\Delta, (z_c \in \mathcal{Z}_c)$ and, according to Proposition 4.8, there exists a homing sequence for the overall model $\bar{A}_\Delta$ of the closed-loop system in (4.14) with respect to any initial uncertainty $\mathcal{Z}^* \subseteq \{z_c\} \times \mathcal{Z}_\Delta, (z_c \in \mathcal{Z}_c)$.

However, the condition in Proposition 4.8 is only sufficient, but not necessary for the existence of the adaptive homing sequences. For every pair $(\bar{z}_1, \bar{z}_2) \in \bar{\mathcal{Z}}_\Delta$ of compatible states there might exist a preset homing sequence $V_H(0 \ldots k_e)$ such that they still can not be distinguished based on the generated output sequences

$$\bar{\Phi}_\Delta(\bar{z}_1, V_H(0 \ldots k_e)) = \bar{\Phi}_\Delta(\bar{z}_2, V_H(0 \ldots k_e)),$$

but they reach the same final state

$$\bar{G}_\Delta^\infty(\bar{z}_1, V_H(0 \ldots k_e)) = \bar{G}_\Delta^\infty(\bar{z}_2, V_H(0 \ldots k_e))$$

such that (4.2) is fulfilled.

An example for this effect is shown in Fig. 4.7. States $z = 1$ and $z = 2$ are compatible, because the only acceptable input $v = v_1$ leads to the same output $w = w_1$ and the same next state $z' = 2$. However, since with this input the next state is always known, the input $v_H(0) = v_1$ is a preset homing sequence with respect to the initial uncertainty $\mathcal{Z}^* = \{1, 2\}$.

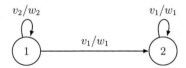

Figure 4.7: Automaton $\mathcal{A}$ with unknown initial state $z_0 \in \{1, 2\}$.

4.3.3 Diagnosability with passive diagnosis

Diagnosability combines the ability to detect and identify a present fault in the plant $\mathcal{P}$. For the case that passive diagnosis shall be used, the reference trajectory $R(0 \ldots k_e)$ that the controller $\mathcal{C}(\mathcal{A}_0)$ receives is generated by the trajectory planning unit $\mathcal{T}(\mathcal{A}_0)$ within the nominal tracking controller $\mathcal{C}_T(\mathcal{A}_0)$. Hence, diagnosability with passive diagnosis has to be analyzed with respect to a given reference trajectory $R(0 \ldots k_e)$.

Definition 4.5 (Diagnosability with passive diagnosis). *The plant $\mathcal{P}$ in the closed-loop system $\mathcal{P} \circ \mathcal{C}_T(\mathcal{A}_0)$ is said to be diagnosable using passive diagnosis with respect to a fault set $\mathcal{F}$ and the reference trajectory $R(0 \ldots k_e)$ generated by the trajectory planning unit $\mathcal{T}(\mathcal{A}_0)$ if every fault $f \in \mathcal{F}$ is detectable and identifiable with passive diagnosis.*

Based on this definition, the following criterion for the diagnosability with passive diagnosis can be stated. It results from combining the detectability condition in Proposition 4.4 and the identifiability condition in Proposition 4.6.

Theorem 4.1 (Diagnosability with passive diagnosis). *Let $z_p(k_f)$ be the state of the plant $\mathcal{P}$ at the fault occurrence time $k_f \geq 0$. Let $V_p(0 \ldots k_e) = \Phi_c(z_{c0}, R(0 \ldots k_e))$ be the input sequence for the plant $\mathcal{P}$ generated by the nominal tracking controller $\mathcal{C}_T(\mathcal{A}_0)$.*

Then the plant $\mathcal{P}$ in the closed-loop system $\mathcal{P} \circ \mathcal{C}_T(\mathcal{A}_0)$ is diagnosable with passive diagnosis with respect to the fault set $\mathcal{F}$ and the reference trajectory $R(0 \ldots k_e)$ generated by the trajectory planning unit $\mathcal{T}(\mathcal{A}_0)$ if and only if the following conditions are fulfilled:

- *(Detectability) For all faults $f \in \mathcal{F}$ there exists a time $k_f^* \leq k_e$ such that*

$$\Phi_f(z_p(k_f), V_p(k_f \ldots k_f^*)) \neq \Phi_0(z_p(k_f), V_p(k_f \ldots k_f^*)). \tag{4.17}$$

- *(Identifiability) The reference trajectory $R(k_f^* \ldots k_e)$ after the fault detection time k_f^* is a preset homing sequence for the overall model $\bar{A}_\Delta$ of the closed-loop system in (4.14) with respect to the initial uncertainty $\mathcal{Z}^* = \{z_c(k_f^*)\} \times \mathcal{Z}_\Delta \subset \bar{\mathcal{Z}}_\Delta$.*

Proof. The first condition is necessary and sufficient for the detectability of all faults $f \in \mathcal{F}$ (Proposition 4.4), while the second condition is necessary and sufficient for the identifiability of all faults with passive diagnosis (Proposition 4.6). Consequently, the fulfillment of both conditions at a time is necessary and sufficient for diagnosability of the plant $\mathcal{P}$ with passive diagnosis according to Definition 4.5. □

Example 4.2 (cont.) *Fault diagnosis of automated warehouse*

The automated warehouse in the closed-loop system with its nominal tracking controller $\mathcal{C}_T(\mathcal{A}_0)$ in Example 3.1 is not diagnosable with passive diagnosis according to Theorem 4.1, because the reference input $r(0) = B$ is not a preset homing sequence for the overall model $\bar{A}_\Delta$ of the closed-loop system. This can be seen, because the input sequence $v_p(0) = AB$ in (3.22) that the tracking controller $\mathcal{C}_T(\mathcal{A}_0)$ generates for the plant $\mathcal{P}$ is not a preset homing sequence for the overall model $\mathcal{A}_\Delta$ of the faulty plant in Fig. 2.7 with respect to the initial uncertainty

$$\mathcal{Z}^* = \mathcal{Z}_\Delta = \left\{ \begin{pmatrix} A \\ 1 \end{pmatrix}, \begin{pmatrix} B \\ 1 \end{pmatrix}, \begin{pmatrix} C \\ 1 \end{pmatrix}, \begin{pmatrix} A \\ 2 \end{pmatrix}, \begin{pmatrix} B \\ 2 \end{pmatrix}, \begin{pmatrix} C \\ 2 \end{pmatrix} \right\}.$$

For example,

$$G_\Delta \left(\begin{pmatrix} A \\ 1 \end{pmatrix}, AB \right) = \begin{pmatrix} A \\ 1 \end{pmatrix} \neq G_\Delta \left(\begin{pmatrix} A \\ 2 \end{pmatrix}, AB \right) = \begin{pmatrix} A \\ 2 \end{pmatrix},$$

$$\text{but } \Phi_\Delta \left(\begin{pmatrix} A \\ 1 \end{pmatrix}, AB \right) = \text{BkA} = \Phi_\Delta \left(\begin{pmatrix} A \\ 2 \end{pmatrix}, AB \right).$$

Hence, for states $\begin{pmatrix} A & 1 \end{pmatrix}^\top, \begin{pmatrix} A & 2 \end{pmatrix}^\top \in \mathcal{Z}^*$ and input $V_H(0 \ldots k_e) = AB$ (4.2) is violated. □

4.3.4 Diagnosability with active diagnosis

In this section, diagnosability with active diagnosis is considered, which is defined similarly to diagnosability with passive diagnosis in Definition 4.5 as follows.

Definition 4.6 (Diagnosability with active diagnosis). *The plant $\mathcal{P}$ in the closed-loop system $\mathcal{P} \circ \mathcal{C}_T(\mathcal{A}_0)$ is said to be diagnosable using active diagnosis with respect to a fault set $\mathcal{F}$ if every fault $f \in \mathcal{F}$ is detectable and identifiable with active diagnosis.*

Combining the detectability condition in Proposition 4.4 and the identifiability condition in Proposition 4.7, the following criterion for diagnosability with active diagnosis results.

Theorem 4.2 (Diagnosability with active diagnosis). *Let $z_p(k_f)$ be the state of the plant $\mathcal{P}$ at the fault occurrence time $k_f \geq 0$. Let $V_p(0 \dots k_e) = \Phi_c(z_{c0}, R(0 \dots k_e))$ be the input sequence for the plant $\mathcal{P}$ generated by the nominal tracking controller $\mathcal{C}_T(\mathcal{A}_0)$. Let $\check{\mathcal{V}}_c(z_c) = \{v_c \in \mathcal{V}_c : G_c(z_c, v_c)!\}$.*

Then the plant $\mathcal{P}$ in the closed-loop system $\mathcal{P} \circ \mathcal{C}_T(\mathcal{A}_0)$ is diagnosable with passive diagnosis with respect to the fault set $\mathcal{F}$ and the reference trajectory $R(0 \dots k_e)$ generated by the trajectory planning unit $\mathcal{T}(\mathcal{A}_0)$ if and only if the following conditions are fulfilled:

- *(Detectability) For all faults $f \in \mathcal{F}$ there exists a time $k_f^* \leq k_e$ such that*

$$\Phi_f(z_p(k_f), V_p(k_f \dots k_f^*)) \neq \Phi_0(z_p(k_f), V_p(k_f \dots k_f^*)). \tag{4.18}$$

- *(Identifiability) For the overall model $\bar{A}_\Delta$ of the closed-loop system in (4.14) there exists an adaptive homing sequence with first input symbol v_c with respect to any initial uncertainty $\mathcal{Z}^* = \{z_c\} \times \mathcal{Z}_\Delta, (z_c \in \mathcal{Z}_c)$ and every $v_c \in \check{\mathcal{V}}_c(z_c)$.*

Proof. The first condition is necessary and sufficient for the detectability of all faults $f \in \mathcal{F}$ (Proposition 4.4), while the second condition is necessary and sufficient for the identifiability of all faults with active diagnosis (Proposition 4.7). Consequently, the fulfillment of both conditions at a time is necessary and sufficient for diagnosability of the plant $\mathcal{P}$ with active diagnosis according to Definition 4.6. □

Of course the identifiability can also be analyzed based on the sufficient condition in Proposition 4.8.

Remark. In this thesis the active diagnosis setting in Fig. 4.1b is considered, in which the diagnostic unit $\mathcal{D}$ generates inputs $v_c(k)$ for the controller $\mathcal{C}(\mathcal{A}_0)$. Therefore, the inputs $v_p(k)$ for the plant $\mathcal{P}$ are restricted by the possible outputs $w_c(k)$ of the controller $\mathcal{C}(\mathcal{A}_0)$ in its current state $z_c(k)$:

$$v_p(k) \in \{w_c \in \mathcal{W}_c : (\exists v_c \in \mathcal{V}_c) \, H_c(z_c(k), v_c) = w_c\}.$$

If the diagnostic unit $\mathcal{D}$ had the possibility to generate the inputs $v_p(k)$ for the plant $\mathcal{P}$ directly, this restriction was overcome and any input symbol $v \in \mathcal{V}_0$ could be applied to the plant $\mathcal{P}$ at any time step k. Therefore, in this setting the plant $\mathcal{P}$ might be diagnosable with active diagnosis even when it is not diagnosable with active diagnosis in the setting including the controller $\mathcal{C}(\mathcal{A}_0)$. The chosen structure reflects the usual technical realization in which no external unit but a controller has direct access of the plant.

Example 4.2 (cont.)   *Fault diagnosis of automated warehouse*

In the following it is shown that the automated warehouse in the closed-loop system with its nominal tracking controller $\mathcal{C}_T(\mathcal{A}_0)$ from Example 3.1 is diagnosable with active diagnosis according to Theorem 4.2. It has already been proved that all faults are detectable. According to Proposition 4.8, identifiability with active diagnosis is given if there are no compatible states within any of the state sets $\{z_c\} \times \mathcal{Z}_\Delta \subset \bar{\mathcal{Z}}_\Delta, (z_c \in \mathcal{Z}_c)$ in the overall model $\bar{A}_\Delta$ of the closed-loop system defined in (4.14). A part of the overall model $\bar{A}_\Delta$ of the closed-loop system for the automated warehouse consisting of the controller $\mathcal{C}(\mathcal{A}_0)$ in Fig. 3.2 and the overall model $\mathcal{A}_\Delta$ of the faulty plant in Fig. 2.7 is shown in Fig. 4.8.

The compatibility partitions of the automated warehouse are shown in Appendix B. It can be seen that there are two pairs of compatible states:

$$\begin{pmatrix} A \\ C \\ 1 \end{pmatrix} \sim \begin{pmatrix} B \\ C \\ 1 \end{pmatrix} \text{ and } \begin{pmatrix} A \\ C \\ 2 \end{pmatrix} \sim \begin{pmatrix} B \\ C \\ 2 \end{pmatrix}.$$

However, in both of these pairs the first elements of the vectors, corresponding to the state z_c of the controller, differ. That is, there are no compatible states within any of the state sets $\{z_c\} \times \mathcal{Z}_\Delta$, $(z_c \in \mathcal{Z}_c)$. Therefore, the automated warehouse is diagnosable with active diagnosis.   □

4.4 General approach for the fault diagnosis

For a diagnosable plant $\mathcal{P}$, the fault diagnosis is performed in two steps, fault detection and fault identification. Both rely on an estimate of the current state $z_p(k)$ of the plant $\mathcal{P}$ based on its I/O-pair $(V_p(0 \ldots k-1), W_p(0 \ldots k-1))$ and will be described in the following.

4.4.1 Fault detection

The aim of fault detection is to decide whether or not a fault occurred. Therefore, the I/O-pairs $(V_p(0 \ldots k), W_p(0 \ldots k))$ of the plant $\mathcal{P}$ in the closed-loop system with the nominal tracking controller $\mathcal{C}_T(\mathcal{A}_0)$ are checked for consistency with the model $\mathcal{A}_0$ of the faultless plant using

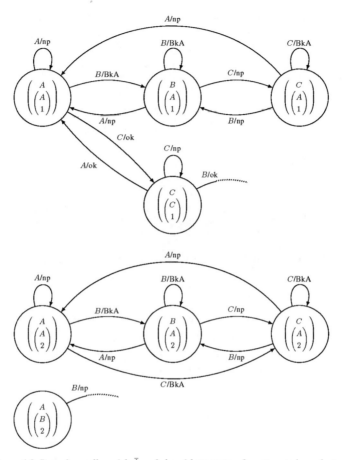

Figure 4.8: Part of overall model $\bar{A}_\Delta$ of closed-loop system for automated warehouse.

(2.23). In order to allow for an incremental update of (2.23) at every time step $k > 0$, the *current state estimate* $z^*(k)$ *of the plant* $\mathcal{P}$ is introduced.

Recall that the initial state $z_\mathrm{p}(0) = z_{00}$ of the plant $\mathcal{P}$ is assumed to be known (Assumption 3). Therefore, the current state estimate of the plant is initialized as

$$z^*(0) = z_{00}. \tag{4.19}$$

At every time step k, the measured I/O-pair $(v_\mathrm{p}(k), w_\mathrm{p}(k))$ of the plant is checked for consistency with the model $\mathcal{A}_0$ of the faultless plant based on the current state estimate $z^*(k)$ by

$$w_{\mathrm{p}}(k) \overset{?}{=} H_0(z^*(k), v_{\mathrm{p}}(k)). \tag{4.20}$$

Based on this equation, the current state estimate $z^*(k)$ is updated by

$$z^*(k+1) = \begin{cases} G_0(z^*(k), v_{\mathrm{p}}(k)) & \text{if } w_{\mathrm{p}}(k) = H_0(z^*(k), v_{\mathrm{p}}(k)), \\ \text{undefined} & \text{if } w_{\mathrm{p}}(k) \neq H_0(z^*(k), v_{\mathrm{p}}(k)). \end{cases} \tag{4.21}$$

A fault is detected at the *fault detection time* k_{f}^* when the current state estimate $z^*(k_{\mathrm{f}}^* + 1)$ becomes undefined, hence as soon as the measured I/O-pair $(V_{\mathrm{p}}(0 \ldots k_{\mathrm{f}}^*), W_{\mathrm{p}}(0 \ldots k_{\mathrm{f}}^*))$ becomes inconsistent with the model $\mathcal{A}_0$:

$$W_{\mathrm{p}}(0 \ldots k_{\mathrm{f}}^*) \neq \Phi_0(z_{00}, V_{\mathrm{p}}(0 \ldots k_{\mathrm{f}}^*)). \tag{4.22}$$

In general, the fault detection time k_{f}^* is not equal to the fault occurrence time k_{f}:

$$k_{\mathrm{f}}^* \geq k_{\mathrm{f}}. \tag{4.23}$$

A delay between fault occurrence and fault detection occurs if the present fault $\bar{f}$ does not change the next output $w_{\mathrm{p}}(k_{\mathrm{f}})$ of the plant $\mathcal{P}$, but only some later one $k_{\mathrm{f}}^* \geq k_{\mathrm{f}}$. For example, this is the case when the fault affects a part of the plant that is not used until a later time step.

The left part of Fig. 4.10 summarizes the described fault detection method. The following proposition proves that this fault detection method detects every fault that is detectable according to Definition 4.3.

Proposition 4.9 (Fault detection). *Using the fault detection method in the left part of Fig. 4.10, every fault $f \in \mathcal{F}$ that is detectable in the closed-loop system $\mathcal{P} \circ \mathcal{C}_T(\mathcal{A}_0)$ is detected at a fault detection time $k_{\mathrm{f}}^* \leq k_{\mathrm{e}}$.*

Proof. If a fault $f \in \mathcal{F}$ occurs at the plant $\mathcal{P}$ at the fault occurrence time k_{f}, the behavior of the plant after time k_{f} is described by the automaton $\mathcal{A}_f$. Therefore, the output sequence of the plant up to time k, $(k \geq k_{\mathrm{f}})$ is given by the concatenation

$$W_{\mathrm{p}}(0 \ldots k) = \Phi_0(z_{00}, V_{\mathrm{p}}(0 \ldots k_{\mathrm{f}} - 1)) \cdot \Phi_f(z_{\mathrm{p}}(k_{\mathrm{f}}), V_{\mathrm{p}}(k_{\mathrm{f}} \ldots k)). \tag{4.24}$$

Since it is assumed that the current state $z_{\mathrm{p}}(k)$ of the plant $\mathcal{P}$ is not affected by the occurrence of a fault until a transition is performed, the plant state at time k_{f} is given by

$$z_{\mathrm{p}}(k_{\mathrm{f}}) = G_0^\infty(z_{00}, V_{\mathrm{p}}(0 \ldots k_{\mathrm{f}} - 1)).$$

According to (4.10), if the fault f is detectable, the last part $\Phi_f(z_p(k_f), V_p(k_f \ldots k))$ of the output sequence $W_p(0 \ldots k)$ in (4.24) differs from the output $\Phi_0(z_p(k_f), V_p(k_f \ldots k))$ of the model $\mathcal{A}_0$ of the faultless plant for $k = k_e$.

Therefore, for some time $k_f^* \in [k_f, \ldots, k_e]$, (4.20) is violated, such that the current state estimate $z^*(k)$ in (4.21) becomes the empty set and the fault is detected at a fault detection time $k_f^* \leq k_e$. $\qquad\qquad\square$

4.4.2 Fault identification

After the detection of the fault, during the fault identification, the current state estimate of the plant $\mathcal{P}$ becomes a set $\mathcal{Z}^*(k) \subseteq \mathcal{Z}_\Delta$, $k \geq k_f^*$ that contains all states from the overall model $\mathcal{A}_\Delta$ of the faulty plant in (2.42) with which the measured I/O-pairs $(V_p(k_f^* \ldots k), W_p(k_f^* \ldots k))$ of the plant are consistent.

At the fault occurrence time k_f the model of the plant changes from $\mathcal{A}_0$ to $\mathcal{A}_{\bar{f}}$, $(\bar{f} \in \mathcal{F})$, but the current state $z_p(k_f)$ of the plant remains unchanged until a transition is performed (cf. Section 2.3.5). Consequently, the current state estimate of the plant at the fault detection time k_f^* under the assumption that the fault is immediately detected (i.e., $k_f^* = k_f$) is given by

$$\mathcal{Z}^*(k_f^* \mid k_f^* = k_f) = \left\{ z_\Delta \in \mathcal{Z}_\Delta : z_\Delta = \begin{pmatrix} z^*(k_f^*) \\ f \end{pmatrix}, f \in \mathcal{F} \right\}. \qquad (4.25)$$

However, as stated in (4.23), it is not guaranteed that the fault detection time k_f^* equals the fault occurrence time k_f. If, for example, the fault occurred at time $k_f = k_f^* - 1$, the current state estimate had to be changed to

$$\mathcal{Z}^*(k_f^* \mid k_f^* = k_f + 1) = \left\{ z_\Delta \in \mathcal{Z}_\Delta : z_\Delta = G_\Delta \left(\begin{pmatrix} z^*(k_f^* - 1) \\ f \end{pmatrix}, v_p(k_f^* - 1) \right), f \in \mathcal{F} \right\}. \qquad (4.26)$$

In the same way the current state estimate of the plant can be computed for all possible fault occurrence times $k_f = k_f^*, \ldots, 0$.

To take the missing information about the fault occurrence time k_f into account, the current state estimate $\mathcal{Z}^*(k_f^*)$ of the plant at the fault detection time k_f^* has to be the union of all current state estimates $\mathcal{Z}^*(k_f^* \mid k_f)$ for all possible fault occurrence times $k_f = 0, \ldots, k_f^*$:

$$\mathcal{Z}^*(k_f^*) = \bigcup_{k_f = 0 \ldots k_f^*} \mathcal{Z}^*(k_f^* \mid k_f). \qquad (4.27)$$

When the fault detection time k_{f}^{*} becomes large, for example because the plant runs fault-free for a very long time before a fault occurs, the evaluation of (4.27) becomes a tedious task. In addition, it can be expected that $\mathcal{Z}^{*}(k_{\mathrm{f}}^{*})$ computed according to (4.27) approaches $\mathcal{Z}_{\Delta}$ when k_{f}^{*} grows.

Therefore, instead of using (4.27), at the fault detection time k_{f}^{*} the current state estimate of the plant is initialized by the entire state set of the overall model $\mathcal{A}_{\Delta}$ of the faulty plant:

$$\mathcal{Z}^{*}(k_{\mathrm{f}}^{*}) = \mathcal{Z}_{\Delta}. \tag{4.28}$$

Then, similar as during fault detection, for every measured I/O-pair $(v_{\mathrm{p}}(k), w_{\mathrm{p}}(k))$, $(k \geq k_{\mathrm{f}}^{*})$, of the plant $\mathcal{P}$ the current state estimate of the plant is updated by the following formula:

$$\mathcal{Z}^{*}(k+1) = \Big\{ z_{\Delta}' \in \mathcal{Z}_{\Delta} : (\exists z_{\Delta} \in \mathcal{Z}^{*}(k))$$
$$\Big[z_{\Delta}' = G_{\Delta}(z_{\Delta}, v_{\mathrm{p}}(k)) \Big] \wedge \Big[w_{\mathrm{p}}(k) = H_{\Delta}(z_{\Delta}, v_{\mathrm{p}}(k)) \Big] \Big\}. \tag{4.29}$$

The following proposition proves that the current state estimate $\mathcal{Z}^{*}(k)$, $(k \geq k_{\mathrm{f}}^{*})$, of the plant iteratively constructed according to (4.29) is indeed correct with respect to the previously measured I/O-pair $(V_{\mathrm{p}}(k_{\mathrm{f}}^{*} \ldots k-1), W_{\mathrm{p}}(k_{\mathrm{f}}^{*} \ldots k-1))$ of the plant $\mathcal{P}$.

Proposition 4.10 (Correctness of current state estimate). *The current state estimate $\mathcal{Z}^{*}(k)$, $(k \geq k_{\mathrm{f}}^{*})$ of the plant contains all states $z_{\Delta} \in \mathcal{Z}_{\Delta}$ of the overall model $\mathcal{A}_{\Delta}$ of the faulty plant with which the measured I/O-pair $(V_{\mathrm{p}}(k_{\mathrm{f}}^{*} \ldots k-1), W_{\mathrm{p}}(k_{\mathrm{f}}^{*} \ldots k-1))$ of the plant $\mathcal{P}$ is consistent.*

Proof. (By induction) According to the consistency definition in (2.24), it needs to be shown for all $k \geq k_{\mathrm{f}}^{*}$ that

$$\mathcal{Z}^{*}(k) = \Big\{ z_{\Delta}' \in \mathcal{Z}_{\Delta} : (\exists z_{0} \in \mathcal{Z}_{\Delta}) \Big[z_{\Delta}' = G_{\Delta}^{\infty}(z_{0}, V_{\mathrm{p}}(k_{\mathrm{f}}^{*} \ldots k-1)) \Big]$$
$$\wedge \Big[W_{\mathrm{p}}(k_{\mathrm{f}}^{*} \ldots k-1) = \Phi_{\Delta}(z_{0}, V_{\mathrm{p}}(k_{\mathrm{f}}^{*} \ldots k-1)) \Big] \Big\}. \tag{4.30}$$

For $k = k_{\mathrm{f}}^{*}$ based on (4.28) the above equation is obviously fulfilled. Given that the equation is fulfilled for some k, for $k+1$ the equivalence of (4.29) and (4.30) can be shown as follows. From the definition of the extended state transition function G^{∞} in (2.10) and the automaton map Φ in (2.8), it follows directly that

$$G_{\Delta}^{\infty}(z_{0}, V_{\mathrm{p}}(k_{\mathrm{f}}^{*} \ldots k)) = G_{\Delta}(G_{\Delta}^{\infty}(z_{0}, V_{\mathrm{p}}(k_{\mathrm{f}}^{*} \ldots k-1)), v_{\mathrm{p}}(k))$$

and the output sequences are concatenated by

$$\Phi_\Delta(z_0, V_p(k_f^* \ldots k)) = \Phi_\Delta(z_0, V_p(k_f^* \ldots k-1)) \cdot \Phi_\Delta(G_\Delta^\infty(z_0, V_p(k_f^* \ldots k-1)), v_p(k))$$
$$= \Phi_\Delta(z_0, V_p(k_f^* \ldots k-1)) \cdot H_\Delta(G_\Delta^\infty(z_0, V_p(k_f^* \ldots k-1)), v_p(k))$$

in (4.30). When substituting $G_\Delta^\infty(z_0, V_p(k_f^* \ldots k-1))$ by z_Δ, the following equations result:

$$G_\Delta^\infty(z_0, V_p(k_f^* \ldots k)) = G_\Delta(z_\Delta, v_p(k)),$$
$$\Phi_\Delta(z_0, V_p(k_f^* \ldots k)) = \Phi_\Delta(z_0, V_p(k_f^* \ldots k-1)) \cdot H_\Delta(z_\Delta, v_p(k)).$$

If z_Δ is included in $\mathcal{Z}^*(k)$, $\Phi_\Delta(z_0, V_p(k_f^* \ldots k-1)) = W_p(k_f^* \ldots k-1)$ holds true. Therefore, given that $z_\Delta \in \mathcal{Z}^*(k)$, it can be seen that, the state z_Δ' is included into the set $\mathcal{Z}^*(k+1)$, both, according to (4.29) and (4.30) if and only if the following two equations are fulfilled:

$$z_\Delta' = G_\Delta(z_\Delta, v_p(k)),$$
$$w_p(k) = H_\Delta(z_\Delta, v_p(k)).$$

Consequently (4.30) is fulfilled for all $k \geq k_f^*$. $\qquad\square$

In the considered case of deterministic I/O automata, the size of the current state estimate $\mathcal{Z}^*(k)$ of the plant is monotonically decreasing as proved in the following proposition.

Proposition 4.11 (Monotony of current state estimate). *The size of the current state estimate* $\mathcal{Z}^*(k)$ *of the plant is monotonically decreasing, that is*

$$|\mathcal{Z}^*(k+1)| \leq |\mathcal{Z}^*(k)|, \quad \forall k \geq k_f^*. \tag{4.31}$$

Proof. For deterministic I/O automata the state transition function G in (2.2) either returns exactly one next state (if it is defined for the given current state and input symbol) or no next state at all (if it is not defined for the given current state and input symbol). Therefore, from (4.29) the above equation follows directly. $\qquad\square$

The time at which the current state estimate of the plant becomes a singleton is called *diagnosis time* k_d. Proposition 4.11 shows that the size of the current state estimate $\mathcal{Z}^*(k)$ of the plant never increases. Therefore, it remains a singleton after the diagnosis time k_d:

$$|\mathcal{Z}^*(k)| = 1, \quad \forall k \geq k_d.$$

The current state estimate $\mathcal{Z}^*(k)$ of the plant never becomes empty ($|\mathcal{Z}^*(k)| = 0$), which would mean that no fault could be identified, because Assumption 3 states that all possible faults are known and correctly described by the models $\mathcal{A}_f$, $(f \in \mathcal{F})$.

The main steps of the fault identification are summarized in the right part of Fig. 4.10.

Current state estimate of the closed-loop system. When additionally the current state $z_c(k)$ of the controller $\mathcal{C}(\mathcal{A}_0)$ is taken into account, the overall model $\bar{\mathcal{A}}_\Delta$ of the closed-loop system in (4.14) instead of the overall model $\mathcal{A}_\Delta$ of the faulty plant is applicable. Then the *current state estimate* $\bar{\mathcal{Z}}^*(k)$ *of the closed-loop system* is defined based on the current state estimate $\mathcal{Z}^*(k)$ of the plant in (4.29) as follows:

$$\bar{\mathcal{Z}}^*(k) = \{z_c(k)\} \times \mathcal{Z}^*(k). \tag{4.32}$$

Therefore, the current state estimate of the closed-loop system at the fault detection time k_f^* is given by

$$\bar{\mathcal{Z}}^*(k_f^*) = \{z_c(k_f^*)\} \times \mathcal{Z}^*(k_f^*) = \{z_c(k_f^*)\} \times \mathcal{Z}_\Delta. \tag{4.33}$$

According to Proposition 3.4, the state $z_c(k)$ of the controller $\mathcal{C}(\mathcal{A}_0)$ can always be reconstructed based on the input sequence $V_p(0 \ldots k-1)$ to the plant $\mathcal{P}$, which can be measured. It can be updated iteratively by

$$z_c(k+1) = G_c(z_c(k), \Phi_c^{-1}(z_c(k), v_p(k))). \tag{4.34}$$

Set of fault candidates. The *set* $\mathcal{F}^*(k)$ *of fault candidates* can be deduced from the current state estimate $\mathcal{Z}^*(k)$ of the plant as follows:

$$\mathcal{F}^*(k) = \left\{ f \in \mathcal{F} : (\exists z \in \mathcal{Z}_f) \begin{pmatrix} z & f \end{pmatrix}^\top \in \mathcal{Z}^*(k) \right\}. \tag{4.35}$$

For the set $\mathcal{F}^*(k)$ of fault candidates the monotony property is even stronger than the one for the current state estimate $\mathcal{Z}^*(k)$ of the plant in Proposition 4.11, because it is not only applicable to the size of the set $\mathcal{F}^*(k)$, but to the set $\mathcal{F}^*(k)$ of fault candidates itself as shown in the following proposition.

Proposition 4.12 (Monotony of set of fault candidates). *For every time* $k \geq k_f^*$ *the following relation for the set of fault candidates holds:*

$$\mathcal{F}^*(k+1) \subseteq \mathcal{F}^*(k). \tag{4.36}$$

Proof. The above relation holds if and only if

$$f \notin \mathcal{F}^*(k) \Rightarrow f \notin \mathcal{F}^*(k+1).$$

This is true, because in the overall model $\mathcal{A}_\Delta$ of the faulty plant in (2.42) there are no transitions between states of different automata $\mathcal{A}_f$, $(f \in \mathcal{F})$ (cf. (2.42d)). Consequently, if all states from an automaton $\mathcal{A}_f$, $(f \in \mathcal{F})$ are excluded from the current state estimate $\mathcal{Z}^*(k)$ of the plant at time k such that the fault f is not contained in the set $\mathcal{F}^*(k)$ of fault candidates, no states from the automaton $\mathcal{A}_f$ will occur in the current state estimate $\mathcal{Z}^*(k+1)$ of the plant. Hence, also the fault f remains excluded from the set $\mathcal{F}^*(k+1)$ of fault candidates. $\qquad\square$

Proposition 4.12 implies that the set $\mathcal{F}^*(k)$ of fault candidates is a monotone sequence

$$\mathcal{F}^*(0) \supseteq \mathcal{F}^*(1) \supseteq \cdots \supseteq \mathcal{F}^*(k). \tag{4.37}$$

A fault $\bar{f} \in \mathcal{F}$ is identified if the set of fault candidates is the singleton $\mathcal{F}(k) = \{\bar{f}\}$.

The evolution of the set $\mathcal{F}^*(k)$ of fault candidates over time is illustrated in Fig. 4.9. At the fault detection time k_f^* it contains all possible faults $f \in \mathcal{F}$, while afterwards it is reduced until it becomes a singleton before or at the diagnosis time k_d.

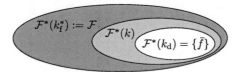

Figure 4.9: Evolution of set $\mathcal{F}^*(k)$ of fault candidates for $k_f^* \leq k \leq k_d$.

4.4.3 Fault diagnosis flowchart

The interaction of the fault detection method in Section 4.4.1 and the fault identification method in Section 4.4.2 are summarized in the flowchart in Fig. 4.10. The fault diagnosis starts with the initialization of the current state estimate $z^*(0)$ of the plant for the fault detection in the upper left part of the picture. This current state estimate is updated in every time step k until it becomes undefined such that a fault is detected.

Then the fault identification, visualized in the right part of the picture, is started with the initialization of the current state estimate $\mathcal{Z}^*(k_f^*)$ of the plant. The current state estimate of the plant is updated in every time step $k \geq k_f^*$ until its size becomes one, such that the present fault $\bar{f} \in \mathcal{F}$ and the current state $z_p(k_d)$ of the plant are identified. The blue blocks labeled "Obtain

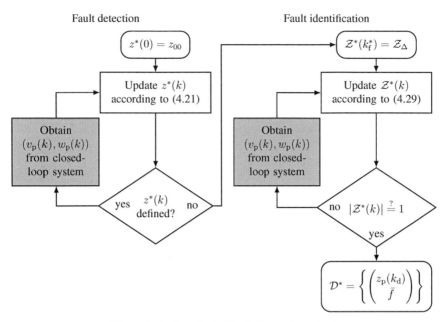

Figure 4.10: Flowchart of fault diagnosis method.

$(v_{\mathrm{p}}(k), w_{\mathrm{p}}(k))$ from closed-loop system" represent the fact that a new input symbol $v_{\mathrm{p}}(k)$ is applied to the plant $\mathcal{P}$ and its output symbol $w_{\mathrm{p}}(k)$ is observed.

In case of passive fault diagnosis the input $v_{\mathrm{p}}(k)$ to the plant $\mathcal{P}$ is generated by the nominal tracking controller $\mathcal{C}_{\mathcal{T}}(\mathcal{A}_0)$. However, in the case of active diagnosis, the diagnostic unit $\mathcal{D}$ has to generate a diagnostic input $r(k)$ for the nominal controller $\mathcal{C}(\mathcal{A}_0)$ such that the controller $\mathcal{C}(\mathcal{A}_0)$ generates the input $v_{\mathrm{p}}(k)$ for the plant $\mathcal{P}$. A method on how to choose this diagnostic input $r(k)$ for every time step $k > k_{\mathrm{f}}^*$ will be described in the next section.

Example 4.2 (cont.)   *Fault diagnosis of automated warehouse*

Consider the automated warehouse, in which the robot M breaks down (see Fig. 2.4), hence fault $f = 2$ occurred and let the fault occurrence time be $k_{\mathrm{f}} = 0$. The warehouse is controlled by the nominal tracking controller $\mathcal{C}_{\mathcal{T}}(\mathcal{A}_0)$ from Example 3.1. The fault diagnosis is performed as summarized in the flowchart in Fig. 4.10.

First, it has to be detected whether a fault occurred or not (left part of Fig. 4.10). Therefore, the current state estimate of the plant for the fault detection in (4.19) is initialized to

$$z^*(0) = z_{00} = A.$$

When the first I/O-pair

$$(v_p(0), w_p(0)) = (AB, BkA) \tag{4.38}$$

of the plant $\mathcal{P}$ arrives at the diagnostic unit $\mathcal{D}$, it is checked for consistency with the model $\mathcal{A}_0$ of the faultless plant in Fig. 2.5 by (4.21). Since

$$w_p(0) = BkA \neq H_0(z^*(0), v_p(0)) = H_0(A, AB) = ok,$$

the current state estimate $z^*(1)$ becomes undefined and a fault is detected at the fault detection time $k_f^* = 0$, which equals the fault occurrence time $k_f = 0$ in this example.

Consequently, the fault identification (right part of Fig. 4.10) starts with the initialization of the current state estimate of the faulty plant according to (4.28):

$$\mathcal{Z}^*(k_f^*) = \mathcal{Z}_\Delta = \left\{ \begin{pmatrix} A \\ 1 \end{pmatrix}, \begin{pmatrix} B \\ 1 \end{pmatrix}, \begin{pmatrix} C \\ 1 \end{pmatrix}, \begin{pmatrix} A \\ 2 \end{pmatrix}, \begin{pmatrix} B \\ 2 \end{pmatrix}, \begin{pmatrix} C \\ 2 \end{pmatrix} \right\}. \tag{4.39}$$

Afterwards, the current state estimate of the faulty plant is updated according to (4.29) using again the I/O-pair in (4.38):

$$\mathcal{Z}^*(1) = \left\{ \begin{pmatrix} A \\ 1 \end{pmatrix}, \begin{pmatrix} A \\ 2 \end{pmatrix} \right\}. \tag{4.40}$$

According to (4.35), the set of fault candidates is

$$\mathcal{F}^*(1) = \{1, 2\}. \tag{4.41}$$

That is, at time $k = 0$ it is known that the current state of the plant $\mathcal{P}$ is $z_p(1) = A$, but the present fault $\bar{f}$ remains unknown. In order to identify the present fault $\bar{f}$, the fault identification has to be continued. Since it has already been shown that the plant $\mathcal{P}$ is not diagnosable using passive diagnosis, active diagnosis has to be used. The example will be continued after the active fault diagnosis method has been introduced.

Avoiding illegal transitions. If the set $\mathcal{E}_{ill}$ of illegal transitions in (2.54) has to be avoided by the plant $\mathcal{P}$, the tracking controller $\mathcal{C}_T(\mathcal{A}_0)$ with trajectory planning unit $\mathcal{T}(\mathcal{A}_0, \mathcal{E}_{ill})$ and controller $\mathcal{C}(\mathcal{A}_0)$ in Fig. 3.2 is used. In this case, the first I/O-pair of the plant is given by

$$(v_p(0), w_p(0)) = (AC, BkA) \tag{4.42}$$

(cf. (3.23)). Again, a fault is detected at the fault detection time $k_f^* = 0$ based on (4.20), because

$$w_p(0) = BkA \neq H_0(z^*(0), v_p(0)) = H_0(A, AC) = ok.$$

When the current state estimate $\mathcal{Z}^*(k_f^*)$ of the faulty plant in (4.39) is updated according to (4.29) based on the I/O-pair in (4.42), the following state set results:

$$\mathcal{Z}^*(1) = \left\{ \begin{pmatrix} A \\ 2 \end{pmatrix} \right\}. \tag{4.43}$$

That is, within this setting, the present fault $\bar{f} = 2$ and the current state $z_p(1) = A$ of the plant $\mathcal{P}$ are already identified at time $k_d = 1$ without any further fault diagnosis. $\qquad\square$

4.5 Generation of input sequences for active diagnosis

In this section the case that active diagnosis is used in order to identify the current state $z_p(k_d)$ of the plant and the present fault $\bar{f} \in \mathcal{F}$ is considered (cf. Fig. 4.1b). More specifically, it is described how the diagnostic trajectory $R(k_f^* + 1 \ldots k_e)$ that the diagnostic unit $\mathcal{D}$ generates for the controller $\mathcal{C}(\mathcal{A}_0)$ in the nominal tracking controller $\mathcal{C}_T(\mathcal{A}_0)$ is chosen after a fault has been detected.

4.5.1 Main idea for the generation of a homing sequence

From the criterion for diagnosability with active diagnosis in Theorem 4.2 it can be deduced that the value $r(k_f^*)$ of the reference trajectory at the fault detection time k_f^* concatenated with the diagnostic trajectory $R(k_f^* + 1 \ldots k_e)$ for the controller $\mathcal{C}(\mathcal{A}_0)$ should be a realization of an adaptive homing sequence for the overall model $\bar{A}_\Delta$ of the closed-loop system in (4.14) with respect to the initial uncertainty $\mathcal{Z}^* = \bar{\mathcal{Z}}^*(k_f^*)$ in (4.33) and the (unknown) current state

$$
\bar{z}_\Delta(k_f^*) = \begin{pmatrix} z_c(k_f^*) \\ \begin{pmatrix} z_p(k_f^*) \\ \bar{f} \end{pmatrix} \end{pmatrix}
\tag{4.44}
$$

of the overall model $\bar{A}_\Delta$ of the closed-loop system at the fault detection time k_f^*.

In order to find an optimal (i.e., shortest) homing sequence, a homing tree as described in Section 4.2.2 could be constructed for the overall model $\bar{A}_\Delta$ of the closed-loop system with respect to the initial uncertainty $\bar{\mathcal{Z}}^*(k_f^*)$ in (4.33). However, as mentioned before, the construction of the homing tree is not practical when dealing with large systems.

In the proof of Proposition 4.8 an alternative idea on how to construct a homing sequence iteratively has been sketched. It avoids the construction of a homing tree at the cost of yielding a homing sequence that is not optimal. The main idea is to perform the following steps alternately:

1. Select a pair (z_1, z_2) of incompatible states $z_1, z_2 \in \bar{\mathcal{Z}}^*(k)$ from the current state estimate $\bar{\mathcal{Z}}^*(k)$ of the closed-loop system.

2. Compute a separating sequence $V_S(0 \ldots k_e)$ for the state pair (z_1, z_2) according to (2.22).

3. Use the separating sequence $V_S(0 \ldots k_e)$ as the next part of the diagnostic trajectory $R(k \ldots k + k_e)$ for the controller $\mathcal{C}(\mathcal{A}_0)$.

4. Update the current state estimate $\bar{\mathcal{Z}}^*(k)$ of the closed-loop system iteratively based on the measured I/O-pairs $(v_p(l), w_p(l))$ of the plant $\mathcal{P}$ in the following steps $l = k, \ldots, k_e$:

a) Update the current state estimate $\mathcal{Z}^*(l)$ of the plant according to (4.29).

b) Update the current state $z_c(l)$ of the controller $\mathcal{C}(\mathcal{A}_0)$ according to (4.34).

c) Update the current state estimate $\bar{\mathcal{Z}}^*(l)$ of the closed-loop system based on (4.32).

By applying separating sequences for pairs of incompatible states in the overall model $\bar{\mathcal{A}}_\Delta$ of the closed-loop system it is guaranteed that the current state estimate $\bar{\mathcal{Z}}^*(k)$ of the closed-loop system and therefore also the current state estimate $\mathcal{Z}^*(k)$ of the plant are reduced by at least one element after the application of an entire separating sequence $V_S(0 \ldots k_e)$. If all of these separating sequences are concatenated, the required homing sequence results.

When the steps described above are integrated into the right part of the fault diagnosis flowchart in Fig. 4.10, the detail shown in Fig. 4.11 results, where the newly occurring blocks have a gray background.

In the following, Steps 1 and 2 are described in more detail, while Steps 3 and 4 have already been presented in Sections 4.3.4 and 4.4.2.

Remark. The method is restricted to plants for which the condition in Proposition 4.8 is fulfilled, such that there are only incompatible states in the current state estimate $\bar{\mathcal{Z}}^*(k)$ of the closed-loop system for any $k \geq k_f^*$. In this case it is always possible to perform Step 1, i.e., to select

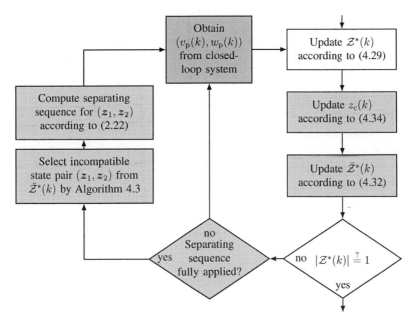

Figure 4.11: Detail of flowchart in Fig. 4.10 for active fault diagnosis.

a pair of incompatible states from $\bar{Z}^*(k)$. If there exist compatible state pairs, the presented active fault diagnosis method can be used anyhow until there do not remain any incompatible states in $\bar{Z}^*(k)$. In this case the fault diagnosis ends with an ambiguous diagnostic result $\mathcal{D}^*$.

4.5.2 Selection of state pairs

In Step 1, a pair (z_1, z_2) of incompatible states $z_1, z_2 \in \bar{Z}^*(k)$ from the current state estimate $\bar{Z}^*(k)$ of the closed-loop system needs to be chosen such that a separating sequence $V_S(0 \ldots k_e)$ can be computed in Step 2. In general, an arbitrary pair of incompatible states $z_1, z_2 \in \bar{Z}^*(k)$ can be used. However, if it is aimed to obtain an unambiguous diagnostic result $\mathcal{D}^*$ as soon as possible, some strategy for selecting the state pairs beneficially should be employed.

Therefore, two heuristic rules for the selection of state pairs are developed. Both rely on partitioning the state sets $\{z_c\} \times \mathcal{Z}_\Delta$, $(z_c \in \mathcal{Z}_c)$ into sets $\mathcal{Z}_i^k$, $(k = 0, \ldots, \bar{n}_\Delta - 2, i = 1, \ldots, m_k)$ of k-compatible states that has been introduced in Section 4.3.4 for the diagnosability analysis.

Rule 1. Choose the state pair (z_1, z_2) such that it can be separated by an input sequence as short as possible.

Rule 2. Choose the state pair (z_1, z_2) such that as many states as possible are removed from the current state estimate $\bar{Z}^*(k)$ of the closed-loop system when a separating sequence for (z_1, z_2) is applied to the controller $\mathcal{C}(\mathcal{A}_0)$.

Algorithm 4.3 summarizes the selection of an incompatible state pair (z_1, z_2) from the current state estimate $\bar{Z}^*(k)$ of the closed-loop system based on Rule 1 and Rule 2, where Rule 1 has a higher priority than Rule 2.

The two rules can be formalized as follows. Rule 1 requires to find the smallest index $l \geq 0$ such that there are at least two states in the current state estimate $\bar{Z}^*(k)$ of the closed-loop system, which are l-incompatible. For two l-incompatible states it is possible to find a separating sequence $V_S(0 \ldots l)$ of length $l + 1$. In terms of the compatibility partitions, the first rule means that not all states in $\bar{Z}^*(k)$ lie in the same set $\mathcal{Z}_i^l$ of l-compatible states, which can be expressed as follows:

$$\min_{l=0,\ldots,\bar{n}_\Delta-2} l : (\nexists i \in \{1, \ldots, m_l\}) \; \bar{Z}^*(k) \cap \mathcal{Z}_i^l = \bar{Z}^*(k). \tag{4.45}$$

The evaluation of this equation is performed in Lines 1–11 of Algorithm 4.3.

The states (z_1, z_2) are chosen from sets $\mathcal{Z}_{i_1}^l, \mathcal{Z}_{i_2}^l$, $(i_1 \neq i_2)$ of l-compatible states. By the application of a separating sequence $V_S(0 \ldots k_e)$ for (z_1, z_2) it is guaranteed that all states from one of these state sets $\mathcal{Z}_{i_1}^l$ or $\mathcal{Z}_{i_2}^l$ are removed from $\bar{Z}^*(k)$. Since Rule 2 requires to choose (z_1, z_2) from sets which contain as many states from the current state estimate $\bar{Z}^*(k)$ of the closed-loop system as possible, as many states as possible are removed from $\bar{Z}^*(k)$. The second

---

Algorithm 4.3: Selection of an incompatible state pair (z_1, z_2) from $\bar{\mathcal{Z}}^*(k)$.

Given: Current state estimate $\bar{\mathcal{Z}}^*(k)$ of the closed-loop system without compatible states
        Sets $\mathcal{Z}_i^k$, $(k = 0, \dots, \bar{n}_\Delta - 2,\ i = 1, 2, \dots, m_k)$ of k-compatible states

1 **if** $|\bar{\mathcal{Z}}^*(k)| > 2$ **then**
2   **for** $l = 0, \dots, \bar{n}_\Delta - 2$
3   **for** $i = 1, \dots, m_l$
4   **if** $\bar{\mathcal{Z}}^*(k) \cap \mathcal{Z}_i^l = \bar{\mathcal{Z}}^*(k)$ **then**
5   Goto Line 2
6   **end**
7   **if** $i = m_l$ **then**
8   Goto Line 12   // l fulfilling (4.45) found
9   **end**
10   **end**
11   **end**
12   Find $i_1 = \arg\max_{i=1,\dots,m_l} |\mathcal{Z}_i^l \cap \bar{\mathcal{Z}}^*(k)|$.
13   Find $i_2 = \arg\max_{i=1,\dots,m_l,\ i \neq i_1} |\mathcal{Z}_i^l \cap \bar{\mathcal{Z}}^*(k)|$.
14   Pick any $z_1 \in \mathcal{Z}_{i_1}^l \cap \bar{\mathcal{Z}}^*(k)$.   // z_1 fulfilling (4.46) found
15   Pick any $z_2 \in \mathcal{Z}_{i_2}^l \cap \bar{\mathcal{Z}}^*(k)$.   // z_2 fulfilling (4.47) found
16 **else**
17   Let z_1 be the first element of $\bar{\mathcal{Z}}^*(k)$.
18   Let z_2 be the second element of $\bar{\mathcal{Z}}^*(k)$.
19 **end**

Result: Incompatible state pair (z_1, z_2) selected from $\bar{\mathcal{Z}}^*(k)$ based on Rule 1 and Rule 2

---

rule can be formalized as follows:

$$z_1 \in \mathcal{Z}_{i_1}^l \cap \bar{\mathcal{Z}}^*(k) \text{ with } i_1 = \underset{i=1,\dots,m_l}{\arg\max} |\mathcal{Z}_i^l \cap \bar{\mathcal{Z}}^*(k)| \tag{4.46}$$

$$z_2 \in \mathcal{Z}_{i_2}^l \cap \bar{\mathcal{Z}}^*(k) \text{ with } i_2 = \underset{i=1,\dots,m_l,\ i \neq i_1}{\arg\max} |\mathcal{Z}_i^l \cap \bar{\mathcal{Z}}^*(k)|. \tag{4.47}$$

These equations are realized in Lines 12–15 of Algorithm 4.3.

Lines 17–18 deal with the case that the current state estimate $\bar{\mathcal{Z}}^*(k)$ contains exactly two states such that no further selection is needed. Since it is required that there are no compatible states in the current state estimate $\bar{\mathcal{Z}}^*(k)$ of the closed-loop system, it is guaranteed that the algorithm always yields a result.

Note again that it is by no means guaranteed that the choice of the state pair is optimal with respect to the aim of finding a shortest homing sequence.

4.5.3 Computation of separating sequences

According to Section 4.5.1, the second step for the iterative construction of a homing sequence is to compute a separating sequence $V_S(0 \ldots k_e)$ for the selected pair of incompatible states $z_1, z_2 \in \bar{Z}^*(k)$. A method for computing a separating sequence $V_S(0 \ldots k_e)$ (called distinguishing input sequence there) for a pair (z_1, z_2) of distinguishable states within a completely defined automaton $\mathcal{A}$ has been proposed in [64]. It relies on the k-equivalence classes Z_i^k, $(k = 0, \ldots, n - 2, \ i = 1, \ldots, m_k)$ introduced for the diagnosability analysis in Section 4.3.4. A separating sequence $V_S(0 \ldots k_e)$ for the state pair (z_1, z_2) can be found by starting from the k-equivalence partition with the smallest value of k for which z_1 and z_2 are in different sets $Z_{i_1}^k, Z_{i_2}^k$, $(i_1 \neq i_2)$, respectively. Then the separating sequence $V_S(0 \ldots k_e)$ can be constructed iteratively by "backtracking" the states z_1 and z_2 through the k-equivalence partitions Z_i^l, $(l = k, \ldots, 0)$. It is guaranteed that the resulting separating sequence $V_S(0 \ldots k_e)$ is the shortest possible one for the state pair (z_1, z_2).

The same procedure can be applied to the sets Z_i^k, $(k = 0, \ldots, \bar{n}_\Delta - 2, \ i = 1, \ldots, m_k)$ of k-compatible states used for the diagnosability analysis in Section 4.3.4. As a result, the shortest possible separating sequence $V_S(0 \ldots k_e)$ for any state pair (z_1, z_2) selected according to Algorithm 4.3 is obtained. Since z_1 and z_2 are l-incompatible with l given by (4.45), the resulting separating sequence $V_S(0 \ldots k_e)$ is known to be of length $l + 1$.

4.5.4 Analysis of the resulting input sequence

The following proposition shows that the presented method for generating the diagnostic trajectory $R(k_f^* + 1 \ldots k_e)$ for the controller $C(\mathcal{A}_0)$ indeed leads to the identification of the present fault $\bar{f}$ and the current state $z_p(k_d)$ of the faulty plant. Note again that the described method is restricted to plants for which the condition in Proposition 4.8 is fulfilled.

Proposition 4.13 (Resulting homing sequence). *If all faults $f \in \mathcal{F}$ are detectable in the closed-loop system $\mathcal{P} \circ C_T(\mathcal{A}_0)$ and there are no compatible states in the set $\{z_c\} \times Z_\Delta \subset \bar{Z}_\Delta$ for any $z_c \in Z_c$, the reference trajectory $R(k_f^* \ldots k_e)$ resulting from the diagnosis method described in Fig. 4.11 is a realization of an adaptive homing sequence for the overall model $\bar{A}_\Delta$ of the closed-loop system with respect to the initial uncertainty $Z^* = \bar{Z}^*(k_f^*)$ in (4.33) and the state $\bar{z}_\Delta(k_f^*)$ in (4.44).*

Proof. According to (4.33), the current state estimate of the closed-loop system at the fault detection time k_f^* is given by

$$\bar{Z}^*(k_f^*) = \{z_c(k_f^*)\} \times Z_\Delta.$$

The value $r(k_f^*)$ of the reference trajectory at the fault detection time is still generated by the trajectory planning unit $\mathcal{T}(\mathcal{A}_0)$ such that after the fault detection time some current state estimate

$$\bar{\mathcal{Z}}^*(k_f^* + 1) \subseteq \{z_c(k_f^* + 1)\} \times \mathcal{Z}_\Delta$$

of the closed-loop system results.

Since there are no compatible states in $\{z_c\} \times \mathcal{Z}_\Delta \subset \bar{\mathcal{Z}}_\Delta$ for any $z_c \in \mathcal{Z}_c$, it is possible to select an incompatible state pair (z_1, z_2) from $\bar{\mathcal{Z}}^*(k_f^* + 1)$ and compute a separating sequence $V_S(0 \ldots k_e)$ for it. By definition, the separating sequence $V_S(0 \ldots k_e)$ leads to different output sequences when applied in state z_1 or in state z_2. Therefore, according to Proposition 4.10, either the successor

$$z_1' = \bar{G}_\Delta^\infty(z_1, V_S(0 \ldots k_e))$$

of the state z_1 with respect to the separating sequence $V_S(0 \ldots k_e)$ or the successor

$$z_2' = \bar{G}_\Delta^\infty(z_2, V_S(0 \ldots k_e))$$

of the state z_2 or both will not be included into the current state estimate $\bar{\mathcal{Z}}^*(k_f^* + k_e + 2)$ updated according to (4.29) if the separating sequence $V_S(0 \ldots k_e)$ is applied to the controller $\mathcal{C}(\mathcal{A}_0)$. That is, the size of the current state estimate $\bar{\mathcal{Z}}^*(k)$ of the closed-loop system is reduced by at least one element.

In the resulting current state estimate $\bar{\mathcal{Z}}^*(k)$ of the closed-loop system there are again no compatible states, such that a new incompatible state pair can be chosen and a new separating sequence can be computed. Consequently, after at most $|\bar{\mathcal{Z}}^*(k_f^* + 1)| - 1$ steps, the current state estimate $\bar{\mathcal{Z}}^*(k)$ of the closed-loop system is reduced to a singleton such that the state of the overall model $\bar{\mathcal{A}}_\Delta$ of the closed-loop system is known. The argumentation is valid for all possible current state $\bar{z}_\Delta(k_f^*)$ of the overall model $\bar{\mathcal{A}}_\Delta$ of the closed-loop system at the fault detection time k_f^*.

Therefore, the input sequence concatenated of all applied separating sequences is a realization of an adaptive homing sequence for the overall model $\bar{\mathcal{A}}_\Delta$ of the closed-loop system with respect to the initial uncertainty $\mathcal{Z}^* = \bar{\mathcal{Z}}^*(k_f^*)$ and the current state $\bar{z}_\Delta(k_f^*)$ of the overall model $\bar{\mathcal{A}}_\Delta$ of the closed-loop system at the fault detection time k_f^*. $\qquad\square$

Example 4.2 (cont.) *Fault diagnosis of automated warehouse*

Since the automated warehouse in the closed-loop system $\mathcal{P} \circ \mathcal{C}_\mathcal{T}(\mathcal{A}_0)$ is not diagnosable with passive diagnosis with respect to the fault set $\mathcal{F} = \{1, 2\}$ and the reference trajectory $R(0 \ldots k_e)$ in (3.20), active

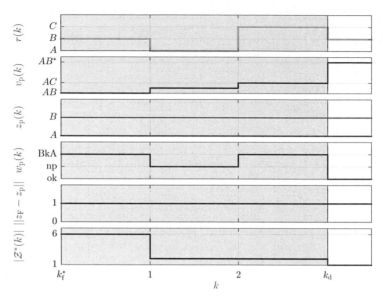

Figure 4.12: Active diagnosis of automated warehouse.

diagnosis has to be used once a fault is detected. The diagnosability analysis revealed that there are no compatible states within any state set $\{z_c\} \times \mathcal{Z}_\Delta$, $(z_c \in \mathcal{Z}_c)$ in this example. Therefore, it is possible to apply the fault diagnosis method described in the previous sections to the automated warehouse example.

Starting from the updated current state estimate $\mathcal{Z}^*(1)$ of the faulty plant after the fault detection time $k_f^* = 0$ in (4.40), the active diagnosis is performed as shown in the flowchart in Fig. 4.11, where the upper right block has just been executed. The course of all important variables during the active diagnosis is displayed in Fig. 4.12. In addition to the previously displayed signals, in the last subplot the size of the current state estimate $\mathcal{Z}^*(k)$ of the plant is displayed.

First, the current state of the controller $\mathcal{C}(\mathcal{A}_0)$ is computed according to (3.12):

$$z_c(1) = G_c^\infty(z_{c0}, \Phi_c^{-1}(z_{c0}, v_p(0))) = G_c^\infty(A, \Phi_c^{-1}(A, AB)) = G_c(A, B) = B.$$

Then the current state estimate of the closed-loop system in Fig. 4.8 is computed based on (4.32):

$$\bar{\mathcal{Z}}^*(1) = \{z_c(1)\} \times \mathcal{Z}^*(1) = \left\{ \left(\begin{array}{c} B \\ A \\ 1 \end{array} \right), \left(\begin{array}{c} B \\ A \\ 2 \end{array} \right) \right\}. \tag{4.48}$$

That is, it is known that the parcel is located at position A, but it is not known whether fault $f = 1$ or fault $f = 2$ occurred.

Since $|\mathcal{Z}^*(1)| = 2 > 1$ and there is no remaining input sequence to be applied to the controller $\mathcal{C}(\mathcal{A}_0)$, an incompatible state pair is selected from $\bar{\mathcal{Z}}^*(1)$ based on Algorithm 4.3. Since there are only two elements in $\bar{Z}^*(1)$, Lines 17 and 18 are applicable, such that the following state pair results:

$$(z_1, z_2) = \left(\begin{pmatrix} B \\ A \\ 1 \end{pmatrix}, \begin{pmatrix} B \\ A \\ 2 \end{pmatrix} \right). \tag{4.49}$$

Next, a separating sequence $V_S(0 \ldots k_e)$ for the state pair in (4.49) is computed as described in Section 4.5.3. A shortest separating sequence is

$$V_S(0 \ldots 1) = (A, C), \tag{4.50}$$

for which the states z_1 and z_2 generate the outputs

$$\bar{W}_1(0 \ldots 1) = \bar{\Phi}_\Delta(z_1, V_S(0 \ldots 1)) = (\text{np}, \text{ok})$$
$$\text{or } \bar{W}_2(0 \ldots 1) = \bar{\Phi}_\Delta(z_2, V_S(0 \ldots 1)) = (\text{np}, \text{BkA}),$$

respectively. This separating sequence is now used as a part of the diagnostic trajectory

$$R(1 \ldots 2) = V_S(0 \ldots 1)$$

for the controller $C(A_0)$.

When the first input symbol $r(1) = v_s(0) = A$ is applied to the controller $C(A_0)$ in Fig. 3.2 in state $z_c(1) = B$, it generates the input $v_p(1) = w_c(1) = BA$ for the faulty plant $\mathcal{P} \models A_2$ in state $z_p(1) = A$. Because the requested transport from position B to position A is not possible when the parcel is at position A, the plant remains in state $z_p(2) = A$ and generates the output $w_p(1) = \text{np}$. Based on the resulting I/O-pair $(v_p(1), w_p(1)) = (BA, \text{np})$ of the plant, the current state estimate of the plant is again updated according to (4.29):

$$Z^*(2) = \left\{ \begin{pmatrix} A \\ 1 \end{pmatrix}, \begin{pmatrix} A \\ 2 \end{pmatrix} \right\}.$$

The current state of the controller is updated according to (4.34):

$$z_c(2) = G_c(B, \Phi_c^{-1}(B, BA)) = G_c(B, A) = A,$$

such that the following current state estimate of the closed-loop system results:

$$\bar{Z}^*(2) = \left\{ \begin{pmatrix} A \\ A \\ 1 \end{pmatrix}, \begin{pmatrix} A \\ A \\ 2 \end{pmatrix} \right\}.$$

Again, there is more than one element in $\bar{Z}^*(2)$. Since the separating sequence in (4.50) has not been fully applied yet, its next element $v_s(1) = C$ is given to the controller $C(A_0)$. Therefore, the controller generates the output $w_c(2) = AC$. In the faulty plant $\mathcal{P} \models A_2$ the requested transport from position A to position C by the broken robot M is not possible. Hence the plant remains in state $z_p(3) = A$ and generates the output $w_p(2) = \text{BkA}$. If the current state estimate of the plant is updated based on this information and (4.29),

$$Z^*(3) = \left\{ \begin{pmatrix} A \\ 2 \end{pmatrix} \right\}$$

results. Hence, $\mathcal{Z}^*(3)$ is a singleton, such that the diagnosis time equals $k_\mathrm{d} = 3$.

In total, the reference trajectory

$$R(0 \ldots 2) = (B, A, C) \tag{4.51}$$

has been applied to the controller $\mathcal{C}(\mathcal{A}_0)$, which is a realization of an adaptive homing sequence for the overall model $\bar{A}_\Delta$ of the closed-loop system with respect to the initial uncertainty

$$\mathcal{Z}^* = \bar{\mathcal{Z}}^*(k_\mathrm{f}^*) = \{z_\mathrm{c}(k_\mathrm{f}^*)\} \times \mathcal{Z}_\Delta$$

$$= \{A\} \times \left\{ \begin{pmatrix} A \\ 1 \end{pmatrix}, \begin{pmatrix} B \\ 1 \end{pmatrix}, \begin{pmatrix} C \\ 1 \end{pmatrix}, \begin{pmatrix} A \\ 2 \end{pmatrix}, \begin{pmatrix} B \\ 2 \end{pmatrix}, \begin{pmatrix} C \\ 2 \end{pmatrix} \right\}.$$

and the state

$$\bar{z}_\Delta(k_\mathrm{f}^*) = \left(\begin{pmatrix} A \\ A \\ 2 \end{pmatrix} \right)$$

(Proposition 4.13). □

4.6 Properties of the diagnostic result

Whenever the fault diagnosis is stopped, the current state estimate $\mathcal{Z}^*(k)$ of the plant is used as the diagnostic result $\mathcal{D}^*$. In this section the properties of the diagnostic result $\mathcal{D}^*$ are analyzed.

4.6.1 Completeness and correctness of the diagnostic result

According to Proposition 4.10, the current state estimate $\mathcal{Z}^*(k)$ of the plant is correct, hence the diagnostic result

$$\mathcal{D}^* = \mathcal{Z}^*(k), \quad \text{where } k \text{ is the time at which the fault diagnosis is stopped} \tag{4.52}$$

is also always correct. The following proposition shows that the diagnostic result $\mathcal{D}^*$ is also complete, because the current state estimate $\mathcal{Z}^*(k)$ of the plant is complete at all times $k \geq k_\mathrm{f}^*$.

Proposition 4.14 (Completeness of diagnostic result). *The current state $z_\mathrm{p}(k)$ of the plant $\mathcal{P}$ and the present fault $\bar{f} \in \mathcal{F}$ are contained in the diagnostic result $\mathcal{D}^*$.*

Proof. (By induction) It is shown that the current state $z_\mathrm{p}(k)$ of the plant $\mathcal{P}$ and the present fault $\bar{f} \in \mathcal{F}$ are contained in the current state estimate $\mathcal{Z}^*(k)$ of the plant $\mathcal{P}$ at all times $k \geq k_\mathrm{f}^*$. Hence, whenever the diagnosis is stopped, the diagnostic result $\mathcal{D}^*$ is complete.

For $k = k_\mathrm{f}^*$, the current state estimate of the plant is initialized according to (4.28). Therefore, it is given by the set

$$\mathcal{Z}^*(k_{\mathrm{f}}^*) = \mathcal{Z}_\Delta = \bigcup_{f \in \mathcal{F}} \mathcal{Z}_f \times \{f\},$$

which obviously contains any possible state-fault vector $\left(z_{\mathrm{p}}(k) \quad \bar{f} \right)^\top$ with $\bar{f} \in \mathcal{F}$, $z_{\mathrm{p}}(k) \in \mathcal{Z}_{\bar{f}}$ corresponding to a faulty plant $\mathcal{P} \vDash \mathcal{A}_{\bar{f}}$ in state $z_{\mathrm{p}}(k)$.

If $z_\Delta(k)$ is included in $\mathcal{Z}^*(k)$, then for any time $k + 1$, $k \geq k_{\mathrm{f}}^*$ it can be proved that also $z_\Delta(k+1) \in \mathcal{Z}^*(k+1)$ as follows. If the fault $\bar{f} \in \mathcal{F}$ is present in the plant $\mathcal{P} \vDash \mathcal{A}_{\bar{f}}$, the current state of the plant is $z_{\mathrm{p}}(k)$ and the plant receives the input symbol $v_{\mathrm{p}}(k)$, the output symbol

$$w_{\mathrm{p}}(k) = H_{\bar{f}}(z_{\mathrm{p}}(k), v_{\mathrm{p}}(k))$$

is generated and the next state of the plant is given by

$$z_{\mathrm{p}}(k + 1) = G_{\bar{f}}(z_{\mathrm{p}}(k), v_{\mathrm{p}}(k)).$$

The current state estimate of the plant $\mathcal{P}$ is updated using (4.29). Hence, for $z_\Delta(k) \in \mathcal{Z}^*(k)$ the next state

$$z_\Delta(k + 1) = G_\Delta(z_\Delta(k), v_{\mathrm{p}}(k)) = \begin{pmatrix} G_{\bar{f}}(z_{\mathrm{p}}(k), v_{\mathrm{p}}(k)) \\ \bar{f} \end{pmatrix}$$

computed according to (2.42) is included in $\mathcal{Z}^*(k + 1)$ if $w_{\mathrm{p}}(k) = H_\Delta(z_\Delta(k), v_{\mathrm{p}}(k))$. Since

$$H_\Delta(z_\Delta(k), v_{\mathrm{p}}(k)) = H_{\bar{f}}(z_{\mathrm{p}}(k), v_{\mathrm{p}}(k))$$

holds, the induction step is completed, which concludes the proof.   □

4.6.2 Termination conditions for the fault identification

Depending on the desired utilization of the diagnostic result $\mathcal{D}^*$, different termination conditions for the fault identification are thinkable, which of course influence the properties of the obtained diagnostic result $\mathcal{D}^*$. Possible termination conditions are, for example,

- the identification of the present fault $\bar{f}$

- the identification of the present fault $\bar{f}$ and the current state $z_{\mathrm{p}}(k_{\mathrm{d}})$ of the faulty plant

- the reduction of the current state estimate $\mathcal{Z}^*(k)$ of the plant to a certain size

- the expiration of a fixed maximal diagnosis time.

An important property of the diagnostic result is the fact whether it contains exactly one element or more than one element, which is reflected by the ambiguity of the diagnostic result defined as follows.

Definition 4.7 (Unambiguous diagnostic result). *The diagnostic result* $\mathcal{D}^*$ *is called unambiguous if* $|\mathcal{D}^*| = 1$. *Otherwise it is called ambiguous.*

From Proposition 4.12 it is known that once the set $\mathcal{F}(k)$ of fault candidates in (4.35) becomes the singleton

$$\mathcal{F}(k) = \{\bar{f}\}, \tag{4.53}$$

it will remain the same for all future times, such that continuing the fault identification does not reveal any new information. Therefore, if only the present fault $\bar{f} \in \mathcal{F}$ needs to be identified, the fault identification can be stopped as soon as the set $\mathcal{F}(k)$ of fault candidates becomes a singleton. In this case, the diagnostic result $\mathcal{D}^*$ is given in form of a set

$$\mathcal{D}^* = \mathcal{Z}^*(k) \subseteq \mathcal{Z}_{\bar{f}} \times \{\bar{f}\}, \tag{4.54}$$

in which each vector has the same second element $\bar{f}$. Hence, if the fault $\bar{f}$ is identified, the diagnostic result might be unambiguous or ambiguous and the current state $z_p(k)$ of the faulty plant might or might not be known.

If, for example in order to allow for a later controller reconfiguration, the current state $z_p(k)$ of the faulty plant $\mathcal{P} \vDash \mathcal{A}_{\bar{f}}$ also needs to be identified, the current state estimate $\mathcal{Z}^*(k_d)$ of the plant has to become a singleton. In Proposition 4.11 it has been shown that the size of the current state estimate of the plant $\mathcal{P}$ does not increase over time. Therefore, the fault identification can be stopped as soon as the current state estimate $\mathcal{Z}^*(k_d)$ of the plant becomes a singleton at the diagnosis time k_d. In this case an unambiguous diagnostic result $\mathcal{D}^*$ is obtained.

Since it has been shown in Proposition 4.14 that the diagnostic result $\mathcal{D}^*$ is always complete, the current state $z_p(k_d)$ of the faulty plant $\mathcal{P} \vDash \mathcal{A}_{\bar{f}}$ and the present fault $\bar{f}$ are identified at the diagnosis time k_d and the corresponding diagnostic result is given by

$$\mathcal{D}^* = \mathcal{Z}^*(k_d) = \left\{ \left(z_p(k_d) \;\; \bar{f} \right)^\top \right\}. \tag{4.55}$$

If one of the other two mentioned termination conditions are used, the diagnostic result $\mathcal{D}^*$ will usually be ambiguous. That is, there are various reasons for the existence of such an ambiguity, namely:

- The plant $\mathcal{P}$ is diagnosable, but the fault diagnosis is stopped before an unambiguous diagnostic result has been found.

- Passive diagnosis is used, even though the plant $\mathcal{P}$ is only diagnosable with active diagnosis but not with passive diagnosis.

- There are compatible states in the overall model $\bar{A}_\Delta$ such that the state of the plant can not be identified using the nominal tracking controller $\mathcal{C}_T(\mathcal{A}_0)$ and the presented active fault diagnosis method.

In the first case it would be possible to keep the diagnosis running to obtain an unambiguous diagnostic result $\mathcal{D}^*$. Nevertheless it might be desirable to stop the diagnosis early, such that the plant does not deviate too much from its desired operating region given by the reference trajectory. In the second case the diagnostic result $\mathcal{D}^*$ remains uncertain unless another reference trajectory $R(0 \ldots k_e)$ is specified if active diagnosis is not available. In the third case it is not possible to obtain an unambiguous diagnostic result $\mathcal{D}^*$ with the presented method when the plant $\mathcal{P}$ happens to be in an unfortunate state when the fault occurs.

4.6.3 Diagnostic result at the diagnosis time

Let the termination condition be to obtain an unambiguous diagnostic result. Then the following theorem regarding the diagnostic result $\mathcal{D}^*$ can be stated for the case that passive diagnosis is used. It states that the proposed passive diagnosis method solves the passive fault diagnosis problem in Problem 4.1 if the plant $\mathcal{P}$ is diagnosable with passive diagnosis.

Theorem 4.3 (Diagnostic result with passive diagnosis). *If the plant $\mathcal{P}$ is diagnosable with passive diagnosis in the closed-loop system $\mathcal{P} \circ \mathcal{C}_T(\mathcal{A}_0)$ with respect to the fault set $\mathcal{F}$ and the reference trajectory $R(0 \ldots k_e)$ generated by the trajectory planning unit $\mathcal{T}(\mathcal{A}_0)$, the present fault $\bar{f} \in \mathcal{F}$ as well as the current state $z_p(k_d)$ of the plant $\mathcal{P}$ are identified at a finite diagnosis time $k_d \leq k_e + 1$ using passive diagnosis.*

Proof. If the plant $\mathcal{P}$ is diagnosable with passive diagnosis in the closed-loop system $\mathcal{P} \circ \mathcal{C}_T(\mathcal{A}_0)$ with respect to the fault set $\mathcal{F}$ and the reference trajectory $R(0 \ldots k_e)$ generated by the trajectory planning unit $\mathcal{T}(\mathcal{A}_0)$, all faults are detectable according to Theorem 4.1. Furthermore, the reference trajectory $R(k_f^* \ldots k_e)$ after the fault detection time k_f^* is a preset homing sequence for the overall model $\bar{A}_\Delta$ of the closed-loop system in (4.14) with respect to the initial uncertainty $\mathcal{Z}^* = \{z_c(k_f^*)\} \times \mathcal{Z}_\Delta \subset \bar{\mathcal{Z}}_\Delta$.

In Proposition 4.9 it has already been proved that the proposed fault detection method detects every detectable fault. Hence, it remains to show that the reference trajectory $R(k_f^* \ldots k_e)$ leads to the identification of the present fault $\bar{f}$ and the current state $z_{\mathrm{p}}(k)$ of the plant $\mathcal{P}$ as required in Problem 4.1. The passive fault diagnosis is performed as described in Section 4.4.2, that is, the current state estimate $\mathcal{Z}^*(k)$ of the plant is initialized according to (4.28) at the fault detection time k_{f}^* and updated according to (4.29) at every time step $k \geq k_{\mathrm{f}}^*$.

From the definition of a preset homing sequence in Definition 4.1 and the proof of Proposition 4.10 on the correctness of the current state estimate it follows that

$$
\mathcal{Z}^*(k_{\mathrm{e}} + 1) = \left\{ z'_{\Delta} \in \mathcal{Z}_{\Delta} : (\exists z_0 \in \mathcal{Z}_{\Delta}) \left[z'_{\Delta} = G_{\Delta}^{\infty}(z_0, V_{\mathrm{p}}(k_{\mathrm{f}}^* \ldots k_{\mathrm{e}})) \right] \right.
$$
$$
\left. \wedge \left[W_{\mathrm{p}}(k_{\mathrm{f}}^* \ldots k_{\mathrm{e}}) = \Phi_{\Delta}(z_0, V_{\mathrm{p}}(k_{\mathrm{f}}^* \ldots k_{\mathrm{e}})) \right] \right\},
$$

where the following implication holds:

$$
W_{\mathrm{p}}(k_{\mathrm{f}}^* \ldots k_{\mathrm{e}}) = \Phi_{\Delta}(z_1, V_{\mathrm{p}}(k_{\mathrm{f}}^* \ldots k_{\mathrm{e}})) = \Phi_{\Delta}(z_2, V_{\mathrm{p}}(k_{\mathrm{f}}^* \ldots k_{\mathrm{e}}))
$$
$$
\Rightarrow G_{\Delta}^{\infty}(z_1, V_{\mathrm{p}}(k_{\mathrm{f}}^* \ldots k_{\mathrm{e}})) = G_{\Delta}^{\infty}(z_2, V_{\mathrm{p}}(k_{\mathrm{f}}^* \ldots k_{\mathrm{e}})) \quad z_1, z_2 \in \mathcal{Z}_{\Delta}.
$$

That is, the current state estimate $\mathcal{Z}^*(k_{\mathrm{e}} + 1)$ of the plant at the diagnosis time $k_{\mathrm{d}} = k_{\mathrm{e}} + 1$ only contains a single element. Together with the completeness of the diagnostic result $\mathcal{D}^* = \mathcal{Z}^*(k_{\mathrm{d}})$ proved in Proposition 4.14 it follows that the present fault $\bar{f} \in \mathcal{F}$ as well as the current state $z_{\mathrm{p}}(k_{\mathrm{d}})$ of the plant are identified at the latest at a finite diagnosis time $k_{\mathrm{d}} = k_{\mathrm{e}} + 1$. $\qquad\square$

For the case of active diagnosis, the following theorem results. It states that the proposed active diagnosis method solves the active fault diagnosis problem in Problem 4.2 if the plant $\mathcal{P}$ is diagnosable with active diagnosis.

Theorem 4.4 (Diagnostic result with active diagnosis). *If the plant $\mathcal{P}$ is diagnosable with active diagnosis in the closed-loop system $\mathcal{P} \circ \mathcal{C}_{\mathcal{T}}(\mathcal{A}_0)$ with respect to the fault set $\mathcal{F}$, the present fault $\bar{f} \in \mathcal{F}$ as well as the current state $z_{\mathrm{p}}(k_{\mathrm{d}})$ of the plant $\mathcal{P}$ are identified at a diagnosis time $k_{\mathrm{d}} \leq k_{\mathrm{f}}^* + (\bar{n}_{\Delta} - 1)(n_{\Delta} - 1)$ using active diagnosis.*

Proof. The proof follows the same argumentation as the proof of Proposition 4.3 above. The only difference is that the reference trajectory $R(k_{\mathrm{f}}^* \ldots k_{\mathrm{e}})$ is now a concatenation of its value $r(k_{\mathrm{f}}^*)$ generated by the trajectory planning unit $\mathcal{T}(\mathcal{A}_0)$ at the fault detection time k_{f}^* and the diagnostic trajectory $R(k_{\mathrm{f}}^* + 1 \ldots k_{\mathrm{e}})$ that is now generated by the diagnostic unit $\mathcal{D}$ according to the method described in Section 4.5. It has to be shown that this reference trajectory $R(k_{\mathrm{f}}^* \ldots k_{\mathrm{e}})$ is indeed a realization of an adaptive homing sequence for the overall model $\bar{\mathcal{A}}_{\Delta}$ of the closed-loop system with respect to the initial uncertainty $\mathcal{Z}^* \subseteq \{z_{\mathrm{c}}(k_{\mathrm{f}}^*)\} \times \mathcal{Z}_{\Delta}$ and the state $\bar{z}_{\Delta}(k_{\mathrm{f}}^*)$ of $\bar{\mathcal{A}}_{\Delta}$ at the fault detection time k_{f}^* and that its length is bounded by $(\bar{n}_{\Delta} - 1)(n_{\Delta} - 1)$. The former has already been proved in Proposition 4.13.

As stated in Section 4.5.3 the length of a shortest separating sequence $V_S(0 \ldots k_e)$ for a pair of l-incompatible states is $l+1$. There are $\bar{n}_\Delta$ (with $\bar{n}_\Delta$ from (4.16)) states in the overall model $\bar{A}_\Delta$ of the closed-loop system. Hence, two incompatible states can be at most $(\bar{n}_\Delta-2)$-incompatible and the length of the separating sequences is bounded by $\bar{n}_\Delta - 1$. If every separating sequence leads to the removal of exactly one element from the current state estimate $\bar{Z}^*(k)$ of the closed-loop system,

$$|\bar{Z}^*(k_f^* + 1)| - 1 \leq |Z_\Delta| - 1 = n_\Delta - 1$$

separating sequences have to be applied until the present fault $\bar{f} \in \mathcal{F}$ as well as the current state $z_p(k_d)$ of the plant $\mathcal{P}$ are identified. That is, after the fault detection time k_f^* at most $(\bar{n}_\Delta - 1)(n_\Delta - 1)$ input symbols have to applied such that the diagnosis time is bounded by

$$k_d \leq k_f^* + (\bar{n}_\Delta - 1)(n_\Delta - 1). \qquad \square$$

Of course the upper bound for the diagnosis time k_d is only a theoretical bound for academic examples that will hardly ever be approached when real systems are considered.

Example 4.2 (cont.) *Fault diagnosis of automated warehouse*

In the automated warehouse example, the current state estimate $\mathcal{Z}^*(k_d)$ became a singleton at the diagnosis time $k_d = 3$. Consistent with Theorem 4.4, the diagnostic result

$$\mathcal{D}^* = \mathcal{Z}^*(k_d) = \left\{ \begin{pmatrix} z_p(k_d) \\ \bar{f} \end{pmatrix} \right\} = \left\{ \begin{pmatrix} A \\ 2 \end{pmatrix} \right\}. \qquad (4.56)$$

reflects correctly that the parcel is located at position A and fault $\bar{f} = 2$ occurred, hence robot M is broken.

The diagnosis time $k_d = 3$ is far below its upper bound

$$k_f^* + (\bar{n}_\Delta - 1)(n_\Delta - 1) = (18 - 1) \cdot (6 - 1) = 85$$

given in Theorem 4.4 $\qquad \square$

4.7 Complexity of the fault diagnosis methods

In order to apply the presented passive or active fault diagnosis methods to real-world systems, it is necessary to analyze their complexity.

4.7.1 Complexity of the diagnosability analysis

Diagnosability with passive diagnosis depends on the reference trajectory $R(0 \ldots k_e)$ generated by the trajectory planning unit $\mathcal{T}(\mathcal{A}_0)$ at runtime. Therefore, it can be analyzed when a fault is detected, but not offline. However, the diagnosability criterion in Theorem 4.1 only requires an analysis whether the remainder $R(k_f^* \ldots k_e)$ of the given reference trajectory is a preset homing sequence for the overall model $\bar{\mathcal{A}}_\Delta$ of the closed-loop system with respect to the initial uncertainty $\mathcal{Z}^* = \{z_c(k_f^*)\} \times \mathcal{Z}_\Delta \subset \bar{\mathcal{Z}}_\Delta$. As a prerequisite, the current state $z_c(k_f^*)$ of the controller $\mathcal{C}(\mathcal{A}_0)$ has to be constructed. Therefore, the complexity of the analysis is linear in the number $n \cdot F$ of states in $\mathcal{A}_\Delta$ and the length $k_e + 1$ of the reference trajectory.

Considering active diagnosis, the diagnosability analysis presented in Section 4.3 can be performed offline. Nevertheless it is desirable to achieve a low time and space complexity in order to guarantee a termination even for large plant models. The main complexity of this diagnosability analysis lies in the test for compatible states in the overall model $\bar{\mathcal{A}}_\Delta$ of the closed-loop system in (4.14), which has been described in Section 4.3.4.

First, the overall model $\bar{\mathcal{A}}_\Delta$ of the closed-loop system has to be constructed from the controller $\mathcal{C}(\mathcal{A}_0)$ and the overall model $\mathcal{A}_\Delta$ of the faulty plant. According to (4.16) it contains $\bar{n}_\Delta = n^2 \cdot F$ states such that the maximal number of transitions is given by

$$\bar{n}_\Delta \cdot |\mathcal{V}_c| = n^2 \cdot F \cdot n = n^3 \cdot F. \tag{4.57}$$

Then, the state set of the overall model $\bar{\mathcal{A}}_\Delta$ of the closed-loop system is partitioned into sets of compatible states considering iteratively input sequences $\bar{V}(0 \ldots k_e)$ with time horizon $k_e = 0, 1, \ldots \bar{n}_\Delta - 1$. In total, at most

$$1 + 2 + \cdots + \bar{n}_\Delta = \frac{\bar{n}_\Delta + 1}{2} = \frac{n^2 \cdot F}{2} \tag{4.58}$$

state sets result.

In summary, the complexity of the diagnosability analysis is at most polynomial in the number of states and linear in the number of faults. Therefore, it is possible to perform the diagnosability analysis for any reasonably sized plant.

4.7.2 Complexity of the online execution of the diagnosis methods

The online execution of the passive fault diagnosis method visualized in Fig. 4.10 requires only the update of the current state estimate $\mathcal{Z}^*(k)$ of the plant in every time step k. According to (4.28) and (4.29) at most $|\mathcal{Z}_\Delta| = n_\Delta$ states need to be considered during the update, hence according to (2.41) the complexity of the method is of order $\mathcal{O}(n \cdot F)$, which makes it well suited for the online application even for large plant models.

The active fault diagnosis is performed according to Fig. 4.11. The update of the current state estimate $\bar{Z}^*(k)$ based on (4.29), (4.32) and (4.34) requires the consideration of at most $|\mathcal{Z}_\Delta| = n_\Delta$ states. In Algorithm 4.3 at most

$$(\bar{n}_\Delta - 1) \cdot \bar{n}_\Delta \leq n^4 \cdot F^2 \tag{4.59}$$

iterations are necessary for the selection of an incompatible state pair. Note that it would also be possible to select the state pair by chance in order to reduce the computation time for this step.

The computationally most expensive step during the online execution of the active diagnosis method is the generation of a separating input sequence for the selected state pair as described in Section 4.5.3. It is performed based on the partitions of the state set of the overall model $\bar{A}_\Delta$ of the closed-loop system resulting from the diagnosability analysis. Consequently, it is of order $\mathcal{O}(n^2 \cdot F)$.

In summary, the complexity of the active diagnosis method has a polynomial complexity with regards to the number n of states in the faultless plant and the number F of faults and is therefore also applicable to large models.

4.8 Active diagnosis avoiding illegal transitions

Neither the passive diagnosis method and the active diagnosis method presented in the previous sections can guarantee the avoidance of illegal transitions $(z, z') \in \mathcal{E}_{\text{ill}}$ by the plant $\mathcal{P}$. Therefore, an alternative method called *active safe diagnosis* is presented in this section in order to solve the active safe diagnosis problem in Problem 4.3.

4.8.1 General approach for the active safe diagnosis

In Section 4.5 it has be described how the active fault diagnosis can be performed if no illegal transitions are present. It has been proposed to choose separating sequences $V_S(0 \ldots k_e)$ for arbitrary state pairs within the current state estimate $\bar{Z}^*(k)$ of the closed-loop system and use them as the next part of the diagnostic trajectory $R(k_f^* + 1 \ldots k_e)$ for the controller $\mathcal{C}(\mathcal{A}_0)$. However, when using this method it can not be guaranteed that the plant $\mathcal{P} \vDash \mathcal{A}_f$ does not execute any illegal transitions $(z_p(k), z_p(k+1)) \in \mathcal{E}_{\text{ill}}$ for any $k_f^* \leq k < k_e$.

Rather, the diagnostic unit $\mathcal{D}$ needs to generate a diagnostic trajectory $R(k_f^* + 1 \ldots k_e)$ for the controller $\mathcal{C}(\mathcal{A}_0)$, such that the concatenation $R(k_f^* \ldots k_e)$ of the value $r(k_f^*)$ of the reference trajectory generated by the trajectory planning unit $\mathcal{T}(\mathcal{A}_0)$ at the fault detection time k_f^* and the diagnostic trajectory

- is a realization of an adaptive homing sequence for the overall model $\bar{A}_\Delta$ of the closed-loop system in (4.14) with respect to the initial uncertainty $\bar{Z}^*(k_f^*)$ in (4.33) and the current state $\bar{z}_\Delta(k_f^*)$ of $\bar{A}_\Delta$ in (4.44) at the fault detection time k_f^*.

- enforces no illegal transitions in the plant $\mathcal{P}$ with inputs from the controller $\mathcal{C}(\mathcal{A}_0)$:

$$(G_{\bar{f}}^\infty(z_p(k_f^*), \Phi_c(z_c(k_f^*), R(k_f^* \ldots k))), G_{\bar{f}}^\infty(z_p(k_f^*), \Phi_c(z_c(k_f^*), R(k_f^* \ldots k+1)))) \notin \mathcal{E}_{ill}$$
$$\forall k_f^* \leq k < k_e. \tag{4.60}$$

Since the present fault $\bar{f}$ and the current state $z_p(k_f^*)$ of the plant $\mathcal{P}$ after the fault detection time k_f^* are unknown, both conditions also have to be fulfilled for all other states

$$\bar{z}_\Delta \in \bar{Z}^*(k_f^*). \tag{4.61}$$

It is unlikely that a preset input sequence $R(k_f^* \ldots k_e)$ fulfilling both properties for all possible current states of the plant $\mathcal{P}$ at the fault detection time k_f^* exists. Such an input sequence would be a preset homing sequence for the overall model $\bar{A}_\Delta$ of the closed-loop system with respect to the initial uncertainty $\bar{Z}^*(k_f^*)$ that at the same time guarantees that no illegal transitions are executed by the plant $\mathcal{P}$ and is therefore called *preset safe homing sequence*.

Even if there is no preset safe homing sequence, it is possible that an adaptive homing sequence exists (see Definition 4.2), which fulfills both properties. Such an input sequence will be called *adaptive safe homing sequence* and will be explained in more detail in the next sections.

4.8.2 Safe diagnosability analysis

Safe diagnosability is closely related to diagnosability with active diagnosis as given in Definition 4.6 with the difference that it has to be ensured that the faulty plant $\mathcal{P}$ does not pass through any illegal transitions $(z_p(k), z_p(k+1)) \in \mathcal{E}_{ill}$ while the diagnostic trajectory $R(k_f^* \ldots k_e)$ is applied to the controller $\mathcal{C}(\mathcal{A}_0)$. However, first of all the fault needs to be safely detectable, which means that no illegal transitions are passed after the fault has occurred but before it is detected, that is for times $k_f \leq k \leq k_f^*$.

Definition 4.8 (Safe detectability). *A fault $f \in \mathcal{F}$ is said to be safely detectable in the closed-loop system $\mathcal{P} \circ \mathcal{C}_T(\mathcal{A}_0)$ with respect to the set $\mathcal{E}_{ill} \in \mathcal{Z}_0 \times \mathcal{Z}_0$ of illegal transitions and the reference trajectory $R(0 \ldots k_e)$ generated by the trajectory planning unit $\mathcal{T}(\mathcal{A}_0)$ if it is possible to decide from the I/O-pair $(V_p(0 \ldots k), W_p(0 \ldots k))$ of the plant $\mathcal{P}$ whether a fault has occurred or not while guaranteeing that the faulty plant $\mathcal{P}$ does not execute any illegal transitions.*

The following condition for safe detectability extends the detectability condition in Proposition 4.4. The additional constraint in (4.63) guarantees the avoidance of illegal transitions by the faulty plant $\mathcal{P} \vDash \mathcal{A}_f$.

Proposition 4.15 (Safe detectability)**.** *Let* $V_p(0 \ldots k_e) = \Phi_c(z_{c0}, R(0 \ldots k_e))$ *be the input sequence for the plant* $\mathcal{P}$ *generated by the nominal tracking controller* $\mathcal{C}_T(\mathcal{A}_0, \mathcal{E}_{ill})$. *Let* $z_p(k_f)$ *be the state of the plant* $\mathcal{P}$ *at the fault occurrence time* $k_f \geq 0$. *Then a fault* $f \in \mathcal{F}$ *is safely detectable in the closed-loop system* $\mathcal{P} \circ \mathcal{C}_T(\mathcal{A}_0, \mathcal{E}_{ill})$ *with respect to the set* $\mathcal{E}_{ill} \in \mathcal{Z}_0 \times \mathcal{Z}_0$ *of illegal transitions and the reference trajectory* $R(0 \ldots k_e)$ *generated by the trajectory planning unit* $\mathcal{T}(\mathcal{A}_0)$ *if there exists a time* $k_f^* \leq k_e$ *such that*

$$\Phi_f(z_p(k_f), V_p(k_f \ldots k_f^*)) \neq \Phi_0(z_p(k_f), V_p(k_f \ldots k_f^*)), \tag{4.62}$$

and

$$(G_f^\infty(z_p(k_f), V_p(k_f \ldots k-1)), G_f^\infty(z_p(k_f), V_p(k_f \ldots k))) \notin \mathcal{E}_{ill}, \quad \forall k_f \leq k < k_f^*. \tag{4.63}$$

Proof. The fulfillment of the condition in (4.62) allows for the decision whether or not a fault occurred (cf. Proposition 4.4). The condition in (4.63) formalizes the requirement that the faulty plant $\mathcal{P} \vDash \mathcal{A}_f$ must not execute any illegal transitions $(z_p(k), z_p(k+1)) \in \mathcal{E}_{ill}$ between the fault occurrence time k_f and the fault detection time k_f^*. □

Additional to safe detectability, safe identifiability has to be given, which is defined as follows.

Definition 4.9 (Safe identifiability)**.** *A fault* $f \in \mathcal{F}$ *is said to be safely identifiable in the closed-loop system* $\mathcal{P} \circ \mathcal{C}_T(\mathcal{A}_0)$ *with respect to a set* $\mathcal{E}_{ill} \subset \mathcal{Z}_0 \times \mathcal{Z}_0$ *of illegal transitions and the reference trajectory* $R(0 \ldots k_e)$ *generated by the trajectory planning unit* $\mathcal{T}(\mathcal{A}_0)$ *if it is possible to decide from the I/O-pair* $(V_p(k_f^* \ldots k), W_p(k_f^* \ldots k))$ *of the plant* $\mathcal{P}$ *whether the fault* f *has occurred or not and which is the current state* $z_p(k)$ *of the faulty plant while guaranteeing that the faulty plant* $\mathcal{P}$ *does not execute any illegal transitions.*

In Proposition 4.7 a criterion for identifiability with active diagnosis which requires the existence of a homing sequence for the overall model $\bar{A}_\Delta$ of the closed-loop system has been presented. Here additionally the avoidance of illegal transitions has to be taken into account. Therefore, now *safe* homing sequences need to exist. In the next Section 4.8.3 the term "safe homing sequence" is explained in more detail for an arbitrary automaton $\mathcal{A}$, whereas afterwards the computation of a safe homing sequence for the overall model $\bar{A}_\Delta$ of the closed-loop system is presented.

Again, the combination of safe detectability and safe identifiability results in safe diagnosability. Analogously to Definition 4.6, diagnosability of a plant $\mathcal{P}$ with respect to a fault set $\mathcal{F}$ and a set $\mathcal{E}_{\text{ill}} \subset \mathcal{Z}_0 \times \mathcal{Z}_0$ of illegal transitions using active safe diagnosis is defined as follows.

Definition 4.10 (Safe diagnosability). *The plant $\mathcal{P}$ in the closed-loop system $\mathcal{P} \circ \mathcal{C}_T(\mathcal{A}_0)$ is said to be safely diagnosable with respect to a fault set $\mathcal{F}$ and a set $\mathcal{E}_{\text{ill}} \subset \mathcal{Z}_0 \times \mathcal{Z}_0$ of illegal transitions if every fault $f \in \mathcal{F}$ is safely detectable and safely identifiable.*

4.8.3 Adaptive safe homing sequence

An adaptive safe homing sequence for an automaton $\mathcal{A}$ with respect to a set $\mathcal{E}_{\text{ill}} \subset \mathcal{Z} \times \mathcal{Z}$ of illegal transitions guarantees that the current state of the automaton $\mathcal{A}$ is known after it has been completely applied and that automaton $\mathcal{A}$ does not pass through any illegal transitions $(z_p(k), z_p(k+1)) \in \mathcal{E}_{\text{ill}}$ in the meantime. Therefore, the term adaptive safe homing sequence is defined as an extension of the adaptive homing sequence in Definition 4.2 as follows.

Definition 4.11 (Adaptive safe homing sequence). *An adaptive safe homing sequence for an automaton $\mathcal{A}$ with respect to an initial uncertainty $\mathcal{Z}^* \subseteq \mathcal{Z}$ and a set $\mathcal{E}_{\text{ill}} \subset \mathcal{Z} \times \mathcal{Z}$ of illegal transitions is an adaptive homing sequence for $\mathcal{A}$ with respect to $\mathcal{Z}^*$ as given in Definition 4.2 for which additionally the following condition holds for every state z in a set $\mathcal{Z}_l \subseteq \mathcal{Z}^*$ corresponding to a leaf in the adaptive safe homing sequence:*

$$(G^\infty(z, V(0 \dots k)), G^\infty(z, V(0 \dots k+1))) \notin \mathcal{E}_{\text{ill}}, \quad \forall 0 \leq k < k_e, \qquad (4.64)$$

where $V(0 \dots k_e)$ is the input sequence formed by the vertex labels on the path from the root to the leaf.

An adaptive safe homing sequence for an automaton $\mathcal{A} = (\mathcal{Z}, \mathcal{V}, \mathcal{W}, G, H, z_0)$ with respect to an initial uncertainty $\mathcal{Z}^* \subseteq \mathcal{Z}$ and a set $\mathcal{E}_{\text{ill}} \subset \mathcal{Z} \times \mathcal{Z}$ of illegal transitions can be found during the construction of a special version of the extended homing tree described in Section 4.2.2 called *safe homing tree*. There is only one difference between the safe homing tree and the extended homing tree that is used in order to find an adaptive homing sequence:

- During the construction of the safe homing tree, a new branch and its corresponding new vertex x' is only added if no illegal transition is enforced by it.

That is, in order to avoid the use of illegal transitions, a new pruning rule is introduced. Starting from a vertex x containing a state set $\mathcal{Z}_x$, a new edge labeled with the I/O-pair $(v, w) \in \mathcal{V} \times \mathcal{W}$ and the corresponding new vertex x' with state set $\mathcal{Z}_{x'}$ given in (4.6) are only added to the safe homing tree if the following relation holds:

$$(z, G(z, v)) \notin \mathcal{E}_{\text{ill}}, \quad \forall z \in \mathcal{Z}_x. \tag{4.65}$$

In Section 4.2.3 it has been described how an adaptive homing sequence can be constructed from a given extended homing tree. This method is now analyzed for its applicability to the construction of an adaptive safe homing sequence based on a safe homing tree.

For the extended homing tree it has been shown in Proposition 4.2 that all states within the set $\mathcal{Z}_x^{\text{pre}}$ of predecessors in (4.8) of a set $\mathcal{Z}_x$ corresponding to a vertex x, fulfill the properties (4.3) and (4.4) of an adaptive homing sequence if $\mathcal{Z}_x$ is a singleton. It turns out that the additional property (4.64) of an adaptive safe homing sequence is also fulfilled in this case if the safe homing tree described above is considered.

Proposition 4.16 (Set corresponding to leaf in adaptive safe homing sequence). *In a safe homing tree, consider a vertex x with corresponding state set $\mathcal{Z}_x = \{z'\}$ and the I/O-pair $(V_x(0 \ldots l-1), W_x(0 \ldots l-1))$ formed by the edge labels on the path from the root vertex to the vertex x.*

Then for all states z from the set $\mathcal{Z}_x^{\text{pre}}$ in (4.8) property (4.64) in Definition 4.11 is fulfilled with $V(0 \ldots k_e) = V_x(0 \ldots l-1)$.

Proof. (By induction) From (4.6) and (4.65) it follows that for every state set $\mathcal{Z}_{x'}$ corresponding to a new vertex x' in the safe homing tree the following equation holds with respect to its parent x:

$$\mathcal{Z}_{x'} = \Big\{ z' \in \mathcal{Z} : \Big((\exists z \in \mathcal{Z}_x) \big[z' = G(z, v) \big] \wedge [w = H(z, v)] \Big)$$
$$\wedge \Big((z, G(z, v)) \notin \mathcal{E}_{\text{ill}} \, \forall z \in \mathcal{Z}_x \Big) \Big\}.$$

That is, if x is the root vertex such that the input sequence $V_{x'}(0 \ldots l-1)$ corresponding to the vertex x' consists of only one element, property (4.64) is directly fulfilled. For all further vertices x' it can be shown that (4.64) is fulfilled if it is fulfilled for the parent x, because the second line of the above equation guarantees the fulfillment of (4.64) for $k = l - 2$. $\qquad \square$

Consequently, an adaptive safe homing sequence can be constructed based on a safe homing tree in exactly the same way as described in Section 4.2.2 for obtaining an adaptive homing sequence from an extended homing tree. Particularly, Algorithm 4.2 can be used for its construction. The following proposition is the correspondent to Proposition 4.3 on the adaptive homing sequence obtained from an extended homing tree.

Proposition 4.17 (Adaptive safe homing sequence). *A tree, which is constructed by the method described above and whose leafs contain state sets Z_l for which*

$$\mathcal{Z}^* = \bigcup_{\text{all leaves}} Z_l \tag{4.66}$$

holds, is an adaptive safe homing sequence for the automaton $\mathcal{A}$ with respect to the initial uncertainty $\mathcal{Z}^* \subseteq \mathcal{Z}$ and the set $\mathcal{E}_{ill} \subset \mathcal{Z} \times \mathcal{Z}$ of illegal states according to Definition 4.11.

Proof. Since the safe homing tree is also an extended homing tree, it is guaranteed by Proposition 4.3 that the tree constructed from it is an adaptive homing sequence. According to Definition 4.11, additionally (4.64) has to be satisfied such that the avoidance of illegal transitions is guaranteed. This has been proved in Proposition 4.16. □

Example 4.3 *Construction of safe homing sequences*

Consider again automaton $\mathcal{A}$ in Fig. 4.3. An adaptive safe homing sequence for automaton $\mathcal{A}$ with respect to the set

$$\mathcal{E}_{ill} = \{(4,3)\} \tag{4.67}$$

of illegal transitions shall be constructed. Therefore, the safe homing tree for automaton $\mathcal{A}$ is defined. The result is shown in Fig. 4.13. Compared to the extended homing tree in Fig. 4.5 the right part is missing, because these vertices correspond to illegal transitions. From the root vertex no edges labeled with input symbol $v = v_1$ emerge, because for $z = 4 \in \mathcal{Z}_x$ (4.65) is not fulfilled.

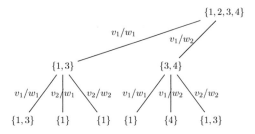

Figure 4.13: Safe homing tree for automaton $\mathcal{A}$ in Fig. 4.3.

From the safe homing tree in Fig. 4.13 with the same method as in Example 4.1 an adaptive safe homing sequence is constructed. It turns out that the adaptive homing sequence in Fig. 4.4 is also an adaptive *safe* homing sequence with respect to the set $\mathcal{E}_{ill}$ of illegal transitions in (4.67). □

4.8.4 Adaptive safe homing sequence for the overall model of the closed-loop system

In the previous sections it has been described how an adaptive safe homing sequence for an arbitrary automaton $\mathcal{A}$ with respect to an initial uncertainty $\mathcal{Z}^* \subseteq \mathcal{Z}$ and a set $\mathcal{E}_{\text{ill}} \in \mathcal{Z} \times \mathcal{Z}$ of illegal transitions can be found based on a safe homing tree. In this section it is shown how these concepts can be used for the active safe diagnosis.

According to Problem 4.3 it is aimed to identify the present fault $\bar{f}$ and current state $z_p(k_d)$ of the plant $\mathcal{P}$ within the overall model $\bar{A}_\Delta$ of the closed-loop system in (4.14). The initial uncertainty for this homing problem is given by the state set $\bar{\mathcal{Z}}^*(k_f^*)$ in (4.33). The first input symbol has already been applied in form of the value $r(k_f^*)$ of the reference sequence generated by the trajectory planning unit $\mathcal{T}(\mathcal{A}_0, \mathcal{E}_{\text{ill}})$ at the fault detection time k_f^*. Additionally it has to be guaranteed that the plant $\mathcal{P}$ with inputs from the controller $\mathcal{C}(\mathcal{A}_0)$ does not execute any illegal transitions $(z_p(k), z_p(k+1)) \in \mathcal{E}_{\text{ill}} \subset \mathcal{Z}_0 \times \mathcal{Z}_0$ during the active safe diagnosis (cf. (4.60)).

With respect to the overall model $\bar{A}_\Delta$ of the closed-loop system, the set $\mathcal{E}_{\text{ill}}$ of illegal transitions for the plant $\mathcal{P}$ in (2.45) can be restated as

$$\bar{\mathcal{E}}_{\text{ill}} = \{(\bar{z}_\Delta, \bar{z}'_\Delta) \in \bar{\mathcal{Z}}_\Delta \times \bar{\mathcal{Z}}_\Delta : (\bar{z}_{\Delta 2}, \bar{z}'_{\Delta 2}) \in \mathcal{E}_{\text{ill}}\} \subset \bar{\mathcal{Z}}_\Delta \times \bar{\mathcal{Z}}_\Delta. \tag{4.68}$$

That is, a transition $(\bar{z}_\Delta, \bar{z}'_\Delta)$ is called illegal in the overall model $\bar{A}_\Delta$ of the closed-loop system if the contained transition of the plant $\mathcal{P}$ is illegal, regardless of the state of the controller $\mathcal{C}(\mathcal{A}_0)$.

Consequently an adaptive safe homing sequence for the overall model $\bar{A}_\Delta$ of the closed-loop system in (4.14) with respect to every initial uncertainty $\{z_c\} \times \mathcal{Z}_\Delta$, $(z_c \in \mathcal{Z}_c)$ starting with every possible input symbol $r(k_f^*)$ such that $G_c(z_c, v_c)!$ and the set $\bar{\mathcal{E}}_{\text{ill}}$ of illegal transitions in (4.68) has to be found. In case that such adaptive safe homing sequences for all $z_c \in \mathcal{Z}_c$ exist, the plant $\mathcal{P}$ is diagnosable with active safe diagnosis as given by the following criterion.

Proposition 4.18 (Safe identifiability). *Let* $\tilde{\mathcal{V}}_c(z_c) = \{v_c \in \mathcal{V}_c : G_c(z_c, v_c)! \wedge (z_c, v_c) \notin \mathcal{E}_{\text{ill}}\}$ *be the set of all possible values of the reference trajectory at the fault detection time. Then all faults $f \in \mathcal{F}$ are safely identifiable in the closed-loop system $\mathcal{P} \circ \mathcal{C}_T(\mathcal{A}_0, \mathcal{E}_{\text{ill}})$ with respect to a set $\mathcal{E}_{\text{ill}} \subset \mathcal{Z}_0 \times \mathcal{Z}_0$ of illegal transitions and the reference trajectory $R(0 \ldots k_e)$ generated by the trajectory planning unit $\mathcal{T}(\mathcal{A}_0)$ if and only if for the overall model $\bar{A}_\Delta$ of the closed-loop system in (4.14) there exists an adaptive safe homing sequence whose first input symbol is v_c with respect to*

- *the set $\bar{\mathcal{E}}_{\text{ill}}$ of illegal transitions in (4.68),*
- *any initial uncertainty $\mathcal{Z}^* = \{z_c\} \times \mathcal{Z}_\Delta$, $(z_c \in \mathcal{Z}_c)$ and*
- *every $v_c \in \tilde{\mathcal{V}}_c(z_c)$.*

Proof. The proof follows the same argumentation as the one of Proposition 4.7 on the identifiability of a fault with active diagnosis. The only difference is that here the adaptive safe homing sequence has to additionally ensure that the plant $\mathcal{P}$ does not pass through any illegal transitions $(z_\mathrm{p}(k), z_\mathrm{p}(k+1)) \in \mathcal{E}_\mathrm{ill}$ for any $k \geq k_\mathrm{f}^*$. This is always guaranteed, because the adaptive safe homing sequence is defined with respect to the set $\bar{\mathcal{E}}_\mathrm{ill}$ of illegal transitions for the closed-loop system, which, according to (4.68), covers all illegal transitions $(z, z') \in \mathcal{E}_\mathrm{ill}$ of the plant $\mathcal{P}$. The additional condition in the set $\tilde{\mathcal{V}}_\mathrm{c}(z_\mathrm{c})$ accounts for the fact that the trajectory planning unit $\mathcal{T}(\mathcal{A}_0, \mathcal{E}_\mathrm{ill})$ will never request any illegal transitions from the controller $\mathcal{C}(\mathcal{A}_0)$ such that some input symbols v_c are never applied in some states z_c. $\qquad\square$

When the conditions for safe detectability in Proposition 4.15 and for safe identifiability in Proposition 4.18 are combined, the following condition for safe diagnosability as defined in Definition 4.10 results.

Theorem 4.5 (Safe diagnosability with active diagnosis). *Let $z_\mathrm{p}(k_\mathrm{f})$ be the state of the plant $\mathcal{P}$ at the fault occurrence time $k_\mathrm{f} \geq 0$. Let $V_\mathrm{p}(0 \ldots k_\mathrm{e}) = \Phi_\mathrm{c}(z_\mathrm{c0}, R(0 \ldots k_\mathrm{e}))$ be the input sequence for the plant $\mathcal{P}$ generated by the nominal tracking controller $C_T(\mathcal{A}_0, \mathcal{E}_\mathrm{ill})$. Let $\tilde{\mathcal{V}}_\mathrm{c}(z_\mathrm{c}) = \{v_\mathrm{c} \in \mathcal{V}_\mathrm{c} : G_\mathrm{c}(z_\mathrm{c}, v_\mathrm{c})! \wedge (z_\mathrm{c}, v_\mathrm{c}) \notin \mathcal{E}_\mathrm{ill}\}$ be the set of all possible values of the reference trajectory at the fault detection time.*

Then the plant $\mathcal{P}$ in the closed-loop system $\mathcal{P} \circ C_T(\mathcal{A}_0, \mathcal{E}_\mathrm{ill})$ is safely diagnosable with active diagnosis with respect to the fault set $\mathcal{F}$ and the set $\mathcal{E}_\mathrm{ill} \subset \mathcal{Z}_0 \times \mathcal{Z}_0$ of illegal transitions if and only if all of the following conditions are fulfilled:

- *(Safe detectability) For all faults $f \in \mathcal{F}$ there exists a time $k_\mathrm{f}^* \leq k_\mathrm{e}$ such that*

$$\Phi_f(z_\mathrm{p}(k_\mathrm{f}), V_\mathrm{p}(k_\mathrm{f} \ldots k_\mathrm{f}^*)) \neq \Phi_0(z_\mathrm{p}(k_\mathrm{f}), V_\mathrm{p}(k_\mathrm{f} \ldots k_\mathrm{f}^*)),$$

   and

$$(G_f^\infty(z_\mathrm{p}(k_\mathrm{f}), V_\mathrm{p}(k_\mathrm{f} \ldots k-1)), G_f^\infty(z_\mathrm{p}(k_\mathrm{f}), V_\mathrm{p}(k_\mathrm{f} \ldots k))) \notin \mathcal{E}_\mathrm{ill}, \quad \forall k_\mathrm{f} \leq k < k_\mathrm{f}^*.$$

- *(Safe identifiability) For the overall model $\bar{A}_\Delta$ of the closed-loop system in (4.14) there exists an adaptive safe homing sequence whose first input symbol is v_c with respect to*
   - *the set $\bar{\mathcal{E}}_\mathrm{ill}$ of illegal transitions in (4.68),*
   - *any initial uncertainty $\mathcal{Z}^* = \{z_\mathrm{c}\} \times \mathcal{Z}_\Delta$, $(z_\mathrm{c} \in \mathcal{Z}_\mathrm{c})$ and*
   - *every $v_\mathrm{c} \in \tilde{\mathcal{V}}_\mathrm{c}(z_\mathrm{c})$.*

Proof. The first condition is necessary and sufficient for the safe detectability of all faults $f \in \mathcal{F}$ (Proposition 4.15), while the second condition is necessary and sufficient for the safe identifiability of all faults (Proposition 4.18). Consequently, the fulfillment of both conditions at a time is necessary and sufficient for safe diagnosability of the plant $\mathcal{P}$ with respect to a set $\mathcal{E}_{ill}$ of illegal transitions according to Definition 4.10. □

Example 4.4 *Active safe diagnosis of automated warehouse*

The safe diagnosability of the automated warehouse example considering the set $\mathcal{E}_{ill}$ of illegal transitions in (2.54) is analyzed based on Theorem 4.5. That is, during the fault diagnosis it has to be ensured that the direct route between position A and position B is avoided.

First safe detectability according to Definition 4.8 is considered. Fault $f = 1$ is not safely detectable, because a transport from position A to position B is never requested by the tracking controller $\mathcal{C}_T(\mathcal{A}_0, \mathcal{E}_{ill})$. However, since the presence of these illegal transitions guarantees that the direct route between position A and position B is never used for the transport of parcels, fault $f = 1$ corresponding to the blocking of the route by an obstacle does not affect the behavior of the controlled plant anyhow and does not violate the safety constraint in (4.63).

Fault $f = 2$ is safely detectable with respect to any input sequence generated by the nominal tracking controller $\mathcal{C}_T(\mathcal{A}_0, \mathcal{E}_{ill})$, because the tracking controller always requests a use of the robot M such that the plant generates an output $w_p \in \{BkA, BkB, BkC\}$ which will never be generated by the faultless plant $\mathcal{P} \vDash \mathcal{A}_0$. The safety constraint in (4.63) is never violated, because the nominal tracking controller $\mathcal{C}_T(\mathcal{A}_0, \mathcal{E}_{ill})$ never generates the input symbols $v_p \in \{AB^*, BA^*\}$ for the plant, which are the only inputs leading to an illegal transition in $\mathcal{A}_2$ in Fig. 2.7.

According to Theorem 4.5 the automated warehouse can therefore not be safely diagnosable with respect to the fault set $\mathcal{F} = \{1, 2\}$ and the set $\mathcal{E}_{ill}$ of illegal transitions in (2.54). However, since the undetectable fault $f = 1$ does not lead to a violation of the safety constraint in (4.63) and does not affect the behavior of the controlled plant anyhow, the existence of the adaptive safe homing sequences for the overall model $\bar{A}_\Delta$ of the closed-loop system in Fig. 4.8 with respect to the set $\bar{\mathcal{E}}_{ill}$ of illegal transitions in (4.68) and any initial uncertainty $\mathcal{Z}^* = \{z_c\} \times \mathcal{Z}_\Delta$ with $\mathcal{Z}_\Delta$ in (2.57) and $z_c \in \mathcal{Z}_c = \{A, B, C\}$ is proved anyhow.

For this example according to (4.68) the set of illegal transitions for the overall model $\bar{A}_\Delta$ of the closed-loop system based on the set $\mathcal{E}_{ill}$ of illegal transitions for the plant $\mathcal{P}$ in (2.54) is given by

$$\bar{\mathcal{E}}_{ill} = \left\{ (\bar{z}_\Delta, \bar{z}'_\Delta) \in \bar{\mathcal{Z}}_\Delta \times \bar{\mathcal{Z}}_\Delta : (\bar{z}_{\Delta 2}, \bar{z}'_{\Delta 2}) \in \{(A, B), (B, A)\} \right\}$$
$$= \left\{ \left(\begin{pmatrix} A \\ A \end{pmatrix}, \begin{pmatrix} A \\ B \end{pmatrix} \right), \left(\begin{pmatrix} A \\ A \end{pmatrix}, \begin{pmatrix} B \\ B \end{pmatrix} \right), \dots, \left(\begin{pmatrix} A \\ B \end{pmatrix}, \begin{pmatrix} A \\ A \end{pmatrix} \right), \dots, \left(\begin{pmatrix} C \\ B \end{pmatrix}, \begin{pmatrix} C \\ A \end{pmatrix} \right) \right\}. \quad (4.69)$$

First, a safe homing tree has to be constructed for every initial uncertainty $\mathcal{Z}^* = \{z_c\} \times \mathcal{Z}_\Delta$, $(z_c \in \{A, B, C\})$ as described in Section 4.8.3. The safe homing trees contain 73, 46 and 53 vertices, respectively. Therefore, in Fig. 4.14 only a part of the safe homing tree for $\bar{A}_\Delta$ with respect to the initial uncertainty $\mathcal{Z}^* = \{A\} \times \mathcal{Z}_\Delta$ and the set $\bar{\mathcal{E}}_{ill}$ of illegal transitions in (4.69) is shown.

Its root vertex is labeled with the initial uncertainty $\mathcal{Z}^* = \{A\} \times \mathcal{Z}_\Delta$. From the root vertex edges labeled with I/O-pairs

$$(v, w) \in \bar{V}_\Delta \times \bar{W}_\Delta = \{A, B, C\} \times \{ok, np, BkA, BkB, BkC\}$$

emerge. The topmost edge labeled (C, BkA) leads to a vertex x with state set

$$
\mathcal{Z}_x = \left\{ \begin{pmatrix} C \\ A \\ 2 \end{pmatrix} \right\}.
$$

Since $\mathcal{Z}_x$ is a singleton, no further edges start from it. Considering that the automaton $\bar{A}_\Delta$ is known to be in one of the states within the initial uncertainty $\mathcal{Z}^*$, this vertex x states that if the input symbol $v(0) = C$ is applied to $\bar{A}_\Delta$ and the output $w(0) = \mathrm{BkA}$ is observed, $\bar{A}_\Delta$ is known to be in state

$$
\bar{z}_\Delta(1) = \begin{pmatrix} C \\ A \\ 2 \end{pmatrix} \in \mathcal{Z}_x.
$$

In contrast, the edge labeled (C, np) leads to a vertex x containing the state set

$$
\mathcal{Z}_x = \left\{ \begin{pmatrix} C \\ B \\ 1 \end{pmatrix}, \begin{pmatrix} C \\ C \\ 1 \end{pmatrix}, \begin{pmatrix} C \\ B \\ 2 \end{pmatrix}, \begin{pmatrix} C \\ C \\ 2 \end{pmatrix} \right\}
$$

such that the state of $\bar{A}_\Delta$ is not known exactly when the output symbol $w(0) = \mathrm{np}$ is observed. Consequently, there emerge some edges from this vertex x which correspond to the application of another input symbol $v(1)$.

Based on the safe homing trees for the three considered initial uncertainties, adaptive safe homing sequences are constructed as described in Section 4.8.3. In the following, the construction of an adaptive safe homing sequence with respect to the initial uncertainty

$$
\mathcal{Z}^* = \{A\} \times \mathcal{Z}_\Delta \tag{4.70}
$$

using Algorithm 4.2 is described. Initially, the set $\mathcal{S}$ of candidates for the adaptive safe homing sequence is empty.

The first level in the safe homing tree which contains a vertex x whose corresponding state set $\mathcal{Z}_x$ is a singleton is the level $l = 1$ (Line 3). This is for example the case for the topmost vertex x in the first level of the safe homing tree in Fig. 4.14, which is labeled with the state set

$$
\mathcal{Z}_x = \left\{ \begin{pmatrix} C \\ A \\ 2 \end{pmatrix} \right\}. \tag{4.71}
$$

The I/O-pair at the edge labels on the path from the root vertex to the vertex x is given by

$$
(V_x(0 \ldots l - 1), W_x(0 \ldots l - 1)) = (C, \mathrm{BkA})
$$

(Line 4). Therefore, the tree T_{new}, which is the topmost one in Fig. 4.15, is added to the set $\mathcal{S}$ (Line 5). Its root vertex corresponds to the input symbol $v_x(0) = C$, its leaf is labeled by the set

$$
\mathcal{Z}_1 = \mathcal{Z}_x^{\mathrm{pre}} = \left\{ \begin{pmatrix} A \\ A \\ 2 \end{pmatrix} \right\}
$$

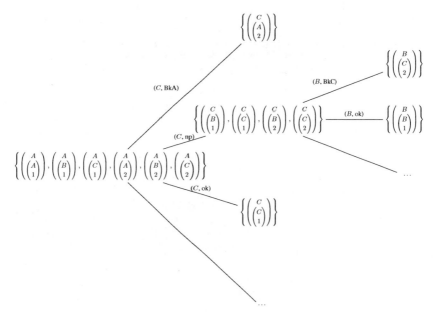

Figure 4.14: Part of safe homing tree for $\bar{A}_\Delta$ with respect to $\mathcal{Z}^*$ in (4.70) and $\bar{\mathcal{E}}_{\text{ill}}$ in (4.69).

according to (4.8) and the state

$$z_1 = \left(\begin{pmatrix} C \\ A \\ 2 \end{pmatrix} \right) \in \mathcal{Z}_x$$

and its edge corresponds to the output symbol $w_x(0) = \text{BkA}$. Since there are no other trees in $\mathcal{S}$, the for-loop in Line 2 goes to the next iteration.

Another vertex with a singleton state set is the vertex x corresponding to the state set

$$\mathcal{Z}_x = \left\{ \begin{pmatrix} C \\ C \\ 1 \end{pmatrix} \right\},$$

whose corresponding I/O-pair is given by

$$(V_x(0 \ldots l-1), W_x(0 \ldots l-1)) = (C, \text{ok}).$$

Again the corresponding linear graph T_{new} (middle graph in Fig. 4.15) is added to the set $\mathcal{S}$. Additionally it is checked whether T_{new} can be united with the graph already in $\mathcal{S}$ (Line 7). Since the root vertices of both trees contain the same input symbol $v = C$ and the output symbols at their branches differ, the trees can be united such that the lowermost graph in Fig. 4.15 results (Line 8).

Since there is no tree in the set $\mathcal{S}$ in whose leaves the state sets $\mathcal{Z}_l$ contain all states $z \in \{A\} \times \mathcal{Z}_\Delta$, no adaptive homing sequence has been found yet (Line 10) and the algorithm continues by considering another vertex within the first level of the safe homing tree (Line 2). When all vertices within the first level of the safe homing tree in Fig. 4.14 have been analyzed, the algorithm proceeds with the following levels $l = 2, 3, \ldots$ until eventually the adaptive safe homing sequence shown in Fig. 4.16 results.

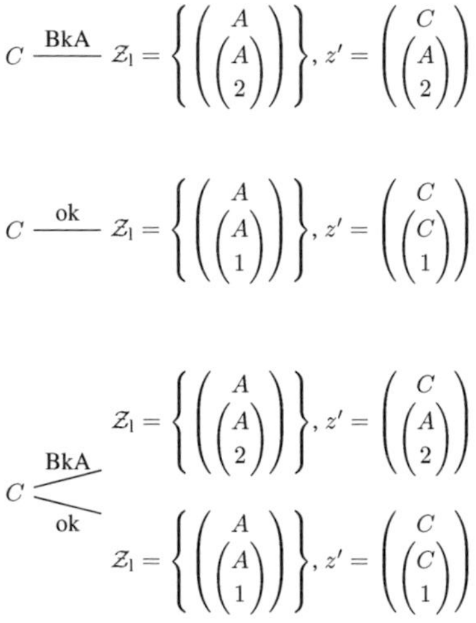

$$
C \xrightarrow{\text{BkA}} \mathcal{Z}_1 = \left\{ \left(\begin{pmatrix} A \\ A \\ 2 \end{pmatrix} \right) \right\}, z' = \left(\begin{pmatrix} C \\ A \\ 2 \end{pmatrix} \right)
$$

$$
C \xrightarrow{\text{ok}} \mathcal{Z}_1 = \left\{ \left(\begin{pmatrix} A \\ A \\ 1 \end{pmatrix} \right) \right\}, z' = \left(\begin{pmatrix} C \\ C \\ 1 \end{pmatrix} \right)
$$

$$
C \begin{array}{c} \xrightarrow{\text{BkA}} \\ \xrightarrow{\text{ok}} \end{array} \begin{array}{l} \mathcal{Z}_1 = \left\{ \left(\begin{pmatrix} A \\ A \\ 2 \end{pmatrix} \right) \right\}, z' = \left(\begin{pmatrix} C \\ A \\ 2 \end{pmatrix} \right) \\[2em] \mathcal{Z}_1 = \left\{ \left(\begin{pmatrix} A \\ A \\ 1 \end{pmatrix} \right) \right\}, z' = \left(\begin{pmatrix} C \\ C \\ 1 \end{pmatrix} \right) \end{array}
$$

Figure 4.15: Trees within set $\mathcal{S}$ after two vertices with singleton sets have been found.

According to Proposition 4.17, the tree in Fig. 4.16 is an adaptive safe homing sequence for the overall model $\bar{A}_\Delta$ of the closed-loop system in Fig. 4.8 with respect to the initial uncertainty $\mathcal{Z}^* = \{A\} \times \mathcal{Z}_\Delta$ and the set $\bar{\mathcal{E}}_{\text{ill}}$ of illegal transitions in (4.69), because all states $z \in \mathcal{Z}^*$ are contained in the state sets $\mathcal{Z}_1$ of its leaves. The adaptive safe homing sequence applies to the case that the value of the reference trajectory at the fault detection time is given by

$$
r(k_{\text{f}}^*) = C.
$$

With respect to the initial uncertainties $\mathcal{Z}^* = \{B\} \times \mathcal{Z}_\Delta$ and $\mathcal{Z}^* = \{C\} \times \mathcal{Z}_\Delta$ and all other possible value of the reference trajectory at the fault detection time, adaptive safe homing sequences can be found as well such that even though according to Theorem 4.5 the plant $\mathcal{P}$ in the closed-loop system $\mathcal{P} \circ \mathcal{C}_{\mathcal{T}}(\mathcal{A}_0, \mathcal{E}_{\text{ill}})$ is not safely diagnosable, whenever it is known that a fault is present in the plant $\mathcal{P}$, its identification is possible.   □

4.8.5 Active safe fault diagnosis

The safe fault diagnosis can be executed if an adaptive safe homing sequence for the overall model $\bar{A}_\Delta$ of the closed-loop system in (4.14) according to Definition 4.11 has been found for every initial uncertainty $\mathcal{Z}^* = \{z_{\text{c}}\} \times \mathcal{Z}_\Delta$, $(z_{\text{c}} \in \mathcal{Z}_{\text{c}})$ using the method described in the previous section. Compared to the active fault diagnosis without illegal transitions described in Section 4.5, no separating sequences are applied to the controller $\mathcal{C}(\mathcal{A}_0)$, but the adaptive safe

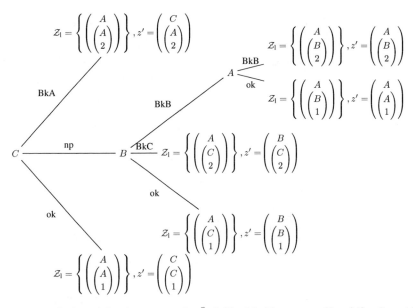

Figure 4.16: Adaptive safe homing sequence for $\bar{A}_\Delta$ in Fig. 4.8 with respect to $\mathcal{Z}^* = \{A\} \times \mathcal{Z}_\Delta$ and $\bar{\mathcal{E}}_{\text{ill}}$ in (4.69).

homing sequence is used to iteratively generate a diagnostic trajectory $R(k_f^* + 1 \ldots k_e)$. The adaptive safe homing sequence corresponding to the state $z_c(k_f^* + 1)$ of the controller $\mathcal{C}(\mathcal{A}_0)$ after the fault detection time k_f^* is applied to the controller $\mathcal{C}(\mathcal{A}_0)$ in accordance with Algorithm 4.1. Therefore, the flowchart of the active fault diagnosis method in Fig. 4.11 is modified such that the flowchart in Fig. 4.17 is obtained.

The following theorem proves that the proposed active safe diagnosis method solves the active safe fault diagnosis problem in Problem 4.3 if the plant $\mathcal{P}$ is diagnosable with active safe diagnosis.

Theorem 4.6 (Diagnostic result with active safe diagnosis). *If the plant $\mathcal{P}$ is safely diagnosable in the closed-loop system $\mathcal{P} \circ \mathcal{C}_\mathcal{T}(\mathcal{A}_0, \mathcal{E}_{\text{ill}})$ with respect to the fault set $\mathcal{F}$ and the set $\mathcal{E}_{\text{ill}} \subset \mathcal{Z}_0 \times \mathcal{Z}_0$ of illegal transitions, the present fault $\bar{f} \in \mathcal{F}$ as well as the current state $z_p(k_d)$ of the plant $\mathcal{P}$ are identified at a finite diagnosis time k_d using active safe diagnosis, while the plant $\mathcal{P}$ does not pass through any illegal transitions $(z_p(k), z_p(k+1)) \in \mathcal{E}_{\text{ill}}$ for any $k_f \leq k < k_d$.*

Proof. The proof follows the same argumentation as the proof of Theorem 4.4 on the diagnostic

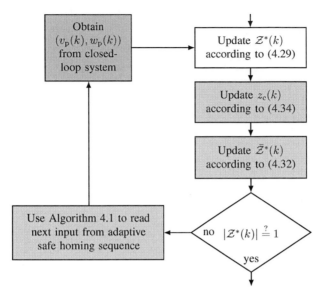

Figure 4.17: Detail of flowchart in Fig. 4.10 for active safe fault diagnosis.

result with active diagnosis. The only difference is that the reference trajectory $R(k_f^* \ldots k_e)$ now has to be a realization of an adaptive *safe* homing sequence instead of a mere adaptive homing sequence and that the plant may not execute any illegal transitions before the fault detection time k_f^*. The latter is guaranteed because the diagnosability condition in Theorem 4.5 requires the safe detectability of all faults $f \in \mathcal{F}$.

The required adaptive safe homing sequence for the overall model $\bar{A}_\Delta$ of the closed-loop system in (4.14) with respect to the set $\bar{\mathcal{E}}_{ill}$ of illegal transitions in (4.68) and any initial uncertainty $\mathcal{Z}^* = \{z_c\} \times \mathcal{Z}_\Delta$ with $z_c \in \mathcal{Z}_c$ always exists if the plant $\mathcal{P}$ in the closed-loop system $\mathcal{P} \circ \mathcal{C}_\mathcal{T}(A_0, \mathcal{E}_{ill})$ is diagnosable using active safe diagnosis with respect to the fault set $\mathcal{F}$ and the set $\mathcal{E}_{ill} \in \mathcal{Z}_0 \times \mathcal{Z}_0$ of illegal transitions (Theorem 4.5).

The application of the adaptive safe homing sequence corresponding to a state $z_c \in \mathcal{Z}_c$ to the controller $\mathcal{C}(A_0)$ guarantees that no illegal transitions $(\bar{z}_\Delta, \bar{z}'_\Delta) \in \bar{\mathcal{E}}_{ill}$ are passed by the overall model $\bar{A}_\Delta$ of the closed-loop system starting in any state $\bar{z}_\Delta \in \mathcal{Z}^* = \{z_c\} \times \mathcal{Z}_\Delta$. From the definition of $\bar{A}_\Delta$ in (4.14) and the definition of $\bar{\mathcal{E}}_{ill}$ in (4.68) it follows that (4.60) is fulfilled if

$$\begin{pmatrix} z_c(k_f^*) \\ \begin{pmatrix} z_p(k_f^*) \\ \bar{f} \end{pmatrix} \end{pmatrix} \in \mathcal{Z}^* = \bar{\mathcal{Z}}^*(k_f^*) = \{z_c\} \times \mathcal{Z}_\Delta, \tag{4.72}$$

hence, the current state $z_p(k_f^* + 1)$ of the plant $\mathcal{P}$ at the fault detection time k_f^* lies in the set $\mathcal{Z}^*(k_f^* + 1) = \mathcal{Z}_\Delta$, which is always true.

In Proposition 4.3 it has been proved that an adaptive homing experiment performed according to Algorithm 4.1 indeed leads to the identification of the current state of the considered automaton if its initial state lies in the initial uncertainty. Since according to (4.72) this condition is fulfilled, it follows that the current state

$$\bar{z}_\Delta(k) = \left(\begin{array}{c} z_c(k) \\ \left(\begin{array}{c} z_c(k) \\ z_p(k) \\ \bar{f} \end{array} \right) \end{array} \right)$$

within the overall model $\bar{A}_\Delta$ of the closed-loop system is identified such that the present fault $\bar{f}$ and the current state $z_p(k)$ of the plant $\mathcal{P}$ are identified. $\qquad \square$

Remark. If the plant $\mathcal{P}$ is not safely diagnosable, no adaptive safe homing sequence which could be used for the fault diagnosis is found. However, in this case it is possible to use some tree from the set $\mathcal{S}$ whose leafs do not contain all states from the initial uncertainty $\mathcal{Z}^*$ for the generation of input symbols. The use of any of these trees guarantees the avoidance of illegal transitions and reduces the current estimate $\mathcal{Z}^*(k)$ of the plant. The only drawback is that it might not be possible to obtain an unambiguous diagnostic result $\mathcal{D}^*$. Further research has to be conducted in order to develop criteria for selecting a tree that is not an adaptive safe homing sequence from the set $\mathcal{S}$ in an "optimal" way.

Example 4.4 (cont.) *Active safe diagnosis of automated warehouse*

Consider again the case that the robot M in the automated warehouse is broken such that fault $\bar{f} = 2$ occurred. For demonstration purposes assume that a parcel is located at position C, hence the initial state of the plant $\mathcal{P}$ is $z_p(0) = C$. Further assume that it is known a priori that a fault $f \in \{1, 2\}$ occurred, even before the tracking controller $\mathcal{C}_T(\mathcal{A}_0, \mathcal{E}_{ill})$ applies a first input $v_p(0)$ to the plant $\mathcal{P}$. Therefore, the current state estimate of the plant is initialized at time $k = 0$ as

$$\mathcal{Z}^*(0) = \mathcal{Z}_\Delta = \left\{ \begin{pmatrix} A \\ 1 \end{pmatrix}, \begin{pmatrix} B \\ 1 \end{pmatrix}, \begin{pmatrix} C \\ 1 \end{pmatrix}, \begin{pmatrix} A \\ 2 \end{pmatrix}, \begin{pmatrix} B \\ 2 \end{pmatrix}, \begin{pmatrix} C \\ 2 \end{pmatrix} \right\}.$$

Since by (3.16) the initial state of the controller $\mathcal{C}(\mathcal{A}_0)$ is known to be $z_c(0) = z_{c0} = A$, the initial uncertainty is given by

$$\bar{\mathcal{Z}}^*(0) = \{A\} \times \mathcal{Z}_\Delta = \left\{ \begin{pmatrix} A \\ \begin{pmatrix} A \\ 1 \end{pmatrix} \end{pmatrix}, \begin{pmatrix} A \\ \begin{pmatrix} B \\ 1 \end{pmatrix} \end{pmatrix}, \begin{pmatrix} A \\ \begin{pmatrix} C \\ 1 \end{pmatrix} \end{pmatrix}, \begin{pmatrix} A \\ \begin{pmatrix} A \\ 2 \end{pmatrix} \end{pmatrix}, \begin{pmatrix} A \\ \begin{pmatrix} B \\ 2 \end{pmatrix} \end{pmatrix}, \begin{pmatrix} A \\ \begin{pmatrix} C \\ 2 \end{pmatrix} \end{pmatrix} \right\}.$$

Consequently, the adaptive safe homing sequence in Fig. 4.16 can be used for the active safe fault diagnosis. The course of all important variables during the active safe diagnosis is displayed in Fig. 4.18.

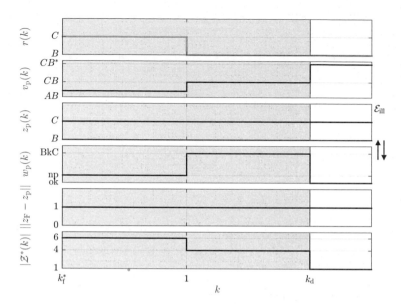

Figure 4.18: Active safe diagnosis of automated warehouse.

The active safe fault diagnosis is executed according to the flowchart in Fig. 4.17. That is, since $|\mathcal{Z}^*(0)| = 6 > 1$, Algorithm 4.1 is used to apply the first input $v_c(0) = C$ from the adaptive safe homing sequence in Fig. 4.16 to the controller $\mathcal{C}(\mathcal{A}_0)$ (Line 1). Therefore, the controller $\mathcal{C}(\mathcal{A}_0)$ in Fig. 3.2 requests a transport from position A to position C by robot M and hence generates the input

$$v_p(0) = w_c(0) = H_c(z_c(0), v_c(0)) = H_c(A, C) = AC$$

for the plant $\mathcal{P}$ and goes to state

$$z_c(1) = G_c(z_c(0), v_c(0)) = G_c(A, C) = C.$$

As a result the plant $\mathcal{P}$ generates the output

$$w_p(0) = H_2(z_p(0), v_p(0)) = H_2(C, AC) = \text{np},$$

which states that the transport is currently not possible, because the parcel is located at position C instead of position A. The plant remains in state

$$z_p(1) = G_2(z_p(0), v_p(0)) = G_2(C, AC) = C.$$

When the current state estimates of the plant and of the closed-loop system are updated according to (4.29) and (4.32), respectively, the following sets result:

$$\mathcal{Z}^*(1) = \left\{ \begin{pmatrix} B \\ 1 \end{pmatrix}, \begin{pmatrix} C \\ 1 \end{pmatrix}, \begin{pmatrix} B \\ 2 \end{pmatrix}, \begin{pmatrix} C \\ 2 \end{pmatrix} \right\} \tag{4.73}$$

$$\bar{\mathcal{Z}}^*(1) = \left\{ \begin{pmatrix} C \\ \begin{pmatrix} B \\ 1 \end{pmatrix} \end{pmatrix}, \begin{pmatrix} C \\ \begin{pmatrix} C \\ 1 \end{pmatrix} \end{pmatrix}, \begin{pmatrix} C \\ \begin{pmatrix} B \\ 2 \end{pmatrix} \end{pmatrix}, \begin{pmatrix} C \\ \begin{pmatrix} C \\ 2 \end{pmatrix} \end{pmatrix} \right\}. \tag{4.74}$$

Again the current state estimate $\mathcal{Z}^*(1)$ of the plant is no singleton such that the next input from the adaptive safe homing sequence in Fig. 4.16 is applied. The next vertex x reached by the edge labeled $w_p(0) = $ np is labeled with the input symbol B (Line 3 in Algorithm 4.1). Since x is not a leaf, the input $v_c(1) = B$ is applied to the controller $\mathcal{C}(\mathcal{A}_0)$ (Lines 4–7).

As a result the controller $\mathcal{C}(\mathcal{A}_0)$ passes into state $z_c(2) = B$ and generates input $v_p(1) = w_c(1) = CB$ for the plant $\mathcal{P}$, which remains in state $z_p(2) = C$ and generates the output $w_p(1) = $ BkC. That is, since the robot M is broken, the parcel can not be transported from position C to position B. The updated current state estimates are given by

$$\mathcal{Z}^*(2) = \left\{ \begin{pmatrix} C \\ 2 \end{pmatrix} \right\} \text{ and } \bar{\mathcal{Z}}^*(2) = \left\{ \begin{pmatrix} B \\ \begin{pmatrix} C \\ 2 \end{pmatrix} \end{pmatrix} \right\},$$

respectively. That is, at the diagnosis time $k_d = 2$, the present fault $\bar{f} = 2$ and the current state $z_p(2) = C$ of the plant are correctly identified as predicted by Theorem 4.6:

$$\mathcal{D}^* = \mathcal{Z}^*(k_d) = \left\{ \begin{pmatrix} C \\ 2 \end{pmatrix} \right\}. \tag{4.75}$$

Note that the adaptive safe homing sequence in Fig. 4.16 also reveals this result, because the leaf reached by the edge labeled BkC is labeled with the current state

$$z' = \begin{pmatrix} B \\ \begin{pmatrix} C \\ 2 \end{pmatrix} \end{pmatrix} = \begin{pmatrix} z_c(k_d) \\ \begin{pmatrix} z_p(k_d) \\ \bar{f} \end{pmatrix} \end{pmatrix}.$$

$\square$

4.8.6 Complexity of the safe fault diagnosis method

Complexity of the diagnosability analysis. The diagnosability analysis and the active safe diagnosis itself require the construction of adaptive safe homing sequences based on a safe homing tree. In the vertices of the safe homing tree all subsets of the state set $\bar{\mathcal{Z}}_\Delta$ of the overall model $\bar{\mathcal{A}}_\Delta$ can occur. Therefore, the number of vertices in a safe homing tree is bounded by $2^{\bar{n}_\Delta}$ with $\bar{n}_\Delta = n^2 \cdot F$ according to (4.16). Since for each state $z_c \in \mathcal{Z}_c$ a separate safe homing tree has to be constructed, the overall computational complexity is of order $\mathcal{O}(n \cdot 2^{n^2 \cdot F})$.

If the adaptive safe homing sequences based on Algorithm 4.2 are constructed during the definition of the safe homing tree, the main complexity lies in the definition of the trees which are candidates for the adaptive safe homing sequence. In the worst case every new tree can be combined with every previously defined tree such that the number of trees in the set $\mathcal{S}$ grows exponentially with the number of leafs in the safe homing tree.

Consequently, the computational complexity for checking the diagnosability with active safe diagnosis is exponential in the number of states in the plant model and the method is restricted to "sufficiently small" plants. However, the construction of adaptive safe homing sequences based on a safe homing tree is only a straightforward method. It is likely that there exist methods for their construction which are, at least on average, more efficient.

Complexity of the online execution of the safe diagnosis method.   Once the adaptive safe homing sequences have been found, the online execution of the active safe diagnosis is of linear complexity, because the update of the current state estimate $\bar{\mathcal{Z}}^*(k)$ is of order $\mathcal{O}(n \cdot F)$ (cf. Section 4.7.2) and the selection of the next input symbol from the adaptive safe homing sequence according to Algorithm 4.1 only requires to find the edge in the adaptive safe homing sequence which is labeled with the observed output symbol. Consequently, the overall complexity is of order $\mathcal{O}(n \cdot F + q)$. That is, the online execution of the active safe diagnosis is a computationally simple task. The challenge only lies in the offline construction of the adaptive safe homing sequences.

5 Reconfiguration of the tracking controller

This chapter deals with the reconfiguration of the tracking controller based on an unambiguous or ambiguous diagnostic result. For the first case it is shown that the controllability of the faulty plant is a necessary and sufficient condition for the reconfigurability of the tracking controller. A method for the reconfiguration of the tracking controller by re-planning the trajectory and modifying part of the transitions of the controller is proposed, which guarantees that the control aim is fulfilled by the faulty plant in the closed-loop system with the reconfigured tracking controller. In case of an ambiguous diagnostic result it is proposed to reconfigure the tracking controller based on a common model of the faulty plant that describes the identical part of the behavior of the plant under all fault candidates in the diagnostic result.

5.1 Reconfiguration setting

The aim of the reconfiguration unit $\mathcal{R}$ is to use the diagnostic result $\mathcal{D}^* \subseteq \mathcal{Z}_\Delta$ provided by the diagnostic unit $\mathcal{D}$ to find a reconfigured tracking controller $\mathcal{C}_{\mathcal{T}}^r$ which guarantees the fulfillment of the control aim (2.44) by the controlled faulty plant $\mathcal{P} \models \mathcal{A}_{\bar{f}}$. The interaction between the reconfiguration unit $\mathcal{R}$ and the tracking controller $\mathcal{C}_{\mathcal{T}}$ is shown in Fig. 5.1. During the reconfiguration, both, the nominal trajectory planning unit $\mathcal{T}(\mathcal{A}_0)$ and the nominal controller $\mathcal{C}(\mathcal{A}_0)$ might need to be modified. The time step at which the diagnostic result $\mathcal{D}^*$ is provided by the diagnostic unit $\mathcal{D}$ and the reconfiguration of the tracking controller $\mathcal{C}_{\mathcal{T}}$ is executed by the reconfiguration unit $\mathcal{R}$ is called *reconfiguration time* k_r.

In general, the reconfiguration problem to be solved by the reconfiguration unit $\mathcal{R}$ can be formalized as follows.

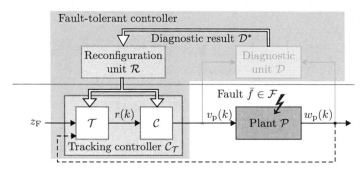

Figure 5.1: Reconfiguration setting.

Problem 5.1 (Reconfiguration problem). *Given a complete and correct diagnostic result $\mathcal{D}^*$, a nominal tracking controller $\mathcal{C}_{\mathcal{T}}(\mathcal{A}_0) = (\mathcal{T}(\mathcal{A}_0), \mathcal{C}(\mathcal{A}_0))$ and the set $\{E_{zf}, E_{vf}, E_{wf}, (f \in \mathcal{F})\}$ of error relations, find a reconfigured tracking controller $\mathcal{C}_{\mathcal{T}}^{\mathrm{r}} = (\mathcal{T}^{\mathrm{r}}, \mathcal{C}^{\mathrm{r}})$ such that the controlled faulty plant $\mathcal{P} \vDash \mathcal{A}_{\bar{f}}$ reaches the desired final state z_{F} at some finite time $k_{\mathrm{F}} \geq k_{\mathrm{r}}$.*

The reconfiguration problem above assumes the completeness and correctness of the diagnostic result $\mathcal{D}^*$, but does not require the diagnostic result $\mathcal{D}^*$ to be unambiguous. That is, it also considers the case that either the present fault $\bar{f}$ or the current state $z_{\mathrm{p}}(k_{\mathrm{r}})$ of the faulty plant at the reconfiguration time k_{r} or both are not known unambiguously. Some reasons for the presence of such an ambiguous diagnostic result $\mathcal{D}^*$ are given in Section 4.6.2.

The reconfigured tracking controller $\mathcal{C}_{\mathcal{T}}^{\mathrm{r}}$ shall be defined such that it can be used to control the faulty plant $\mathcal{P} \vDash \mathcal{A}_{\bar{f}}$ for all times $k \geq k_{\mathrm{r}}$. That is, it is *not* aimed to find a reconfigured tracking controller that is only capable of steering the faulty plant $\mathcal{P}$ into the currently valid desired final state z_{F}. Rather, similar to the nominal tracking controller $\mathcal{C}_{\mathcal{T}}(\mathcal{A}_0)$, the reconfigured tracking controller shall be flexible enough to also deal with other desired final states z_{F} given at runtime.

The ability to find a solution to the reconfiguration problem in Problem 5.1 is denoted reconfigurability of the tracking controller $\mathcal{C}_{\mathcal{T}}$, which is defined as follows.

Definition 5.1 (Reconfigurability). *A nominal tracking controller $\mathcal{C}_{\mathcal{T}}(\mathcal{A}_0)$ is said to be reconfigurable with respect to a complete and correct diagnostic result $\mathcal{D}^*$ and a desired final state z_{F} if it is possible to find a reconfigured tracking controller $\mathcal{C}_{\mathcal{T}}^{\mathrm{r}}$ based on the diagnostic result $\mathcal{D}^*$, the nominal tracking controller $\mathcal{C}_{\mathcal{T}}(\mathcal{A}_0)$ and the set $\{E_{zf}, E_{vf}, E_{wf}, (f \in \mathcal{F})\}$ of error relations, which steers the faulty plant $\mathcal{P} \vDash \mathcal{A}_{\bar{f}}$ from its current state $z_{\mathrm{p}}(k_{\mathrm{d}})$ into the state z_{F}.*

Avoiding illegal transitions.   Analogously, the corresponding safe reconfiguration problem can be stated as follows. It considers the existence of a set $\mathcal{E}_{\mathrm{ill}}$ of illegal transitions that have to be avoided by the controlled faulty plant after the reconfiguration of the tracking controller C_T.

Problem 5.2 (Safe reconfiguration problem). *Given a complete and correct diagnostic result $\mathcal{D}^*$, a nominal tracking controller $C_T(\mathcal{A}_0, \mathcal{E}_{\text{ill}})$, a set $\mathcal{E}_{\text{ill}}$ of illegal transitions and the set $\{E_{zf}, E_{vf}, E_{wf}, (f \in \mathcal{F})\}$ of error relations, find a reconfigured tracking controller C_T^r such that the controlled faulty plant $\mathcal{P} \vDash \mathcal{A}_{\bar{f}}$ reaches the desired final state z_F at some finite time $k_F \geq k_r$ while not executing any illegal transitions $(z_p(k), z_p(k+1)) \in \mathcal{E}_{\text{ill}}, (k_r \leq k < k_F)$.*

The ability to find a solution to the safe reconfiguration problem in Problem 5.2 is denoted as safe reconfigurability of the tracking controller C_T, which is defined as follows. It extends the definition of reconfigurability in Definition 5.1 by the claim that the faulty plant must not execute any illegal transitions.

Definition 5.2 (Safe reconfigurability). *A nominal tracking controller $C_T(\mathcal{A}_0, \mathcal{E}_{\text{ill}})$ is said to be safely reconfigurable with respect to a complete and correct diagnostic result $\mathcal{D}^*$ and a desired final state z_F if it is possible to find a reconfigured tracking controller C_T^r based on the diagnostic result $\mathcal{D}^*$, the nominal tracking controller $C_T(\mathcal{A}_0, \mathcal{E}_{\text{ill}})$, the set $\mathcal{E}_{\text{ill}}$ of illegal transitions and the set $\{E_{zf}, E_{vf}, E_{wf}, (f \in \mathcal{F})\}$ of error relations, which steers the faulty plant $\mathcal{P} \vDash \mathcal{A}_{\bar{f}}$ from its current state $z_p(k_r)$ into the desired final state z_F such that does not execute any illegal transitions $(z_p(k), z_p(k+1)) \in \mathcal{E}_{\text{ill}}, (k_r \leq k < k_F)$.*

5.2 Reconfiguration with unambiguous diagnostic result

This section deals with the reconfiguration problem in Problem 5.1 for the special case that the given diagnostic result $\mathcal{D}^*$ is unambiguous. That is, the reconfiguration of the tracking controller C_T shall be performed at the reconfiguration time $k_r = k_d$ based on a diagnostic result $\mathcal{D}^*$ that contains only the present fault $\bar{f}$ and the current state $z_p(k_d)$ of the faulty plant $\mathcal{P} \vDash \mathcal{A}_{\bar{f}}$.

5.2.1 Reconfigurability analysis

The following theorem states a necessary and sufficient condition under which the tracking controller C_T is reconfigurable according to Definition 5.1 when an unambiguous diagnostic result $\mathcal{D}^*$ is provided by the diagnostic unit $\mathcal{D}$.

Theorem 5.1 (Reconfigurability with unambiguous diagnostic result). *A nominal tracking controller $\mathcal{C}_T(\mathcal{A}_0)$ is reconfigurable with respect to the complete, correct and unambiguous diagnostic result* $\mathcal{D}^* = \left\{ \left(z_p(k_d) \quad \bar{f} \right)^\top \right\}$ *and the desired final state z_F if and only if the automaton $\mathcal{A}_{\bar{f}} = (\mathcal{Z}_{\bar{f}}, \mathcal{V}_{\bar{f}}, \mathcal{W}_{\bar{f}}, G_{\bar{f}}, H_{\bar{f}}, z_p(k_d))$ is controllable with respect to the state z_F.*

Proof. When the diagnostic result $\mathcal{D}^* = \left\{ \left(z_p(k_d) \quad \bar{f} \right)^\top \right\}$ is correct, it is known that the plant $\mathcal{P}$ is modeled by the automaton $\mathcal{A}_{\bar{f}}$ and its current state at the reconfiguration time k_r is given by $z_p(k_r) = z_p(k_d)$. When the set $\{E_{zf}, E_{vf}, E_{wf}, (f \in \mathcal{F})\}$ of error relations and the nominal tracking controller $\mathcal{C}_T(\mathcal{A}_0)$ and therefore the model $\mathcal{A}_0$ of the faulty plant are known, the automaton $\mathcal{A}_{\bar{f}} = (\mathcal{Z}_{\bar{f}}, \mathcal{V}_{\bar{f}}, \mathcal{W}_{\bar{f}}, G_{\bar{f}}, H_{\bar{f}}, z_{\bar{f}0})$ can be computed according to (2.32)–(2.35) and (2.37) or (2.38) with $z_{\bar{f}0} = z_p(k_d)$.

If the automaton $\mathcal{A}_{\bar{f}}$ with initial state $z_{\bar{f}0} = z_p(k_d)$ is controllable with respect to z_F, a plant $\mathcal{P} \vDash \mathcal{A}_{\bar{f}}$ with $z_{\bar{f}0} = z_p(k_d)$ in the closed-loop system $\mathcal{P} \circ \mathcal{C}_T(\mathcal{A}_{\bar{f}})$ reaches the desired final state z_F at a finite time k_F (Theorem 3.3). That is, since the faulty plant is modeled by the automaton $\mathcal{A}_{\bar{f}}$ with current state $z_p(k_r) = z_p(k_d)$ and the automaton $\mathcal{A}_{\bar{f}}$ can be constructed based on the available information, the tracking controller $\mathcal{C}_T(\mathcal{A}_0)$ is reconfigurable according to Definition 5.1 with $\mathcal{C}_T^r = \mathcal{C}_T(\mathcal{A}_{\bar{f}})$.

If the automaton $\mathcal{A}_{\bar{f}}$ with initial state $z_{\bar{f}0} = z_p(k_d)$ is not controllable with respect to z_F, there exists no admissible state sequence $Z(0 \ldots k_e + 1) \in \mathcal{Z}_{\bar{f}}^\infty$ with $z(0) = z_p(k_d)$ and $z(k_e + 1) = z_F$ for it (Theorem 3.1). That is, in the automaton graph of $\mathcal{A}_{\bar{f}}$ there exists no path from $z_p(k_d)$ to z_F. Consequently no input sequence $V(0 \ldots k_e) \in \mathcal{V}_{\bar{f}}^\infty$ which steers the plant $\mathcal{P} \vDash \mathcal{A}_{\bar{f}}$ from its current state $z_p(k_d)$ into the desired final state z_F exists, such that it is not possible to find any reconfigured tracking controller $\mathcal{C}_T^r$ fulfilling this task. Therefore, the tracking controller $\mathcal{C}_T(\mathcal{A}_0)$ is not reconfigurable in this case, which completes the proof. $\qquad\square$

In order to check the reconfigurability of the nominal tracking controller $\mathcal{C}_T(\mathcal{A}_0)$ based on the above theorem, the diagnostic result $\mathcal{D}^*$ has to be known. That is, the criterion can only be evaluated at runtime. However, if every automaton $\mathcal{A}_f$, $(f \in \mathcal{F})$ with every possible initial state $z_{f0} \in \mathcal{Z}_f$, $(f \in \mathcal{F})$ is controllable with respect to a desired final state z_F, the tracking controller $\mathcal{C}_T(\mathcal{A}_0)$ is reconfigurable with respect to the state z_F regardless of which unambiguous diagnostic result $\mathcal{D}^*$ is obtained.

Avoiding illegal transitions.   If additionally a set $\mathcal{E}_{ill}$ of illegal transitions has to be avoided by the faulty plant in the closed-loop system with the reconfigured tracking controller, Theorem 5.1 is modified such that the following theorem results.

Theorem 5.2 (Safe reconfigurability with unambiguous diagnostic result). *A nominal tracking controller $C_T(\mathcal{A}_0, \mathcal{E}_{\text{ill}})$ is safely reconfigurable with respect to a complete, correct and unambiguous diagnostic result $\mathcal{D}^* = \left\{ \left(z_{\text{p}}(k_{\text{d}}) \quad \bar{f} \right)^{\top} \right\}$, a set $\mathcal{E}_{\text{ill}}$ of illegal transitions and a desired final state z_{F} if and only if the automaton $\mathcal{A}_{\bar{f}}$ with initial state $z_{\bar{f}0} = z_{\text{p}}(k_{\text{d}})$ is safely controllable with respect to the state z_{F} and the set $\mathcal{E}_{\text{ill}}$.*

Proof. The proof is analogous to the one of Theorem 5.1 with the difference that the condition for safe controllability is given in Theorem 3.2 instead of Theorem 3.1, the reconfigured tracking controller C_T^{r} is here given by $C_T(\mathcal{A}_{\bar{f}}, \mathcal{E}_{\text{ill}})$ and the avoidance of illegal transitions by the faulty plant is guaranteed by Theorem 3.4 instead of Theorem 3.3. □

The above theorem shows that the safe controllability of the model $\mathcal{A}_{\bar{f}}$ of the model of the faulty plant is a necessary and sufficient condition for the safe reconfigurability of the tracking controller based on an unambiguous diagnostic result.

Example 5.1 *Reconfiguration of tracking controller for automated warehouse*

For the automated warehouse example, the reconfigurability of the nominal tracking controller $C_T(\mathcal{A}_0)$ from Example 3.1 with respect to the complete, correct and unambiguous diagnostic result

$$\mathcal{D}^* = \left\{ \left(\frac{z_{\text{p}}(k_{\text{d}})}{\bar{f}} \right) \right\} = \left\{ \begin{pmatrix} A \\ 2 \end{pmatrix} \right\}$$

in (4.56) and every possible desired final state $z_{\text{F}} \in \{A, B, C\}$ is analyzed using Theorem 5.1. That is, the controllability of the automaton $\mathcal{A}_{\bar{f}} = \mathcal{A}_2$ In Fig. 2.7 with initial state

$$z_{\bar{f}0} = z_{20} = z_{\text{p}}(k_{\text{d}}) = A$$

is checked using Theorem 3.1. In the automaton graph of $\mathcal{A}_2$ there exists a path from the state $z_{20} = A$ to every other state $z \in \{A, B, C\}$ such that an admissible state sequence $Z(0 \ldots k_{\text{e}} + 1) \in \mathcal{Z}_2^{\infty}$ with $z(0) = z_{20}$ and $z(k_{\text{e}} + 1) = z_{\text{F}}$ for every state $z_{\text{F}} \in \{A, B, C\}$ exists. Consequently, the automaton $\mathcal{A}_2$ is controllable with respect to every state $z_{\text{F}} \in \{A, B, C\}$. Therefore, according to Theorem 5.1, the nominal tracking controller $C_T(\mathcal{A}_0)$ is reconfigurable with respect to the diagnostic result $\mathcal{D}^*$ in (4.56) and any desired final state $z_{\text{F}} \in \{A, B, C\}$.

Avoiding illegal transitions. Considering additionally the set $\mathcal{E}_{\text{ill}}$ of illegal transitions in (2.54), the safe reconfigurability of the nominal tracking controller $C_T(\mathcal{A}_0, \mathcal{E}_{\text{ill}})$ from Example 3.1 with respect to the complete, correct and unambiguous diagnostic result

$$\mathcal{D}^* = \mathcal{Z}^*(k_{\text{d}}) = \left\{ \begin{pmatrix} C \\ 2 \end{pmatrix} \right\}$$

in (4.75), the set $\mathcal{E}_{\text{ill}}$ of illegal transitions in (2.54) and every possible desired final state $z_{\text{F}} \in \{A, B, C\}$ is analyzed using Theorem 5.2. That is, the controllability of the legal part $\mathcal{A}_{\bar{f}, \text{leg}} = \mathcal{A}_{2, \text{leg}}$ of the plant

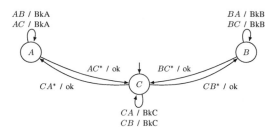

Figure 5.2: Legal part $\mathcal{A}_{2,\text{leg}}$ of model $\mathcal{A}_2$ of faulty automated warehouse.

model $\mathcal{A}_2$ in Fig. 2.7 with initial state

$$z_{\bar{f}0} = z_{20} = z_p(k_d) = C,$$

which is shown in Fig. 5.2, is checked using Theorem 3.2.

It can be seen that all states A, B and C are reachable from the state $z_{20} = C$. Hence, according to Theorem 5.2, the nominal tracking controller $\mathcal{C}_T(\mathcal{A}_0, \mathcal{E}_{\text{ill}})$ is safely reconfigurable with respect to the diagnostic result $\mathcal{D}^*$ in (4.75) and any desired final state $z_F \in \{A, B, C\}$. $\qquad\square$

5.2.2 Reconfiguration of the tracking controller by complete redefinition

When the nominal tracking controller $\mathcal{C}_T(\mathcal{A}_0)$ is reconfigurable with respect to the given diagnostic result $\mathcal{D}^*$, a reconfigured tracking controller $\mathcal{C}_T^{\text{r}}$ shall be computed by the reconfiguration unit $\mathcal{R}$ (Problem 5.1). The available information to fulfill this task includes the complete, correct and unambiguous diagnostic result $\mathcal{D}^* = \left\{ \left(z_p(k_d) \quad \bar{f} \right)^{\top} \right\}$ and the set $\{E_{zf}, E_{vf}, E_{wf}, (f \in \mathcal{F})\}$ of error relations. Additionally, the nominal tracking controller $\mathcal{C}_T(\mathcal{A}_0)$ is given.

As already sketched in the proof of Theorem 5.1, one possibility to find a reconfigured tracking controller $\mathcal{C}_T^{\text{r}}$ is to define a completely new tracking controller $\mathcal{C}_T(\mathcal{A}_{\bar{f}})$ based on the model $\mathcal{A}_{\bar{f}} = (\mathcal{Z}_{\bar{f}}, \mathcal{V}_{\bar{f}}, \mathcal{W}_{\bar{f}}, G_{\bar{f}}, H_{\bar{f}}, z_{\bar{f}0})$ of the faulty plant using the method described in Chapter 3. The model $\mathcal{A}_{\bar{f}}$ of the faulty plant $\mathcal{P} \vDash \mathcal{A}_{\bar{f}}$ is obtained by applying the error relations $E_{z\bar{f}}, E_{v\bar{f}}$ and $E_{w\bar{f}}$ to the model $\mathcal{A}_0$ of the faultless plant according to (2.32)–(2.35) and (2.37) or (2.38) and setting its initial state to $z_{\bar{f}0} = z_p(k_d)$. The following proposition shows that with this procedure the reconfigured tracking controller $\mathcal{C}_T^{\text{r}} = \mathcal{C}_T(\mathcal{A}_{\bar{f}})$ is a solution to the reconfiguration problem in Problem 5.1 given that the diagnostic result $\mathcal{D}^*$ is unambiguous.

Proposition 5.1 (Reconfiguration based on unambiguous diagnostic result by complete redefinition). *Consider a nominal tracking controller $\mathcal{C}_T(\mathcal{A}_0)$ that is reconfigurable with respect to the complete, correct and unambiguous diagnostic result $\mathcal{D}^* = \left\{ \left(z_{\mathrm{p}}(k_{\mathrm{d}}) \quad \bar{f} \right)^\top \right\}$ and the desired final state z_{F}. Let $\mathcal{C}_T^{\mathrm{r}} = (T^{\mathrm{r}}, \mathcal{C}^{\mathrm{r}}) = \mathcal{C}_T(\mathcal{A}_{\bar{f}})$ be the reconfigured tracking controller based on $\mathcal{D}^*$. Let $k_{\mathrm{e}}^{\mathrm{r}}$ be the time horizon of the reference trajectory generated by the reconfigured trajectory planning unit T^{r}. Then the faulty plant $\mathcal{P}$ in the closed-loop system $\mathcal{P} \circ \mathcal{C}_T^{\mathrm{r}}$ reaches the desired final state z_{F} at time $k_{\mathrm{F}} = k_{\mathrm{e}}^{\mathrm{r}} + 1$.*

Proof. The reconfigurability of $\mathcal{C}_T(\mathcal{A}_0)$ with respect to $\mathcal{D}^*$ and z_{F} implies the controllability of the automaton $\mathcal{A}_{\bar{f}}$ with initial state $z_{\bar{f}0} = z_{\mathrm{p}}(k_{\mathrm{d}})$ with respect to z_{F} (Theorem 5.1). The correctness of the unambiguous diagnostic result $\mathcal{D}^*$ guarantees that the faulty plant is modeled by the automaton $\mathcal{A}_{\bar{f}}$ and is currently in state $z_{\mathrm{p}}(k_{\mathrm{d}})$. According to Theorem 3.3, the plant $\mathcal{P}$ in the closed-loop system $\mathcal{P} \circ \mathcal{C}_T^{\mathrm{r}}$ therefore reaches the desired final state z_{F} at time $k_{\mathrm{F}} = k_{\mathrm{e}}^{\mathrm{r}} + 1$. $\quad\square$

Avoiding illegal transitions. When during the complete redefinition of the tracking controller additionally a set $\mathcal{E}_{\mathrm{ill}}$ of illegal transitions is taken into account and the given diagnostic result $\mathcal{D}^*$ is unambiguous, the resulting reconfigured tracking controller solves the safe reconfiguration problem in Problem 5.2 as shown in the following proposition.

Proposition 5.2 (Safe reconfiguration based on unambiguous diagnostic result by complete redefinition). *Consider a nominal tracking controller $\mathcal{C}_T(\mathcal{A}_0, \mathcal{E}_{\mathrm{ill}})$ that is safely reconfigurable with respect to the desired final state z_{F}, the set $\mathcal{E}_{\mathrm{ill}}$ of illegal transitions and the complete, correct and unambiguous diagnostic result $\mathcal{D}^* = \left\{ \left(z_{\mathrm{p}}(k_{\mathrm{d}}) \quad \bar{f} \right)^\top \right\}$. Let*

$$\mathcal{C}_T^{\mathrm{r}} = (T^{\mathrm{r}}, \mathcal{C}^{\mathrm{r}}) = \mathcal{C}_T(\mathcal{A}_{\bar{f}}, \mathcal{E}_{\mathrm{ill}})$$

be the reconfigured tracking controller based on $\mathcal{D}^*$ and $\mathcal{E}_{\mathrm{ill}}$. Let $k_{\mathrm{e}}^{\mathrm{r}}$ be the time horizon of the reference trajectory generated by the reconfigured trajectory planning unit T^{r}. Then the faulty plant $\mathcal{P}$ in the closed-loop system $\mathcal{P} \circ \mathcal{C}_T^{\mathrm{r}}$ reaches the desired final state z_{F} at time $k_{\mathrm{F}} = k_{\mathrm{e}}^{\mathrm{r}} + 1$, while not executing any illegal transitions $(z_{\mathrm{p}}(k), z_{\mathrm{p}}(k+1)) \in \mathcal{E}_{\mathrm{ill}}$, $(k_{\mathrm{r}} \leq k < k_{\mathrm{F}})$.

Proof. The proof follows directly from the proof of Proposition 5.1 and Theorem 3.4 on the avoidance of illegal transitions by any plant $\mathcal{P} \models \mathcal{A}$ controlled by its corresponding tracking controller $\mathcal{C}_T(\mathcal{A}, \mathcal{E}_{\mathrm{ill}})$. $\quad\square$

The above propositions show that the reconfiguration of the tracking controller $\mathcal{C}_T$ by defining a completely new tracking controller for the model $\mathcal{A}_{\bar{f}}$ of the faulty plant is a successful strategy. However, since a fault usually only affects a part of the plant, only certain parts of the newly defined tracking controller C^r_T differ from the nominal tracking controller C_T as well. Therefore, it is desirable to solve the reconfiguration problems in Problem 5.1 and Problem 5.2, respectively, by *modifying* the nominal tracking controller $C_T(\mathcal{A}_0)$ or $C_T(\mathcal{A}_0, \mathcal{E}_{\text{ill}})$ instead of defining a completely new tracking controller. Therefore, the next section presents a method for the reconfiguration of the nominal tracking controller that requires only a partial redefinition.

For the reconfiguration by partial redefinition, it has to be analyzed which parts of the trajectory planning unit $T(\mathcal{A}_0)$ (or $T(\mathcal{A}_0, \mathcal{E}_{\text{ill}})$) in (3.18) or (3.19), respectively, and the controller $C(\mathcal{A}_0)$ in (3.7) can be kept in use and which parts need to be redefined in which way based on the available information in the diagnostic result $\mathcal{D}^*$ and the set $\{E_{zf}, E_{vf}, E_{wf}, (f \in \mathcal{F})\}$ of error relations.

5.2.3 Partial redefinition of the trajectory planning unit

This section deals with the reconfiguration of the trajectory planning unit T. It is aimed to keep as much of the nominal trajectory planning unit $T(\mathcal{A}_0)$ and modify only those parts which are affected by the present fault $\bar{f} \in \mathcal{F}$.

The nominal trajectory planning unit $T(\mathcal{A}_0)$ plans a reference trajectory $R(0 \ldots k_e)$ for the plant $\mathcal{P} \vDash \mathcal{A}_0$ by searching for a path from the initial state $z_p(0) = z_{00}$ to the desired final state z_F in the model $\mathcal{A}_0$ of the faultless plant. In the reconfigured tracking controller C^r_T, the trajectory planning unit T^r shall again plan a reference trajectory for the plant $\mathcal{P}$. However, since now the fault $\bar{f}$ is present and the plant is currently in state $z_p(k_d)$, a path from the state $z_p(k_d)$ to the desired final state z_F has to be searched in some modified automaton $\mathcal{A}^r \neq \mathcal{A}_0$ not necessarily equal to the model $\mathcal{A}_{\bar{f}}$ of the faulty plant. The necessary modifications are summarized in Table 5.1.

	Nominal $T(\mathcal{A}_0)$	Reconfigured T^r
Find reference trajectory for plant	$\mathcal{P} \vDash \mathcal{A}_0$	$\mathcal{P} \vDash \mathcal{A}_{\bar{f}}$
Find reference trajectory	$R(0 \ldots k_e)$	$R(k_r \ldots k^r_e)$
Search for path in	$\mathcal{A}_0$	$\mathcal{A}^r \neq \mathcal{A}_{\bar{f}}$
Search for path from	z_{00}	$z_p(k_d)$
Search for path to	z_F	z_F

Table 5.1: Necessary modifications in trajectory planning unit T.

In order to change the trajectory planning unit as little as possible, the reconfigured trajectory planning unit T^r uses the same search algorithm as the nominal trajectory planning unit $T(\mathcal{A}_0)$. Furthermore, the automaton graph of the model $\mathcal{A}_0$ of the faulty plant on which the search is performed is kept, except for transitions that are modified due to the error relations $E_{z\bar{f}}$ and $E_{v\bar{f}}$

according to (2.35). Finally, the state z_{00} from which the search for a path started is replaced by the current state $z_p(k_d)$ of the faulty plant. Hence, the following automaton $\mathcal{A}_0^r$ results:

$$
\mathcal{A}_0^r : \begin{cases} \mathcal{Z}_0^r = \mathcal{Z}_0, \\ \mathcal{V}_0^r = \mathcal{V}_0, \\ \mathcal{W}_0^r = \mathcal{W}_0, \\ G_0^r(z,v) = \begin{cases} G_0(z,v) & \text{if } \big[v \text{ faultless input}\big] \wedge \big[(z,v) \text{ faultless transition}\big], \\ E_{z\bar{f}}(z, E_{v\bar{f}}(v)) & \text{otherwise,} \end{cases} \\ H_0^r(z,v) = H_0(z,v), \\ z_{00}^r = z_p(k_d). \end{cases}
$$

(5.1)

In the automaton $\mathcal{A}_0^r$ possible changes of the output, e.g., by the error relation $E_{w\bar{f}}$, are neglected. Consequently, the output function of the automaton $\mathcal{A}_0^r$ and the output function of the model $\mathcal{A}_{\bar{f}}$ of the faulty plant differ from each other. Nevertheless, the following proposition shows that both automata can be used interchangeably for the purpose of trajectory planning. More specifically, it states that any reference trajectory $R(k_r \ldots k_e^r)$ ending in the desired final state z_F that the reconfigured trajectory planning unit T^r finds, would also be found by the trajectory planning unit $T(\mathcal{A}_{\bar{f}})$ defined for the faulty plant $\mathcal{P} \models \mathcal{A}_{\bar{f}}$ and vice versa.

Proposition 5.3 (Equivalence of reference trajectories). *A reference trajectory $R(k_r \ldots k_e^r)$ with $r(k_r) = z_p(k_d)$ and $r(k_e^r) = z_F$ is found by the partially redefined trajectory planning unit $T^r = T(\mathcal{A}_0^r)$ with $\mathcal{A}_0^r$ in (5.1) if and only if the same reference trajectory $R(k_r \ldots k_e^r)$ is found by the completely redefined trajectory planning unit $T(\mathcal{A}_{\bar{f}})$.*

Proof. Since the same search algorithm is used by the trajectory planning units $T^r = T(\mathcal{A}_0^r)$ and $T(\mathcal{A}_{\bar{f}})$ to find a path from $z_p(k_d)$ to z_F, it only has to be shown that the search spaces are identical. The trajectory planning units search for a path in the automata graphs of $\mathcal{A}_0^r$ in (5.1) and of the model $\mathcal{A}_{\bar{f}}$ of the faulty plant, respectively. According to Definition 3.2 a path in the graph of an automaton $\mathcal{A}$ always corresponds to an admissible state sequence $Z(0 \ldots k_e + 1) \in \mathcal{Z}^\infty$ for which with (3.2)

$$
\forall\, 0 \le k \le k_e : (\exists v \in \mathcal{V})\, G(z(k),v) = z(k+1)
$$

follows. From the definition of the state transition function $G_{\bar{f}}$ in (2.35) and the definition of the error relations $E_{v\bar{f}}$ and $E_{z\bar{f}}$ in (2.29) and (2.31), respectively, it follows that

$$
G_0^r(z,v) = G_{\bar{f}}(z,v) = \begin{cases} G_0(z,v) & \text{if } \big[v \text{ faultless input}\big] \wedge \big[(z,v) \text{ faultless transition}\big], \\ E_{z\bar{f}}(z, E_{v\bar{f}}(v)) & \text{otherwise.} \end{cases}
$$

Consequently, every path in the automaton graph of $\mathcal{A}_0^r$ is also a path in the automaton graph of $\mathcal{A}_{\bar{f}}$ and vice versa.

Hence, even though the automaton $\mathcal{A}_0^r$ in (5.1) and the automaton $\mathcal{A}_{\bar{f}}$ may obviously differ from each other in the output function $H_0^r \equiv H_0 \not\equiv H_{\bar{f}}$, the same reference trajectories are found anyhow, because the trajectory planning units only search for a path in the automata graphs but neglect the I/O labels at their edges. $\square$

The above proposition shows that a partial reconfiguration of the trajectory planning unit by changing only some transitions in the automaton in which the search is performed according to (5.1) leads to the same reconfigured trajectory planning unit T^r as a complete redefinition based on the model $\mathcal{A}_{\bar{f}}$ of the faulty plant would.

Avoiding illegal transitions.   When a set $\mathcal{E}_{\text{ill}}$ of illegal transitions is considered, the nominal trajectory planning unit $T(\mathcal{A}_0, \mathcal{E}_{\text{ill}})$ performs a search for a path from the initial state z_{00} to the desired final state z_{F} in the legal part $\mathcal{A}_{\text{leg}}$ of the model $\mathcal{A}_0$ of the faultless plant (cf. (3.19)). That is, now the legal part $\mathcal{A}_{\text{leg}}$ of the model $\mathcal{A}_0$ of the faultless plant is modified during the reconfiguration of the trajectory planning unit. First, faulty transitions are modified based on the error relations $E_{z\bar{f}}$ and $E_{v\bar{f}}$ according to (2.35). Afterwards, modified transitions that belong to the set $\mathcal{E}_{\text{ill}}$ of illegal transitions are removed. Finally, the initial state z_{leg0} of the automaton $\mathcal{A}_{\text{leg}}$ from which the search for a path starts is replaced by the current state $z_{\text{p}}(k_{\text{d}})$ of the faulty plant. As a result, the following automaton is obtained:

$$
\mathcal{A}_{\text{leg}}^r : \begin{cases}
\mathcal{Z}_{\text{leg}}^r = \mathcal{Z}_{\text{leg}}, \\
\mathcal{V}_{\text{leg}}^r = \mathcal{V}_{\text{leg}}, \\
\mathcal{W}_{\text{leg}}^r = \mathcal{W}_{\text{leg}}, \\
G_{\text{leg}}^r(z,v) = \begin{cases}
G_{\text{leg}}(z,v) & \text{if } \big[v \text{ faultless input}\big] \wedge \big[(z,v) \text{ faultless transition}\big] \\
& \qquad\qquad\qquad\qquad\qquad \wedge \big[G_{\text{leg}}(z,v)!\big], \\
E_{z\bar{f}}(z, E_{v\bar{f}}(v)) & \text{if } \big(\big[v \text{ faulty input}\big] \vee \big[(z,v) \text{ faulty transition}\big]\big) \\
& \qquad\qquad\qquad \wedge \big[(z, E_{z\bar{f}}(z, E_{v\bar{f}}(v))) \notin \mathcal{E}_{\text{ill}}\big], \\
\text{undefined} & \text{otherwise,}
\end{cases} \\
H_{\text{leg}}^r(z,v) = \begin{cases}
H_{\text{leg}}(z,v) & \text{if } \big[H_{\text{leg}}(z,v)!\big] \wedge \big[G_{\text{leg}}^r(z,v)!\big], \\
\text{undefined} & \text{otherwise,}
\end{cases} \\
z_{\text{leg0}}^r = z_{\text{p}}(k_{\text{d}}).
\end{cases}
\tag{5.2}
$$

Just as during the definition of the nominal tracking controller $C_T(\mathcal{A}_0, \mathcal{E}_{\text{ill}})$, the use of the automaton $\mathcal{A}_{\text{leg}}^{\text{r}}$ in the reconfigured trajectory planning unit T^{r} guarantees that the generated reference trajectory $R(k_{\text{r}} \ldots k_{\text{e}}^{\text{r}})$ after the reconfiguration time k_{r} does not contain any illegal transitions $(r(k), r(k+1)) \in \mathcal{E}_{\text{ill}}$, $(k_{\text{r}} \leq k < k_{\text{e}}^{\text{r}})$. In combination with Proposition 5.3 it can be proved that any reference trajectory $R(k_{\text{r}} \ldots k_{\text{e}}^{\text{r}})$ that the reconfigured trajectory planning unit T^{r} finds, would also be found by the trajectory planning unit $T(\mathcal{A}_{\bar{f}}, \mathcal{E}_{\text{ill}})$ specifically defined for the faulty plant $\mathcal{P} \vDash \mathcal{A}_{\bar{f}}$ with respect to the set $\mathcal{E}_{\text{ill}}$ of illegal transitions and vice versa.

Proposition 5.4 (Equivalence of safe reference trajectories)**.** *The partially redefined trajectory planning unit* $T^{\text{r}} = T(\mathcal{A}_{\text{leg}}^{\text{r}})$ *with* $\mathcal{A}_{\text{leg}}^{\text{r}}$ *in (5.2) finds a reference trajectory* $R(k_{\text{r}} \ldots k_{\text{e}}^{\text{r}})$ *with* $r(k_{\text{r}}) = z_{\text{p}}(k_{\text{d}})$ *and* $r(k_{\text{e}}^{\text{r}}) = z_{\text{F}}$ *if and only if the completely redefined trajectory planning unit* $T(\mathcal{A}_{\bar{f}}, \mathcal{E}_{\text{ill}})$ *finds the same reference trajectory* $R(k_{\text{r}} \ldots k_{\text{e}}^{\text{r}})$.

Proof. In the proof of Proposition 5.3 it has been shown that the reconfigured trajectory planning unit $T(\mathcal{A}_0^{\text{r}})$ and the redefined trajectory planning unit $T(\mathcal{A}_{\bar{f}})$ generate the same reference trajectories, because the state transition functions of the automata $\mathcal{A}_0^{\text{r}}$ and $\mathcal{A}_{\bar{f}}$ are identical. Following the same argumentation, in the presence of a set $\mathcal{E}_{\text{ill}}$ of illegal transitions it therefore only has to be proved that the automaton $\mathcal{A}_{\text{leg}}^{\text{r}}$ in (5.2) and the legal part of the model $\mathcal{A}_{\bar{f}}$ of the faulty plant with respect to the set $\mathcal{E}_{\text{ill}}$ of illegal transitions have the same state transition function.

According to Definition 2.4, the state transition function of the legal part of the model $\mathcal{A}_{\bar{f}}$ of the faulty plant is given by

$$
G_{\bar{f},\text{leg}}(z,v) = \begin{cases} G_{\bar{f}}(z,v) & \text{if } (z, G_{\bar{f}}(z,v)) \notin \mathcal{E}_{\text{ill}}, \\ \text{undefined} & \text{otherwise,} \end{cases}
$$

$$
= \begin{cases} G_0(z,v) & \text{if } \Big[v \text{ faultless input}\Big] \wedge \Big[(z,v) \text{ faultless transition}\Big] \\ & \qquad\qquad\qquad\qquad \wedge \Big[(z, G_0(z,v)) \notin \mathcal{E}_{\text{ill}}\Big], \\ E_{\text{z}\bar{f}}(z, E_{\text{v}\bar{f}}(v)) & \text{if } \Big(\Big[v \text{ faulty input}\Big] \vee \Big[(z,v) \text{ faulty transition}\Big] \Big) \\ & \qquad\qquad\qquad \wedge \Big[(z, E_{\text{z}\bar{f}}(z, E_{\text{v}\bar{f}}(v))) \notin \mathcal{E}_{\text{ill}}\Big], \\ \text{undefined} & \text{otherwise.} \end{cases}
$$

With the definition of G_{leg} in (2.48) it follows that $G_{\bar{f},\text{leg}} \equiv G_{\text{leg}}^{\text{r}}$. □

The above proposition shows that a partial reconfiguration of the trajectory planning unit by changing only some transitions in the automaton in which the search is performed according to (5.2) leads to the same reconfigured trajectory planning unit T^{r} as a complete redefinition based on the model $\mathcal{A}_{\bar{f}}$ of the faulty plant and the set $\mathcal{E}_{\text{ill}}$ of illegal transitions would.

Example 5.1 (cont.)   *Reconfiguration of tracking controller for automated warehouse*

The nominal trajectory planning unit $T(\mathcal{A}_0)$ of the automated warehouse example described in Example 3.1 is reconfigured by modifying the stored model $\mathcal{A}_0$ of the faultless plant according to (5.1) such that the automaton $\mathcal{A}_0^r$ results. That is, based on the diagnostic result $\mathcal{D}^*$ in (4.56), the initial state of the automaton is changed to $z_0^r = z_p(k_d) = A$ and the transitions corresponding to the state-input pairs

$$(z, v) \in \{A, B, C\} \times \{AB, AC, BA, BC, CA, CB\} \tag{5.3}$$

are modified based on the error relations E_{z2} and E_{v2} in (2.56c) and (2.56a), respectively. Thereby the automaton $\mathcal{A}_0^r$ in Fig. 5.3 results.

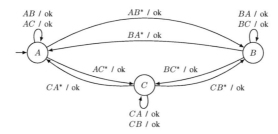

Figure 5.3: Automaton $\mathcal{A}_0^r$ in reconfigured trajectory planning unit T^r.

It can be seen that in contrast to the model $\mathcal{A}_2$ of the faulty plant in Fig. 2.7, the self-loops are still labeled with the nominal output ok instead of the faulty outputs BkA, BkB or BkC. Nevertheless, as proved in Proposition 5.3, the reconfigured trajectory planning unit T^r will always generate a reference trajectory that also would have been generated by the completely redefined trajectory planning unit $T(\mathcal{A}_2)$ defined for the model $\mathcal{A}_2$. In case of the desired final state $z_F = B$, the reference trajectory is given by

$$R(k_r \ldots k_e^r) = r(3) = B. \tag{5.4}$$

Avoiding illegal transitions. Taking the set $\mathcal{E}_{ill}$ of illegal transitions in (2.54) into account, the legal part of the model $\mathcal{A}_0$ of the faultless plant in Fig. 2.6 stored in the nominal trajectory planning unit $T(\mathcal{A}_0, \mathcal{E}_{ill})$ is modified based on the diagnostic result $\mathcal{D}^*$ in (4.75) according to (5.2). First, its initial state is replaced by $z_{leg0}^r = z_p(k_d) = C$. Then, the transitions corresponding to the state-input pairs in (5.3) are modified based on the error relations E_{z2} and E_{v2} in (2.56c) and (2.56a), respectively. It turns out that none of these modified transitions lies in the set $\mathcal{E}_{ill}$ of illegal transitions:

$$(z, E_{z2}(z, E_{v2}(v))) \notin \mathcal{E}_{ill} = \{(A, B), (B, A)\}, \quad \forall (z, v) \text{ from (5.3)}.$$

Figure 5.4 shows the graph of the resulting automaton $\mathcal{A}_{leg}^r$. Note that the outputs of the modified transitions are undefined.

In case of the desired final state $z_F = B$, the reconfigured trajectory planning unit T^r generates the reference trajectory

$$R(k_r \ldots k_e^r) = r(2) = B. \tag{5.5}$$

$\square$

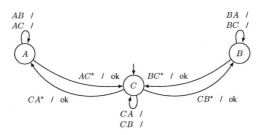

Figure 5.4: Automaton $\mathcal{A}_{\mathrm{leg}}^{\mathrm{r}}$ in reconfigured trajectory planning unit T^{r}.

5.2.4 Partial redefinition of the controller

The main idea for the partial redefinition of the controller $\mathcal{C}$ is to first remove transitions affected by the fault from the nominal controller $\mathcal{C}(\mathcal{A}_0)$ and afterwards newly define relevant transitions and include them into the pruned nominal controller such that the reconfigured controller C^{r} results. According to (2.32)–(2.34) the state set, input set and output set of the plant $\mathcal{P}$ are not changed by any fault $f \in \mathcal{F}$. Therefore, the sets $\mathcal{Z}_{\mathrm{c}}$, $\mathcal{V}_{\mathrm{c}}$ and $\mathcal{W}_{\mathrm{c}}$ of the nominal controller $\mathcal{C}(\mathcal{A}_0)$ deduced from these sets also remain constant and it is sufficient to only change some transitions in the nominal controller $\mathcal{C}(\mathcal{A}_0)$.

Since the controller is completely independent from the set $\mathcal{E}_{\mathrm{ill}}$ of illegal transitions, the method described in this section is applicable, both, in the case that there is a set $\mathcal{E}_{\mathrm{ill}}$ of illegal transitions or not.

According to (3.7d) and (3.7e) the transitions in the nominal controller $\mathcal{C}(\mathcal{A}_0)$ and the transitions in the model $\mathcal{A}_0$ of the faultless plant are related as follows:

$$[G_{\mathrm{c}}(z_{\mathrm{c}}, v_{\mathrm{c}}) = z_{\mathrm{c}}'] \wedge [H_{\mathrm{c}}(z_{\mathrm{c}}, v_{\mathrm{c}}) = w_{\mathrm{c}}] \Rightarrow G_0(z_{\mathrm{c}}, w_{\mathrm{c}}) = v_{\mathrm{c}}. \tag{5.6}$$

That is, a transition from state z_{c} with input v_{c} to the next state z_{c}' with the output w_{c} in the nominal controller $\mathcal{C}(\mathcal{A}_0)$ corresponds to a transition from state z_{c} to state v_{c} with input w_{c} in the model $\mathcal{A}_0$ of the faultless plant. After the reconfiguration, the implication in (5.6) shall again be valid for the reconfigured controller C^{r} and the model $\mathcal{A}_{\bar{f}}$ of the faulty plant:

$$[G_{\mathrm{c}}^{\mathrm{r}}(z_{\mathrm{c}}, v_{\mathrm{c}}) = z_{\mathrm{c}}'] \wedge [H_{\mathrm{c}}^{\mathrm{r}}(z_{\mathrm{c}}, v_{\mathrm{c}}) = w_{\mathrm{c}}] \Rightarrow G_{\bar{f}}(z_{\mathrm{c}}, w_{\mathrm{c}}) = v_{\mathrm{c}}. \tag{5.7}$$

Remove transitions affected by the fault. Transitions in the nominal controller $\mathcal{C}(\mathcal{A}_0)$ that are affected by the present fault $\bar{f}$ have to be identified and removed such that a pruned automaton

$$\mathcal{C}(\mathcal{A}_0)^- = (\mathcal{Z}_c^-, \mathcal{V}_c^-, \mathcal{W}_c^-, G_c^-, H_c^-, z_{c0}^-) \tag{5.8}$$

results. The state set, input set and output set of the pruned automaton $\mathcal{C}(\mathcal{A}_0)^-$ are identical to the respective sets in the nominal controller $\mathcal{C}(\mathcal{A}_0)$ in (3.7a)–(3.7c):

$$\mathcal{Z}_c^- = \mathcal{Z}_c \tag{5.9}$$
$$\mathcal{V}_c^- = \mathcal{V}_c \tag{5.10}$$
$$\mathcal{W}_c^- = \mathcal{W}_c. \tag{5.11}$$

Considering (5.6) and (5.7), all transitions for which

$$[G_c(z_c, v_c) = z_c'] \wedge [H_c(z_c, v_c) = w_c] \Rightarrow G_{\bar{f}}(z_c, w_c) \neq v_c$$

holds have to be removed from the nominal controller $\mathcal{C}(\mathcal{A}_0)$. According to the definition of the state transition function of the faulty plant in (2.35), the right hand side of the implication can be expressed as follows:

$$G_{\bar{f}}(z_c, w_c) \neq v_c \Leftrightarrow E_{z\bar{f}}(z_c, E_{v\bar{f}}(w_c)) \neq v_c.$$

Consequently, the following state transition function and output function for the pruned automaton $\mathcal{C}(\mathcal{A}_0)^-$ result:

$$G_c^-(z_c, v_c) = \begin{cases} G_c(z_c, v_c) & \text{if } \Big[H_c(z_c, v_c) \text{ faultless input} \Big] \\ & \qquad \wedge \Big[(z_c, H_c(z_c, v_c)) \text{ faultless transition} \Big], \\ \text{undefined} & \text{otherwise,} \end{cases} \tag{5.12}$$

$$H_c^-(z_c, v_c) = \begin{cases} H_c(z_c, v_c) & \text{if } \Big[H_c(z_c, v_c) \text{ faultless input} \Big] \\ & \qquad \wedge \Big[(z_c, H_c(z_c, v_c)) \text{ faultless transition} \Big], \\ \text{undefined} & \text{otherwise.} \end{cases} \tag{5.13}$$

The above equations describe that compared to the nominal controller $\mathcal{C}(\mathcal{A}_0)$ only transitions that are not affected by the present fault $\bar{f}$ remain in the pruned automaton $C(\mathcal{A}_0)^-$ in (5.8). Its initial state equals the current state of the faulty plant:

$$z_{c0}^- = z_p(k_d). \tag{5.14}$$

The pruned automaton $\mathcal{C}(\mathcal{A}_0)^-$ in (5.8) could be used as a controller for the faulty plant, because it does not try to enforce any transitions that are not possible due to the present fault $\bar{f}$. However, the pruned automaton $\mathcal{C}(\mathcal{A}_0)^-$ is too restrictive, because it does not make use of possible alternatives in the faulty plant, for example caused by redundancies. Therefore, in the following some additional transitions are defined and included into the pruned automaton $\mathcal{C}(\mathcal{A}_0)^-$.

Define new transitions.   There are two classes of transitions, which might enrich the possibilities of the pruned automaton $\mathcal{C}(\mathcal{A}_0)^-$. They are described by the sets $\mathcal{E}_{\text{new},-}$ and $\mathcal{E}_{\text{new},+}$, respectively.

- **Removed transitions:** The set $\mathcal{E}_{\text{new},-}$ contains all state pairs $(z_c, z_c') \in \mathcal{Z}_c \times \mathcal{Z}_c$ whose corresponding transitions have been removed from the nominal controller $\mathcal{C}(\mathcal{A}_0)$ when defining the pruned automaton $\mathcal{C}(\mathcal{A}_0)^-$. Thereby, it is possible to find redundant inputs for the faulty plant $\mathcal{P}$.

- **New transitions:** The set $\mathcal{E}_{\text{new},+}$ contains all state pairs $(z_c, z_c') \in \mathcal{Z}_c \times \mathcal{Z}_c$ for which new transitions from state z_c to state z_c' occur in the plant $\mathcal{P}$ due to the present fault $\bar{f}$.

The sets $\mathcal{E}_{\text{new},-}$ and $\mathcal{E}_{\text{new},+}$ are illustrated in Fig. 5.5. On the left, a model $\mathcal{A}_0$ of a faultless plant is shown, while the automaton on the right represents the model $\mathcal{A}_{\bar{f}}$ of the faulty plant. The set $\mathcal{E}_{\text{new},-}$ of removed transitions here contains only the transition from state $z = 2$ to state $z' = 3$. The only newly occurring transition, which is included in the set $\mathcal{E}_{\text{new},+}$, is the self-loop at state $z = 2$.

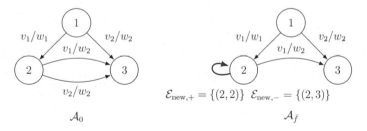

Figure 5.5: Illustration of sets $\mathcal{E}_{\text{new},-}$ and $\mathcal{E}_{\text{new},+}$.

Note that according to (3.7d) for all transitions in the controller $\mathcal{C}$, $z_c' = v_c$ holds. Therefore, it is possible to specify a set of transitions $(z_c, z_c') \in \mathcal{Z}_c \times \mathcal{Z}_c$ to be redefined, even though eventually a set of state-input pairs $(z_c, v_c) \in \mathcal{Z}_c \times \mathcal{V}_c$ has to be considered for the redefinition.

The set $\mathcal{E}_{\text{new},-}$ of transitions to be newly defined because of their removal from the nominal controller $\mathcal{C}(\mathcal{A}_0)$ can be computed based on (5.12) as follows:

$$
\begin{aligned}
\mathcal{E}_{\text{new},-} &= \Big\{ (z_c, z_c') \in \mathcal{Z}_c \times \mathcal{Z}_c : \big[G_c(z_c, z_c') \text{ defined} \big] \wedge \big[G_c^-(z_c, z_c') \text{ undefined} \big] \Big\} \\
&= \Big\{ (z_c, z_c') \in \mathcal{Z}_c \times \mathcal{Z}_c : (\exists w \in \mathcal{W}_c) \big(\big[H_c(z_c, z_c') = w \big] \wedge \big[E_{zf}(z_c, E_{vf}(w)) \neq z_c' \big] \big) \Big\}.
\end{aligned}
\tag{5.15}
$$

Newly appearing transitions from a state z to a next state z' in the model $\mathcal{A}_{\bar{f}}$ of the faulty plant are characterized by the fact that

$$
\big[\mathcal{V}_{a0}(z', z) = \emptyset \big] \wedge \big[\mathcal{V}_{a\bar{f}}(z', z) \neq \emptyset \big].
$$

From the definition of the nominal controller $\mathcal{C}(\mathcal{A}_0)$ in (3.7) it is known that the first literal of the above formula can be rewritten as follows:

$$
\mathcal{V}_{a0}(z', z) = \emptyset \Leftrightarrow \big[G_c(z, z')! \big] \wedge \big[H_c(z, z')! \big] \Leftrightarrow (\nexists w \in \mathcal{W}_c) \, H_c(z, z') = w. \tag{5.16}
$$

The second literal of the same formula can be rewritten based on the definition of the active input set in (2.11) and the definition of the state transition function of the faulty plant in (2.35):

$$
\begin{aligned}
\mathcal{V}_{a\bar{f}}(z', z) \neq \emptyset &\Leftrightarrow (\exists v \in \mathcal{V}_{\bar{f}}) \, G_{\bar{f}}(z, v) = z' \\
&\Leftrightarrow (\exists v \in \mathcal{V}_{\bar{f}}) \, E_{zf}(z, E_{vf}(v)) = z'.
\end{aligned}
\tag{5.17}
$$

Based on (5.16) and (5.17) the set $\mathcal{E}_{\text{new},+}$ of transitions to be newly defined because of newly appearing transitions in the model $\mathcal{A}_{\bar{f}}$ of the faulty plant is given by

$$
\begin{aligned}
\mathcal{E}_{\text{new},+} &= \Big\{ (z_c, z_c') \in \mathcal{Z}_c \times \mathcal{Z}_c : \big[\mathcal{V}_{a0}(z_c', z_c) = \emptyset \big] \wedge \big[\mathcal{V}_{a\bar{f}}(z_c', z_c) \neq \emptyset \big] \Big\} \\
&= \Big\{ (z_c, z_c') \in \mathcal{Z}_c \times \mathcal{Z}_c : \big[(\nexists w \in \mathcal{W}_c) H_c(z_c, z_c') = w \big] \\
&\qquad\qquad \wedge \big[(\exists w \in \mathcal{W}_c) E_{zf}(z_c, E_{vf}(w)) = z_c' \big] \Big\}.
\end{aligned}
\tag{5.18}
$$

Combining the set $\mathcal{E}_{\text{new},-}$ in (5.15) and the set $\mathcal{E}_{\text{new},+}$ in (5.18), the set $\mathcal{E}_{\text{new}}$ of all transitions to be newly defined results to

$$\mathcal{E}_{new} = \mathcal{E}_{new,-} \cup \mathcal{E}_{new,+}$$
$$= \Big\{ (z_c, z_c') \in \mathcal{Z}_c \times \mathcal{Z}_c :$$
$$\Big[(\exists w \in \mathcal{W}_c) \Big(\big[H_c(z_c, z_c') = w \big] \wedge \big[E_{zf}(z_c, E_{vf}(w)) \neq z_c' \big] \Big) \Big]$$
$$\vee \Big[\big((\not\exists w \in \mathcal{W}_c) H_c(z_c, z_c') = w \big) \wedge \big((\exists w \in \mathcal{W}_c) E_{zf}(z_c, E_{vf}(w)) = z_c' \big) \Big] \Big\}.$$
$$(5.19)$$

In order to obtain the reconfigured controller C^r, the state transition function G_c^- and the output function H_c^- of the pruned automaton $C(\mathcal{A}_0)^-$ in (5.12) and (5.13), respectively, are additionally defined for all state pairs in the set $\mathcal{E}_{new}$ in (5.19). That is, the definitions for the state transition function and output function of a controller in (3.7d) and (3.7e), respectively, are applied to all elements in $\mathcal{E}_{new}$. As a result, the reconfigured controller C^r is given by

$$\mathcal{Z}_c^r = \mathcal{Z}_c^- = \mathcal{Z}_c, \tag{5.20a}$$

$$\mathcal{V}_c^r = \mathcal{V}_c^- = \mathcal{V}_c, \tag{5.20b}$$

$$\mathcal{W}_c^r = \mathcal{W}_c^- = \mathcal{W}_c, \tag{5.20c}$$

$$G_c^r(z_c, v_c) = \begin{cases} G_c^-(z_c, v_c) & \text{if } G_c^-(z_c, v_c)!, \\ v_c & \text{if } (z_c, v_c) \in \mathcal{E}_{new} \wedge \mathcal{V}_{a\bar{f}}(v_c, z_c) \neq \emptyset, \\ \text{undefined} & \text{otherwise,} \end{cases} \tag{5.20d}$$

$$H_c^r(z_c, v_c) = \begin{cases} H_c^-(z_c, v_c) & \text{if } H_c^-(z_c, v_c)!, \\ w_c \in \mathcal{V}_{a\bar{f}}(v_c, z_c) & \text{if } (z_c, v_c) \in \mathcal{E}_{new} \wedge \mathcal{V}_{a\bar{f}}(v_c, z_c) \neq \emptyset, \\ \text{undefined} & \text{otherwise,} \end{cases} \tag{5.20e}$$

$$z_{c0}^r = z_{c0}^-. \tag{5.20f}$$

From the definition of the set $\mathcal{E}_{new,-}$ in (5.15) contained in the set $\mathcal{E}_{new}$ in (5.19), it follows that

$$(z_c, v_c) \in \mathcal{E}_{new} \Rightarrow G_c^-(z_c, v_c) \text{ undefined.}$$

Therefore, the conditions for the different cases of the state transition function G_c^r and the output function H_c^r are mutually exclusive.

Algorithm 5.1 summarizes the necessary steps to obtain the reconfigured controller C^r by a partial redefinition of the nominal controller $C(\mathcal{A}_0)$.

The following proposition shows that the controller $C(\mathcal{A}_{\bar{f}})$ that is defined completely newly for the faulty plant $\mathcal{P} \vDash \mathcal{A}_{\bar{f}}$ as described in Section 5.2.2 is identical to the reconfigured con-

---

Algorithm 5.1: Partial redefinition of controller based on unambiguous diagnostic result.

Given: Complete, correct and unambiguous diagnostic result $\mathcal{D}^* = \left\{ \left(z_{\mathrm{p}}(k_{\mathrm{d}}) \quad \bar{f} \right)^{\mathsf{T}} \right\}$.

          Nominal controller $\mathcal{C}(\mathcal{A}_0) = (\mathcal{Z}_{\mathrm{c}}, \mathcal{V}_{\mathrm{c}}, \mathcal{W}_{\mathrm{c}}, G_{\mathrm{c}}, H_{\mathrm{c}}, z_{\mathrm{c}0})$.

          Set $\{ E_{zf}, E_{vf}, E_{wf}, (f \in \mathcal{F}) \}$ of error relations.

1   $\mathcal{Z}_{\mathrm{c}}^{\mathrm{r}} = \mathcal{Z}_{\mathrm{c}}$.   // State set of reconfigured controller based on (5.20a)

2   $\mathcal{V}_{\mathrm{c}}^{\mathrm{r}} = \mathcal{V}_{\mathrm{c}}$.   // Input set of reconfigured controller based on (5.20b)

3   $\mathcal{W}_{\mathrm{c}}^{\mathrm{r}} = \mathcal{W}_{\mathrm{c}}$.   // Output set of reconfigured controller based on (5.20c)

4   $z_{\mathrm{c}0}^{\mathrm{r}} = z_{\mathrm{p}}(k_{\mathrm{d}})$.   // Initial state of reconfigured controller based on (5.20f)

5   Compute the state transition function G_{c}^- and the output function H_{c}^- of the pruned automaton $\mathcal{C}(\mathcal{A}_0)^-$ using (5.12) and (5.13), respectively.

6   Compute the set $\mathcal{E}_{\mathrm{new}}$ of transitions to be newly defined using (5.19).

7   Complete the state transition function and the output function of the pruned automaton $\mathcal{C}(\mathcal{A}_0)^-$ by (5.20d) and (5.20e), respectively, in order to obtain the state transition function $G_{\mathrm{c}}^{\mathrm{r}}$ and the output function $H_{\mathrm{c}}^{\mathrm{r}}$ of the reconfigured controller.

Result: Reconfigured controller $\mathcal{C}^{\mathrm{r}} = (\mathcal{Z}_{\mathrm{c}}^{\mathrm{r}}, \mathcal{V}_{\mathrm{c}}^{\mathrm{r}}, \mathcal{W}_{\mathrm{c}}^{\mathrm{r}}, G_{\mathrm{c}}^{\mathrm{r}}, H_{\mathrm{c}}^{\mathrm{r}}, z_{\mathrm{c}0}^{\mathrm{r}})$.

---

troller $\mathcal{C}^{\mathrm{r}}$ in (5.20).

Proposition 5.5 (Identity of controllers). *The reconfigured controller $\mathcal{C}^{\mathrm{r}}$ resulting from Algorithm 5.1 and the controller $\mathcal{C}(\mathcal{A}_{\bar{f}})$ as defined in (3.7) with $z_{\mathrm{c}0} = z_{\mathrm{p}}(k_{\mathrm{d}})$ are identical.*

Proof. Throughout this proof, without loss of generality, assume that in the model $\mathcal{A}_{\bar{f}}$ of the faulty plant there is only one input symbol v_{p} associated with each transition, i. e.

$$|\mathcal{V}_{a\bar{f}}(z', z)| \leq 1, \quad \forall (z', z) \in \mathcal{Z}_{\bar{f}} \times \mathcal{Z}_{\bar{f}}. \tag{5.21}$$

If (5.21) does not hold, the transitions with the additional input symbols can always be removed from $\mathcal{A}_{\bar{f}}$ without changing the structure of the automaton. Otherwise it would be possible that during the definition of the controller $\mathcal{C}(\mathcal{A}_{\bar{f}})$ and the reconfigured controller $\mathcal{C}^{\mathrm{r}}$ different output symbols w_{c} are chosen such that the two automata can not be identical.

It has to be shown that $\mathcal{L}_{\mathrm{c},\bar{f}} = \mathcal{L}_{\mathrm{c}}^{\mathrm{r}}$, where $\mathcal{L}_{\mathrm{c},\bar{f}}$ and $\mathcal{L}_{\mathrm{c}}^{\mathrm{r}}$ are the behavioral relations of $\mathcal{C}(\mathcal{A}_{\bar{f}})$ and $\mathcal{C}^{\mathrm{r}}$, respectively. The behavioral relation of the controller $\mathcal{C}(\mathcal{A}_{\bar{f}})$ defined for the model $\mathcal{A}_{\bar{f}}$ of the faulty plant can be computed according to (2.7) and simplified using the definition of the controller in (3.7) and the definition of the state transition function of automaton $\mathcal{A}_{\bar{f}}$ in (2.35):

$$\mathcal{L}_{\mathrm{c},\bar{f}} = \left\{ (z_{\mathrm{c}}', w_{\mathrm{c}}, z_{\mathrm{c}}, v_{\mathrm{c}}) \in \mathcal{Z}_{\mathrm{c}} \times \mathcal{W}_{\mathrm{c}} \times \mathcal{Z}_{\mathrm{c}} \times \mathcal{V}_{\mathrm{c}} : \left[G_{\mathrm{c},\bar{f}}(z_{\mathrm{c}}, v_{\mathrm{c}}) = z_{\mathrm{c}}' \right] \wedge \left[H_{\mathrm{c},\bar{f}}(z_{\mathrm{c}}, v_{\mathrm{c}}) = w_{\mathrm{c}} \right] \right\}$$

$$= \left\{ (z_{\mathrm{c}}', w_{\mathrm{c}}, z_{\mathrm{c}}, v_{\mathrm{c}}) \in \mathcal{Z}_{\mathrm{c}} \times \mathcal{W}_{\mathrm{c}} \times \mathcal{Z}_{\mathrm{c}} \times \mathcal{V}_{\mathrm{c}} : \left[v_{\mathrm{c}} = z_{\mathrm{c}}' \right] \wedge \left[G_{\bar{f}}(z_{\mathrm{c}}, w_{\mathrm{c}}) = v_{\mathrm{c}} \right] \right\}$$

$$= \left\{ (z_{\mathrm{c}}', w_{\mathrm{c}}, z_{\mathrm{c}}, v_{\mathrm{c}}) \in \mathcal{Z}_{\mathrm{c}} \times \mathcal{W}_{\mathrm{c}} \times \mathcal{Z}_{\mathrm{c}} \times \mathcal{V}_{\mathrm{c}} : \left[v_{\mathrm{c}} = z_{\mathrm{c}}' \right] \wedge \left[E_{z\bar{f}}(z_{\mathrm{c}}, E_{v\bar{f}}(w_{\mathrm{c}})) = v_{\mathrm{c}} \right] \right\}.$$

The behavioral relation of the reconfigured controller $\mathcal{C}^{\mathrm{r}}$ is also defined according to (2.7) and simplified based on the definition of the reconfigured controller $\mathcal{C}^{\mathrm{r}}$ in (5.20), the state transition function and the output function of the pruned automaton $\mathcal{C}(\mathcal{A}_0)^-$ in (5.12) and (5.13), respectively, and the state transition function of the automaton $\mathcal{A}_{\bar{f}}$ in (2.35):

$$
\begin{aligned}
\mathcal{L}_{\mathrm{c}}^{\mathrm{r}} &= \Big\{ (z_{\mathrm{c}}', w_{\mathrm{c}}, z_{\mathrm{c}}, v_{\mathrm{c}}) \in \mathcal{Z}_{\mathrm{c}} \times \mathcal{W}_{\mathrm{c}} \times \mathcal{Z}_{\mathrm{c}} \times \mathcal{V}_{\mathrm{c}} : \Big[G_{\mathrm{c}}^{\mathrm{r}}(z_{\mathrm{c}}, v_{\mathrm{c}}) = z_{\mathrm{c}}' \Big] \wedge \Big[H_{\mathrm{c}}^{\mathrm{r}}(z_{\mathrm{c}}, v_{\mathrm{c}}) = w_{\mathrm{c}} \Big] \Big\} \\
&= \Big\{ (z_{\mathrm{c}}', w_{\mathrm{c}}, z_{\mathrm{c}}, v_{\mathrm{c}}) \in \mathcal{Z}_{\mathrm{c}} \times \mathcal{W}_{\mathrm{c}} \times \mathcal{Z}_{\mathrm{c}} \times \mathcal{V}_{\mathrm{c}} : \Big(\Big[G_{\mathrm{c}}^{-}(z_{\mathrm{c}}, v_{\mathrm{c}}) = z_{\mathrm{c}}' \Big] \wedge \Big[H_{\mathrm{c}}^{-}(z_{\mathrm{c}}, v_{\mathrm{c}}) = w_{\mathrm{c}} \Big] \Big) \\
&\qquad \vee \Big(\Big[(z_{\mathrm{c}}, z_{\mathrm{c}}') \in \mathcal{E}_{\mathrm{new}} \Big] \wedge \Big[v_{\mathrm{c}} = z_{\mathrm{c}}' \Big] \wedge \Big[G_{\bar{f}}(z_{\mathrm{c}}, w_{\mathrm{c}}) = v_{\mathrm{c}} \Big] \Big) \Big\} \\
&= \Big\{ (z_{\mathrm{c}}', w_{\mathrm{c}}, z_{\mathrm{c}}, v_{\mathrm{c}}) \in \mathcal{Z}_{\mathrm{c}} \times \mathcal{W}_{\mathrm{c}} \times \mathcal{Z}_{\mathrm{c}} \times \mathcal{V}_{\mathrm{c}} : \Big(\Big[G_{\mathrm{c}}(z_{\mathrm{c}}, v_{\mathrm{c}}) = z_{\mathrm{c}}' \Big] \wedge \Big[H_{\mathrm{c}}(z_{\mathrm{c}}, v_{\mathrm{c}}) = w_{\mathrm{c}} \Big] \\
&\qquad \wedge \Big[E_{\mathrm{z}\bar{f}}(z_{\mathrm{c}}, E_{\mathrm{v}\bar{f}}(w_{\mathrm{c}})) = v_{\mathrm{c}} \Big] \wedge \Big[H_{\mathrm{c}}(z_{\mathrm{c}}, v_{\mathrm{c}}) = w_{\mathrm{c}} \Big] \Big) \\
&\qquad \vee \Big(\Big[v_{\mathrm{c}} = z_{\mathrm{c}}' \Big] \wedge \Big[G_{\bar{f}}(z_{\mathrm{c}}, w_{\mathrm{c}}) = v_{\mathrm{c}} \Big] \wedge \Big[(z_{\mathrm{c}}, z_{\mathrm{c}}') \in \mathcal{E}_{\mathrm{new}} \Big] \Big) \Big\} \\
&= \Big\{ (z_{\mathrm{c}}', w_{\mathrm{c}}, z_{\mathrm{c}}, v_{\mathrm{c}}) \in \mathcal{Z}_{\mathrm{c}} \times \mathcal{W}_{\mathrm{c}} \times \mathcal{Z}_{\mathrm{c}} \times \mathcal{V}_{\mathrm{c}} : \\
&\qquad \Big[v_{\mathrm{c}} = z_{\mathrm{c}}' \Big] \wedge \Big[E_{\mathrm{z}\bar{f}}(z_{\mathrm{c}}, E_{\mathrm{v}\bar{f}}(w_{\mathrm{c}})) = v_{\mathrm{c}} \Big] \wedge \Big(\Big[H_{\mathrm{c}}(z_{\mathrm{c}}, v_{\mathrm{c}}) = w_{\mathrm{c}} \Big] \vee \Big[(z_{\mathrm{c}}, z_{\mathrm{c}}') \in \mathcal{E}_{\mathrm{new}} \Big] \Big) \Big\}.
\end{aligned}
$$

In the following it is proved that the two behavioral relations are identical, that is,

$$
\Big((z_{\mathrm{c}}', w_{\mathrm{c}}, z_{\mathrm{c}}, v_{\mathrm{c}}) \in \mathcal{L}_{\mathrm{c}, \bar{f}} \Big) \Leftrightarrow \Big((z_{\mathrm{c}}', w_{\mathrm{c}}, z_{\mathrm{c}}, v_{\mathrm{c}}) \in \mathcal{L}_{\mathrm{c}}^{\mathrm{r}} \Big).
$$

The above formula is equivalent to the statement that

$$
\Big(\Big[v_{\mathrm{c}} = z_{\mathrm{c}}' \Big] \wedge \Big[E_{\mathrm{z}\bar{f}}(z_{\mathrm{c}}, E_{\mathrm{v}f}(w_{\mathrm{c}})) = v_{\mathrm{c}} \Big] \Big) \Leftrightarrow \Big(\Big[v_{\mathrm{c}} = z_{\mathrm{c}}' \Big] \wedge \Big[E_{\mathrm{z}\bar{f}}(z_{\mathrm{c}}, E_{\mathrm{v}\bar{f}}(w_{\mathrm{c}})) = v_{\mathrm{c}} \Big] \\
\wedge \Big(\Big[H_{\mathrm{c}}(z_{\mathrm{c}}, v_{\mathrm{c}}) = w_{\mathrm{c}} \Big] \vee \Big[(z_{\mathrm{c}}, z_{\mathrm{c}}') \in \mathcal{E}_{\mathrm{new}} \Big] \Big) \Big).
$$

A formula of the form $A \Leftrightarrow (A \wedge B)$ like the one above is true if and only if $A \Rightarrow B$ (that is, A implies B). Therefore, it has to be checked whether

$$
A := \Big(\Big[v_{\mathrm{c}} = z_{\mathrm{c}}' \Big] \wedge \Big[E_{\mathrm{z}\bar{f}}(z_{\mathrm{c}}, E_{\mathrm{v}\bar{f}}(w_{\mathrm{c}})) = v_{\mathrm{c}} \Big] \Big)
$$

implies

$$
B := \Big(\Big[H_{\mathrm{c}}(z_{\mathrm{c}}, v_{\mathrm{c}}) = w_{\mathrm{c}} \Big] \vee \Big[(z_{\mathrm{c}}, z_{\mathrm{c}}') \in \mathcal{E}_{\mathrm{new}} \Big] \Big).
$$

Denote the first part of B by b_1, i.e.,

$$
b_1 := \Big[H_{\mathrm{c}}(z_{\mathrm{c}}, v_{\mathrm{c}}) = w_{\mathrm{c}} \Big].
$$

The second part of B can be rewritten using the definition of the set $\mathcal{E}_{new}$ in (5.19) as follows:

$$(z_c, z_c') \in \mathcal{E}_{new} \Leftrightarrow \Big((\exists w \in \mathcal{W}_c) \Big(\big[H_c(z_c, z_c') = w \big] \wedge \big[E_{z\bar{f}}(z_c, E_{v\bar{f}}(w)) \neq z_c' \big] \Big) \Big)$$
$$\vee \Big(\big[(\not\exists w \in \mathcal{W}_c) H_c(z_c, z_c') = w \big] \wedge \big[(\exists w \in \mathcal{W}_c) E_{z\bar{f}}(z_c, E_{v\bar{f}}(w)) = z_c' \big] \Big).$$

Denote the two parts of this logical formula by

$$b_2 := \Big((\exists w \in \mathcal{W}_c) \Big(\big[H_c(z_c, z_c') = w \big] \wedge \big[E_{z\bar{f}}(z_c, E_{v\bar{f}}(w)) \neq z_c' \big] \Big) \Big)$$

and

$$b_3 := \Big(\big[(\not\exists w \in \mathcal{W}_c) H_c(z_c, z_c') = w \big] \wedge \big[(\exists w \in \mathcal{W}_c) E_{z\bar{f}}(z_c, E_{v\bar{f}}(w)) = z_c' \big] \Big),$$

respectively, such that $B = b_1 \vee b_2 \vee b_3$. From A it follows directly that $w \neq w_c$ in b_2. With the assumption in (5.21), the second part of b_2 is therefore always true. Furthermore, from A it also follows that the second part of b_3 is a tautology. Hence, the implication $A \Rightarrow B$ can be simplified to

$$A \Rightarrow \big[H_c(z_c, z_c') = w_c \big] \vee \Big[(\exists w \in \mathcal{W}_c) \Big(\big[H_c(z_c, z_c') = w \big] \wedge \big[w \neq w_c \big] \Big) \Big]$$
$$\vee \Big[(\not\exists w \in \mathcal{W}_c) H_c(z_c, z_c') = w \Big].$$

This statement is always true, because it describes the three possible values of the output function H_c. For a given state-input pair $(z_c, v_c) \in \mathcal{Z}_c \times \mathcal{V}_c$ with $v_c = z_c'$, the output function $H_c(z_c, v_c)$ either equals the output symbol w_c or some other output symbol $w \in \mathcal{W}$, $(w \neq w_c)$ or is undefined. Consequently, it has been formally shown that $\mathcal{L}_{c,\bar{f}} = \mathcal{L}_c^r$ and the proof is completed.   □

Example 5.1 (cont.)   *Reconfiguration of tracking controller for automated warehouse*

The nominal controller $\mathcal{C}(\mathcal{A}_0)$ in Fig. 3.2 is reconfigured based on the diagnostic result $\mathcal{D}^*$ in (4.56) using Algorithm 5.1. The state set, input set, output set and current state of the reconfigured controller $\mathcal{C}^r$ are given by

$$\mathcal{Z}_c^r = \mathcal{Z}_c = \{A, B, C\},$$
$$\mathcal{V}_c^r = \mathcal{V}_c = \{A, B, C\},$$
$$\mathcal{W}_c^r = \mathcal{W}_c = \{AB, AC, BA, BC, CA, CB, AB^*, AC^*, BA^*, BC^*, CA^*, CB^*\},$$
$$z_{c0}^r = z_p(k_d) = A,$$

respectively, (Lines 1–4). From Line 5 of Algorithm 5.1 the state transition function and the output function of the pruned automaton $\mathcal{C}(\mathcal{A}_0)^-$ result, whose automaton graph is visualized in Fig. 5.6. Only

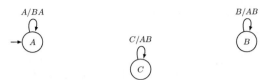

Figure 5.6: Pruned automaton $\mathcal{C}(\mathcal{A}_0)^-$.

the self-loops remain in the pruned automaton $\mathcal{C}(\mathcal{A}_0)^-$, because for all other transitions

$$E_{z2}(z_c, E_{v2}(v_c)) = E_{z2}(z_c, \varepsilon) = z_c \neq v_c$$

holds. Physically, this means that since the nominal controller $\mathcal{C}(\mathcal{A}_0)$ always uses the robot M, which is now broken, for the transport of parcels all transitions requesting any transport need to be deleted.

Next, as specified in Line 6 of Algorithm 5.1, the set of transitions to be newly defined is computed using (5.19). Since no completely new transitions occur in the model $\mathcal{A}_2$ of the faulty plant, only transitions which have been removed from the nominal controller $\mathcal{C}(\mathcal{A}_0)$ are considered in the set

$$\mathcal{E}_{new} = \Big\{(A, B), (A, C), (B, A), (B, C), (C, A), (C, B)\Big\}. \tag{5.22}$$

Then, Line 7 of Algorithm 5.1 is executed. That is, the state transition function and output function of the pruned automaton $\mathcal{C}(\mathcal{A}_0)^-$ in Fig. 5.6 are completed using (5.20d) and (5.20e), respectively. As a result the reconfigured controller $\mathcal{C}^r$ in Fig. 5.7 is obtained. For all elements in the set $\mathcal{E}_{new}$ in (5.22)

$$\mathcal{V}_{a2}(v_c, z_c) \neq \emptyset$$

holds, such that new transitions are added to the pruned automaton. That is, there always exists a redundant input under which a transport is still possible when fault $f = 2$ occurs. It can be seen that these redundant inputs correspond to the usage of robot M^* instead of robot M.

Avoiding illegal transitions. Taking the set $\mathcal{E}_{ill}$ of illegal transitions in (2.54) into account, the same reconfigured controller $\mathcal{C}^r$ as in Fig. 5.7 results, because the reconfiguration method for the controller is independent from the set $\mathcal{E}_{ill}$ of illegal transitions. The only difference is that in this case the initial state of the reconfigured controller $\mathcal{C}^r$ is given by

$$z_{c0}^r = C, \tag{5.23}$$

because of the different current state $z_p(k_d)$ of the plant in the diagnostic result $\mathcal{D}^*$ in (4.75). □

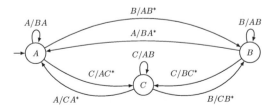

Figure 5.7: Reconfigured controller $\mathcal{C}^r$ corresponding to diagnostic result $\mathcal{D}^*$ in (4.56).

5.2.5 Reconfiguration of the tracking controller by partial redefinition

The reconfiguration of the nominal tracking controller $C_T(A_0)$ based on a partial redefinition of the nominal trajectory planning unit $T(A_0)$ and the nominal controller $C(A_0)$ as described in the previous sections is summarized in Algorithm 5.2.

The following proposition shows that the tracking controller $C_T(A_{\bar{f}})$ that is defined completely newly for the faulty plant $P \models A_{\bar{f}}$ as described in Section 5.2.2 is identical to the reconfigured tracking controller C_T^r resulting from applying Algorithm 5.2 to the nominal tracking controller $C_T(A_0)$.

Proposition 5.6 (Identity of tracking controllers). *The tracking controller $C_T(A_{\bar{f}})$ defined for the model $A_{\bar{f}}$ of the faulty plant is identical to the reconfigured tracking controller C_T^r resulting from Algorithm 5.2.*

Proof. In Proposition 5.3 it has already been shown that the trajectory planning unit $T(A_{\bar{f}})$ and the trajectory planning unit T^r yield the same reference trajectory $R(k_r \ldots k_e^r)$. Since the controllers $C(A_{\bar{f}})$ and C^r in (5.20) have the same I/O behavior (Proposition 5.5), the equivalence of $C_T(A_{\bar{f}})$ and C_T^r is proved. ☐

Based on the result of the above proposition and the fact that the tracking controller $C_T(A_{\bar{f}})$ specifically defined for the faulty plant $P \models A_{\bar{f}}$ steers the faulty plant into the desired final state z_F, the following theorem can be stated. It shows that the reconfigured tracking controller C_T^r that results from a partial redefinition of the trajectory planning unit $T(A_0)$ and the nominal controller $C(A_0)$ is a solution to the reconfiguration problem in Problem 5.1 if the given diagnostic result D^* is unambiguous.

Algorithm 5.2: Reconfiguration of tracking controller with unambiguous diagnostic result.

Given: Complete, correct and unambiguous diagnostic result $D^* = \left\{ \left(z_p(k_d) \quad \bar{f} \right)^\top \right\}$.

Nominal tracking controller $C_T(A_0) = (T(A_0), C(A_0))$.
Set $\{E_{zf}, E_{vf}, E_{wf}, (f \in \mathcal{F})\}$ of error relations.

1 Modify the automaton graph of A_0 stored in $T(A_0)$ according to (5.1) such that the reconfigured trajectory planning unit T^r results.

2 Apply Algorithm 5.1 to $C(A_0)$ such that the reconfigured controller C^r results.

Result: Reconfigured tracking controller $C_T^r = (T^r, C^r)$.

Theorem 5.3 (Fulfillment of control aim after reconfiguration based on unambiguous diagnostic result). *Consider a nominal tracking controller $C_T(A_0)$ that is reconfigurable with respect to a desired final state z_F and a complete, correct and unambiguous diagnostic result $D^* = \left\{ \left(z_p(k_d) \quad \bar{f} \right)^\top \right\}$. Let C_T^r be the reconfigured tracking controller resulting from Algorithm 5.2. Let k_e^r be the time horizon of the reference trajectory generated by the reconfigured trajectory planning unit T^r. Then the faulty plant $P \vDash A_{\bar{f}}$ in the closed-loop system $P \circ C_T^r$ reaches the desired final state z_F at time $k_F = k_e^r + 1$.*

Proof. Since the faulty plant P in the closed-loop system $P \circ C_T(A_{\bar{f}})$ reaches the desired final state z_F at a finite time k_F if the tracking controller is reconfigurable (Proposition 5.1), and the reconfigured tracking controller C_T^r is identical to $C_T(A_{\bar{f}})$ (Proposition 5.6), the proof follows directly. □

Avoiding illegal transitions. Taking into account a set $\mathcal{E}_{ill}$ of illegal transitions for the plant, Line 2 of Algorithm 5.2 describing the reconfiguration procedure has to be changed. Now, instead of the model A_0 of the faultless plant itself, the legal part A_{leg} of the model A_0 of the faultless plant stored in the nominal trajectory planning unit $T(A_0, \mathcal{E}_{ill})$ has to be modified according to (5.2). The remainder of the algorithm remains unchanged. As a result, the following theorem shows that the reconfigured tracking controller C_T^r that results from a partial redefinition of the trajectory planning unit $T(A_0, \mathcal{E}_{ill})$ and the nominal controller $C(A_0)$ under consideration of the set $\mathcal{E}_{ill}$ of illegal transitions is a solution to the safe reconfiguration problem in Problem 5.2 if the given diagnostic result D^* is unambiguous.

Theorem 5.4 (Safe fulfillment of control aim after reconfiguration based on unambiguous diagnostic result). *Consider a nominal tracking controller $C_T(A_0, \mathcal{E}_{ill})$ that is reconfigurable with respect to the desired final state z_F, the set $\mathcal{E}_{ill}$ of illegal transitions and the complete, correct and unambiguous diagnostic result $D^* = \left\{ \left(z_p(k_d) \quad \bar{f} \right)^\top \right\}$. Let C_T^r be the reconfigured tracking controller resulting from Algorithm 5.2 under consideration of $\mathcal{E}_{ill}$. Let k_e^r be the time horizon of the reference trajectory generated by the reconfigured trajectory planning unit T^r. Then the faulty plant $P \vDash A_{\bar{f}}$ in the closed-loop system $P \circ C_T^r$ reaches the desired final state z_F at time $k_F = k_e^r + 1$ while not executing any illegal transitions $(z_p(k), z_p(k+1)) \in \mathcal{E}_{ill}$, $(k_r \leq k < k_F)$.*

Proof. The trajectory planning unit $T(A_{\bar{f}}, \mathcal{E}_{ill})$ completely newly defined for the faulty plant and the reconfigured trajectory planning unit T^r are identical (Proposition 5.4) and the controllers $C(A_{\bar{f}})$ and C^r in (5.20) have the same I/O behavior (Proposition 5.5). Therefore, the tracking controller $\mathcal{C}_T(A_{\bar{f}}, \mathcal{E}_{\text{ill}})$ defined for the model $A_{\bar{f}}$ of the faulty plant under consideration of the set $\mathcal{E}_{\text{ill}}$ of illegal transitions is identical to the reconfigured tracking controller $\mathcal{C}_T^r$ resulting from Algorithm 5.2 when the automaton graph in $\mathcal{T}(A_0, \mathcal{E}_{\text{ill}})$ is modified according to (5.2). With this result the proof is the same as the one of Theorem 5.3 on the fulfillment of the control aim after the reconfiguration in the absence of illegal transitions. $\square$

Example 5.1 (cont.)   *Reconfiguration of tracking controller for automated warehouse*

The nominal tracking controller $\mathcal{C}_T(A_0)$ from Example 3.1 is reconfigured based on the diagnostic result $\mathcal{D}^*$ in (4.56) and the set $\{E_{zf}, E_{vf}, E_{wf}, (f \in \mathcal{F})\}$ of error relations in (2.55a)–(2.55c) and (2.56a)–(2.56c) according to Algorithm 5.2. That is, the nominal trajectory planning unit $\mathcal{T}(A_0)$ is reconfigured as described in Section 5.2.3 and the nominal controller is reconfigured as described in Section 5.2.4. The resulting reconfigured trajectory planning unit $\mathcal{T}^r$ contains the automaton A^r in Fig. 5.3, while the automaton graph of the reconfigured controller $\mathcal{C}^r$ is shown in Fig. 5.7.

After the reconfiguration time $k_r = 3$, the reconfigured tracking controller $\mathcal{C}_T^r = (\mathcal{T}^r, \mathcal{C}^r)$ is used to control the faulty plant A_2 with current state $z_p(k_d = 3) = A$. That is, a parcel lies at position A and the robot M is broken. The trajectory planning unit $\mathcal{T}^r$ generates the reference input $r(k_r = k_e^r = 3) = B$ in (5.4) for the controller $\mathcal{C}^r$, which generates the input $v_p(3) = w_c(3) = AB^*$ for the plant and goes to state $z_c(4) = B$. That is, the reconfigured tracking controller requests the transport of a parcel from position A to position B using the robot M^*. This transport can be executed by the faulty plant such that the plant goes to state $z_p(4) = B$ and generates the output $w_p(3) = \text{ok}$. Consequently, as stated by Theorem 5.3, the faulty plant reaches the desired final state $z_F = B$ at time $k_F = 4 = k_e^r + 1$.

Avoiding illegal transitions. In the presence of illegal transitions, the tracking controller reconfigured based on the diagnostic result $\mathcal{D}^*$ in (4.75) and the set $\mathcal{E}_{\text{ill}}$ of illegal transitions in (2.54) is considered. That is the reconfigured trajectory planning unit contains the automaton A_{leg}^r in Fig. 5.4 and the reconfigured controller is shown in Fig. 5.7 with initial state $z_{c0}^r = C$.

When the trajectory planning unit $\mathcal{T}^r$ generates the reference input $r(k_r = k_e^r = 2) = B$ in (5.5), the controller $\mathcal{C}^r$ generates the input $v_p(2) = CB^*$ for the faulty plant and passes to state $z_c(3) = B$. That is, it requests the transport of a parcel from position C to position B using the robot M^*. This transport can be executed by the faulty plant such that the plant goes to state $z_p(3) = B$ and generates the output $w_p(2) = \text{ok}$. Consequently, as stated by Theorem 5.4, the faulty plant reaches the desired final state $z_F = B$ at time $k_F = 3 = k_e^r + 1$ and no illegal transition between state A and state B has been used. $\square$

5.2.6 Complexity of the reconfiguration method

Complexity of the reconfigurability analysis.   According to Theorem 5.5 and Theorem 5.2 a necessary and sufficient condition for the reconfigurability of the tracking controller is the controllability of the model $A_{\bar{f}}$ of the faulty plant. If the reconfigurability shall be checked in advance such that the present fault $\bar{f} \in \mathcal{F}$ and the current state $z_p(k_d)$ of the faulty plant $\mathcal{P}$ are unknown, the controllability of all models $A_f, (f \in \mathcal{F})$ with respect to all possible initial states $z_{f0} \in \mathcal{Z}_f, (f \in \mathcal{F})$ has to be analyzed. According to Section 3.6.1 the complexity of one of these controllability analyses is of order $\mathcal{O}(n \cdot p)$ such that an overall complexity of order $\mathcal{O}(F \cdot n \cdot n \cdot p) = \mathcal{O}(F \cdot n^2 \cdot p)$ results. Hence, the reconfigurability analysis is quadratic in the number n of states in the model A_0 of the faultless plant, but linear in the number F of faults and p of input symbols.

If the reconfigurability analysis is executed at runtime with respect to a given diagnostic result $\mathcal{D}^*$, its complexity is reduced to $\mathcal{O}(n \cdot p)$. In any case the complexity of the reconfigurability analysis is low enough to apply it to real-world systems.

Complexity of the online execution of the reconfiguration method. The reconfiguration of the nominal tracking controller $\mathcal{C}_{\mathcal{T}}(\mathcal{A}_0)$ according to Algorithm 5.2 has to be executed online, as soon as the diagnostic unit $\mathcal{D}$ provides the diagnostic result $\mathcal{D}^*$. The reconfiguration of the nominal trajectory planning unit $\mathcal{T}(\mathcal{A}_0)$ requires the correction of the initial state of the stored model $\mathcal{A}_0$ of the faultless plant, as well as the replacement of all faulty transitions in its state transition function. Therefore, at most $n \cdot p + 1$ operations are executed in Line 1 of Algorithm 5.2 if all transitions are faulty.

The reconfiguration of the nominal controller $\mathcal{C}(\mathcal{A}_0)$ is conducted according to Algorithm 5.1. In Lines 1–4 of this algorithm, only the initial state of $\mathcal{C}(\mathcal{A}_0)$ is changed, while its state set, input set and output set are kept. During the removal of faulty transitions in Line 5, every faulty transition has to be deleted from the state transition function and the output function, for which at most $2 \cdot n \cdot p$ operations are necessary if all transitions are faulty. In the set $\mathcal{E}_{\text{new}}$ computed in Line 6 of Algorithm 5.1 every faulty transition might occur twice, once for sure in the set $\mathcal{E}_{\text{new},-}$ of removed transitions, but possibly also in the set $\mathcal{E}_{\text{new},+}$ of new transitions. That is, for the construction of the set $\mathcal{E}_{\text{new}}$ at most $n \cdot p$ operations are necessary and it contains at most $2 \cdot n \cdot p$ elements if all transitions are changed by the fault such that completely new transitions occur. The completion of the state transition function and the output function of the pruned automaton based on the set $\mathcal{E}_{\text{new}}$ in Line 7 of Algorithm 5.1 therefore requires at most $2 \cdot 2 \cdot n \cdot p$ operations.

In total, the number of operations for the reconfiguration of the nominal tracking controller $\mathcal{C}_{\mathcal{T}}(\mathcal{A}_0)$ is of order $\mathcal{O}(n \cdot p)$. That is, the computational complexity of the reconfiguration is at most linear in the number n of states and p of input symbols of the model $\mathcal{A}_0$ of the faultless plant. Usually the bound will be even lower, because not all $n \cdot p$ transitions are affected by a fault. It is application-dependent whether the required time before the reconfigured tracking controller $\mathcal{C}_{\mathcal{T}}^{\text{r}}$ is put into action can be tolerated or not.

5.3 Reconfiguration with ambiguous diagnostic result

In this section the method for the reconfiguration of the nominal tracking controller $\mathcal{C}_{\mathcal{T}}(\mathcal{A}_0)$ based on an unambiguous diagnostic result $\mathcal{D}^*$ described in the previous section is generalized in order to obtain a reconfiguration method that can also deal with ambiguous diagnostic results. That is, a solution to the reconfiguration problem in Problem 5.1 and the safe reconfiguration problem in Problem 5.2 is proposed.

5.3.1 Main idea for the reconfiguration with ambiguous diagnostic result

Problem 5.1 and Problem 5.2 both require to find a reconfigured tracking controller $C_{\mathcal{T}}^{r}$ that steers the faulty plant $\mathcal{P} \models \mathcal{A}_{\bar{f}}$ from its current state $z_p(k_d)$ into the desired final state z_F. However, neither the present fault $\bar{f} \in \mathcal{F}$ nor the current state $z_p(k_d)$ of the faulty plant are known unambiguously. Rather, only a diagnostic result $\mathcal{D}^*$ in which multiple fault candidates $f \in \mathcal{F}$ and/or multiple possible current states $z_p \in \mathcal{Z}_f$ might be contained is given. Since the diagnostic result $\mathcal{D}^*$ is required to be complete and correct, it is known that

$$\begin{pmatrix} z_p(k_d) \\ \bar{f} \end{pmatrix} \in \mathcal{D}^* \subseteq \mathcal{Z}_\Delta = \bigcup_{f \in \mathcal{F}} \mathcal{Z}_f \times \{f\} \tag{5.24}$$

holds (cf. (2.42a)). Therefore, in order to guarantee that the reconfigured tracking controller $C_{\mathcal{T}}^{r}$ succeeds in controlling the faulty plant, a tracking controller $C_{\mathcal{T}}^{r}$ which works for *all* fault candidates and possible current states of the faulty plant contained in the diagnostic result $\mathcal{D}^*$ has to be found.

The main idea for finding a reconfigured tracking controller $C_{\mathcal{T}}^{r}$ that solves the reconfiguration problem simultaneously for all fault candidates and possible current states of the faulty plant in the diagnostic result $\mathcal{D}^*$ is to construct a common model of the faulty plant describing the common behavior of the faulty plant in all possible situations contained in the diagnostic result $\mathcal{D}^*$. For the case that exactly two fault candidates are included in the diagnostic result $\mathcal{D}^*$, Fig. 5.8 illustrates the behavior that the common model of the faulty plant shall describe. It needs to contain all transitions contained in both models of the faulty plant.

Like in the case of an unambiguous diagnostic result, the common model shall be expressed in form of error relations, such that the reconfiguration method described in the previous section, possibly with some modifications, can be applied.

In the remainder of the section, first, a definition for such a common model is proposed and it is proved that a tracking controller newly defined based on this common model already solves the reconfiguration problems in Problem 5.1 or Problem 5.2. Afterwards, an equivalent defi-

Figure 5.8: Illustration of common model of the faulty plant for two faults in $\mathcal{D}^*$.

nition of the common model in form of error relations is given. Finally, it is shown how the method for the reconfiguration of the tracking controller by partial redefinition from the previous section has to be adapted in order to use it for the reconfiguration based on an ambiguous diagnostic result.

5.3.2 Construction of the common model of the faulty plant

For the construction of the common model of the faulty plant the operation of intersecting deterministic I/O automata is used. Given a set $\{\mathcal{A}_i, (i \in \mathcal{N})\}$ of deterministic I/O automata, an intersection automaton $\mathcal{A}_\cap = (\mathcal{Z}_\cap, \mathcal{V}_\cap, G_\cap, z_{\cap 0})$ shall be constructed that contains all transitions $z \xrightarrow{v} z'$ that are present in all automata $\mathcal{A}_i, (i \in \mathcal{N})$. Note that the outputs of the automata $\mathcal{A}_i, (i \in \mathcal{N})$ are neglected during the construction of $\mathcal{A}_\cap$. The intersection operation is formally defined as follows.

Definition 5.3 (Intersection operation). *The automaton*

$$
\mathcal{A}_\cap : \begin{cases}
\mathcal{Z}_\cap = \displaystyle\bigcap_{i \in \mathcal{N}} \mathcal{Z}_i \cup \{z_{\cap 0}\} & \text{(5.25a)} \\[2ex]
\mathcal{V}_\cap = \displaystyle\bigcap_{i \in \mathcal{N}} \mathcal{V}_i & \text{(5.25b)} \\[2ex]
G_\cap(z, v) = \begin{cases}
z' & \text{if } \left(\left[G_i(z_{i0}, v) = z' \,\forall i \in \mathcal{N}\right] \wedge \left[z = z_{\cap 0}\right]\right) \\
& \vee \left(\left[G_i(z, v) = z' \,\forall i \in \mathcal{N}\right] \wedge \left[z \neq z_{\cap 0}\right]\right), \\
\text{undefined} & \text{otherwise,}
\end{cases} & \text{(5.25c)} \\[5ex]
z_{\cap 0} = \begin{cases}
z_0 & \text{if } z_{i0} = z_0 \in \mathcal{Z}_i, \quad \forall i \in \mathcal{N}, \\
z_{\text{new}} \notin \displaystyle\bigcup_{i \in \mathcal{N}} \mathcal{Z}_i & \text{otherwise,}
\end{cases} & \text{(5.25d)}
\end{cases}
$$

is called intersection automaton of the set $\{\mathcal{A}_i, (i \in \mathcal{N})\}$ of deterministic I/O automata and is denoted by

$$
\mathcal{A}_\cap = \bigcap_{i \in \mathcal{N}} \mathcal{A}_i. \tag{5.26}
$$

The new initial state $z_{\cap 0} = z_{\text{new}}$ that is introduced if the initial states z_{i0} of the automata $\mathcal{A}_i \ (i \in \mathcal{N})$ differ, is an "artificial state" that is assumed to not belong to any other automaton. Note that if no artificial initial state z_{new} is introduced, both parts of the formula for computing the next state z' in (5.25c) become identical.

In the following the interesting property that every input sequence leading to a state sequence in an intersection automaton $\mathcal{A}_\cap = \cap_{i \in \mathcal{N}} \mathcal{A}_i$ leads to the same state sequence in every intersected automaton $\mathcal{A}_i$, $(i \in \mathcal{N})$ is proved.

Proposition 5.7 (Common state sequence). *Given an intersection automaton $\mathcal{A}_\cap = \bigcap_{i \in \mathcal{N}} \mathcal{A}_i$, every input sequence $V(0 \ldots k_e) \in \mathcal{V}_\cap^\infty$ leading to a state sequence $Z_\cap(0 \ldots k_e + 1) \in \mathcal{Z}_\cap^\infty$ with $z_\cap(0) = z_{\cap 0}$ in the intersection automaton $\mathcal{A}_\cap$ leads to a state sequence $Z_i(0 \ldots k_e + 1) \in \mathcal{Z}_i^\infty$ with $z_i(0) = z_{i0}$ and $Z_i(1 \ldots k_e + 1) = Z_\cap(1 \ldots k_e + 1)$ in any automaton $\mathcal{A}_i$, $(i \in \mathcal{N})$.*

Proof. (By induction) According to (5.25c) and with $z_\cap(0) = z_{\cap 0}$ and $z_i(0) = z_{i0}$, $(i \in \mathcal{N})$ the following equation holds for all $i \in \mathcal{N}$:

$$z_\cap(1) = G_\cap(z_\cap(0), v(0)) = G_i(z_{i0}, v(0)) = G_i(z_i(0), v(0)) = z_i(1).$$

Given $z_\cap(k) = z(k)$, $(0 < k \leq k_e)$, equation (5.25c) leads to

$$z_\cap(k + 1) = G_\cap(z_\cap(k), v(k)) = G_i(z_\cap(k), v(k)) = z_i(k + 1), \quad \forall i \in \mathcal{N}.$$

Therefore, except for the initial state, the traversed state sequence is the same in all intersected automata $\mathcal{A}_i$, $(i \in \mathcal{N})$ and the intersection automaton $\mathcal{A}_\cap$. □

When the intersection operation is applied to the set $\{\mathcal{A}_f, (f \in \mathcal{F}^*(k_r))\}$ of automata corresponding to all fault candidates $f \in \mathcal{F}^*(k_r)$ contained in the diagnostic result $\mathcal{D}^*$, the common model of the faulty plant results, which is defined as follows.

Definition 5.4 (Common model of the faulty plant). *The common model of a faulty plant $\mathcal{P}$ with respect to a diagnostic result $\mathcal{D}^*$ is an intersection automaton given by*

$$\mathcal{A}_\cap = \bigcap_{(z_p \ f)^\top \in \mathcal{D}^*} \left(\mathcal{A}_f = (\mathcal{Z}_f, \mathcal{V}_f, \mathcal{W}_f, G_f, H_f, z_p) \right). \tag{5.27}$$

The common model $\mathcal{A}_\cap$ of the faulty plant describes the common behavior of the plant under all fault candidates $f \in \mathcal{F}^*(k_r)$ in the diagnostic result $\mathcal{D}^*$. If the diagnostic result $\mathcal{D}^*$ contains more than one state from the same automaton $\mathcal{A}_f$, $(f \in \mathcal{F})$ or different states from different automata $\mathcal{A}_f$, $(f \in \mathcal{F})$, a new initial state $z_{\cap 0} := z_{new}$ is introduced during the construction of the common model $\mathcal{A}_\cap$ of the faulty plant (cf. (5.25d)). Note that if the diagnostic result $\mathcal{D}^*$ is unambiguous, the common model of the faulty plant is given by $\mathcal{A}_\cap = (\mathcal{Z}_{\bar{f}}, \mathcal{V}_{\bar{f}}, G_{\bar{f}}, z_p(k_d))$, that is, by the model $\mathcal{A}_{\bar{f}}$ of the faulty plant with removed outputs.

Example 5.2 *Reconfiguration of tracking controller for automated warehouse*

Consider the automated warehouse example introduced in Section 2.5 with the difference that now an additional state $A_\#$ is present. The state $A_\#$ corresponds to the situation that a robot with a parcel gets stuck closely to position A, because the direct route between position A and position B is blocked due to fault $f = 1$. Therefore, for the fault $f = 1$ the error relation E_{zf} is changed to

$$E_{z1}(z,v) = \begin{cases} A_\# & \text{if } (z,v) \in \{(A,AB),(A,AB^*),(A_\#,AB),(A_\#,AB^*)\}, \\ G_0(z,v) & \text{otherwise,} \end{cases} \tag{5.28}$$

while all other error relations remain the same as before (cf. (2.55a), (2.55b) and (2.56a)–(2.56c)). The state $A_\#$ is not reachable in automata $\mathcal{A}_0$ and $\mathcal{A}_2$, because a robot with a parcel can only get stuck if fault $f = 1$ occurs (gray state and transitions in Fig. 5.9b).

When the robot M breaks down, hence fault $f = 2$ occurs, the diagnostic result at time $k = 1$ is given by

$$\mathcal{D}^* = \left\{ \begin{pmatrix} A_\# \\ 1 \end{pmatrix}, \begin{pmatrix} A \\ 2 \end{pmatrix} \right\}. \tag{5.29}$$

This ambiguous diagnostic result reflects that either the route between position A and position B is blocked ($f = 1$) and a robot with a parcel is stuck closely to position A ($z_\mathrm{p} = A_\#$) or robot M failed ($f = 2$) and the parcel still lies at position A ($z_\mathrm{p} = A$). That is, based on the currently available measurements it can not be distinguished between the two faults. At this time $k_\mathrm{r} = 1$ the fault diagnosis is stopped and the reconfiguration shall be executed based on the ambiguous diagnostic result in (5.29).

The common model of the faulty plant with respect to the diagnostic result $\mathcal{D}^*$ in (5.29) is constructed according to Definition 5.4. That is, the automata $\mathcal{A}_1$ and $\mathcal{A}_2$ with initial states $z_{10} = A_\#$ and $z_{20} = A$ in Fig. 5.9a and Fig. 5.9b, respectively, are intersected using the intersection operation from Definition 5.3. As a result the common model $\mathcal{A}_\cap$ of the faulty plant shown in Fig. 5.10 is obtained.

Since the initial states z_{10} and z_{20} differ, based on (5.25d) the common model $\mathcal{A}_\cap$ of the faulty plant obtains an artificial initial state

$$z_{\cap 0} = z_{\mathrm{new}}.$$

The artificial initial state z_{new} represents a combination of the state $A_\#$ in the automaton $\mathcal{A}_1$ in Fig. 5.9a and the state A in the automaton $\mathcal{A}_2$ in Fig. 5.9b. Therefore, according to (5.25a) the state set of the common model $\mathcal{A}_\cap$ of the faulty plant is given by

$$\mathcal{Z}_\cap = \mathcal{Z}_1 \cap \mathcal{Z}_2 \cup \{z_{\mathrm{new}}\} = \{A,B,C,A_\#,z_{\mathrm{new}}\},$$

while from (5.25b) the input set

$$\mathcal{V}_\cap = \mathcal{V}_1 \cap \mathcal{V}_2 = \{AB,AC,BA,BC,CA,CB,AB^*,AC^*,BA^*,BC^*,CA^*,CB^*\}$$

results. The only transition starting at the initial state $z_{\cap 0} = z_{\mathrm{new}}$ of the common model $\mathcal{A}_\cap$ of the faulty plant is

$$G_\cap(z_{\mathrm{new}},AC^*) = C,$$

because

$$G_1(z_{10} = A_\#,AC^*) = C = G_2(z_{20} = A,AC^*)$$

(cf. first part of (5.25c)). Another transition included in $\mathcal{A}_\cap$ is, for example,

$$G_\cap(B,BA^*) = A = G_1(B,BA^*) = G_2(B,BA^*)$$

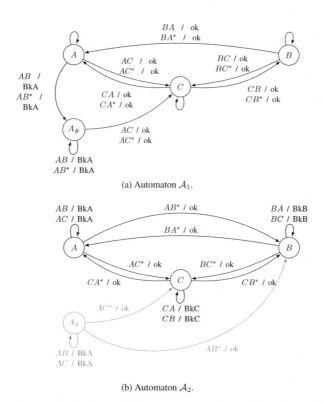

(a) Automaton $\mathcal{A}_1$.

(b) Automaton $\mathcal{A}_2$.

Figure 5.9: Models $\mathcal{A}_1$ and $\mathcal{A}_2$ of faulty automated warehouse with additional state.

(cf. second part of (5.25c)). Conversely, for example, no transition from state A to state B is contained in the common model $\mathcal{A}_\cap$ of the faulty plant, because

$$G_2(A, AB^*) = B \neq G_1(A, AB^*) = A_\#.$$

The common model $\mathcal{A}_\cap$ of the faulty plant in Fig. 5.10 shows that for some inputs the plant $\mathcal{P}$ shows an identical behavior, regardless whether fault $f = 1$ or fault $f = 2$ occurred. In particular, the robot M^* can be used for the transport of parcels in both fault cases.   □

5.3.3 Reconfigurability analysis

The following theorem states a condition under which the nominal tracking controller $\mathcal{C}_\mathcal{T}(\mathcal{A}_0)$ is reconfigurable according to Definition 5.1. In contrast to the condition for the reconfigurability based on an unambiguous diagnostic result $\mathcal{D}^*$ in Theorem 5.1, the following condition

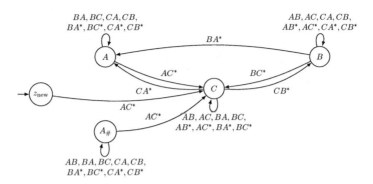

Figure 5.10: Common model $\mathcal{A}_\cap$ of the faulty plant with respect to diagnostic result $\mathcal{D}^*$ in (5.29).

applicable also in the case of an ambiguous diagnostic result $\mathcal{D}^*$ is sufficient, but not necessary for the reconfigurability.

Theorem 5.5 (Reconfigurability with ambiguous diagnostic result). *A nominal tracking controller $\mathcal{C}_T(\mathcal{A}_0)$ is reconfigurable with respect to the complete and correct diagnostic result $\mathcal{D}^*$ and the final state z_F if the common model $\mathcal{A}_\cap$ of the faulty plant in (5.27) is controllable with respect to the state z_F.*

Proof. If the common model $\mathcal{A}_\cap$ of the faulty plant is controllable with respect to z_F, the tracking controller $\mathcal{C}_T(\mathcal{A}_\cap)$ generates an input sequence $V(0\ldots k_e)$ such that the automaton $\mathcal{A}_\cap$ reaches the state z_F at time $k_e + 1$, that is,

$$G_\cap^\infty(z_{\cap 0}, V(0\ldots k_e)) = z_F$$

(Theorem 3.3). From Proposition 5.7 on the traversal of a common state sequence by all intersected automata it follows that

$$G_f^\infty(z_p, V(0\ldots k_e)) = z_F, \quad \forall \begin{pmatrix} z_p & f \end{pmatrix}^\top \in \mathcal{D}^*.$$

That is, the tracking controller $\mathcal{C}_T(\mathcal{A}_\cap)$ steers the faulty plant $\mathcal{P} \models \mathcal{A}_{\bar{f}}$ from its current state $z_p(k_d)$ into the desired final state z_F if $\begin{pmatrix} z_p(k_d) & \bar{f} \end{pmatrix}^\top \in \mathcal{D}^*$, which is true because of the completeness of the diagnostic result $\mathcal{D}^*$.

When the set $\{E_{zf}, E_{vf}, E_{wf}, (f \in \mathcal{F})\}$ of error relations and the nominal tracking controller $\mathcal{C}_T(\mathcal{A}_0)$ and therefore the model $\mathcal{A}_0$ of the faultless plant are known, the automata $\mathcal{A}_f = (\mathcal{Z}_f, \mathcal{V}_f, \mathcal{W}_f, G_f, H_f, z_{f0})$ can be computed according to (2.32)–(2.35) and (2.37) or

(2.38) with $z_{\bar{f}0} = z_p$ for all $\begin{pmatrix} z_p & f \end{pmatrix}^\mathsf{T} \in \mathcal{D}^*$. Hence, the common model $\mathcal{A}_\cap$ of the faulty plant can be constructed based on the available information using (5.27). Consequently, the tracking controller $\mathcal{C}_\mathcal{T}(\mathcal{A}_0)$ is reconfigurable according to Definition 5.1 with $\mathcal{C}_\mathcal{T}^r = \mathcal{C}_\mathcal{T}(\mathcal{A}_\cap)$. □

Compared to the reconfigurability condition in the presence of an unambiguous diagnostic result $\mathcal{D}^*$ in Theorem 5.1, now the controllability of the common model $\mathcal{A}_\cap$ of the faulty plant instead of controllability of the model $\mathcal{A}_{\bar{f}}$ of the faulty plant needs to be given. The reason for this is that the present fault $\bar{f}$ is not known such that the reconfigured tracking controller $\mathcal{C}_\mathcal{T}^r$ has to guarantee the fulfillment of the control aim (2.44) for all possible faulty situations contained in the diagnostic result $\mathcal{D}^*$.

The reconfigurability condition above is not a necessary one, because there might exist a reconfigured tracking controller $\mathcal{C}_\mathcal{T}^r$ which steers any faulty plant $\mathcal{P} \vDash \mathcal{A}_f$ from its current state z_p, where $\begin{pmatrix} z_p & f \end{pmatrix}^\mathsf{T} \in \mathcal{D}^*$, into the desired final state z_F, but does not only use common transitions included in the common model $\mathcal{A}_\cap$ of the faulty plant. However, the reconfigured tracking controller $\mathcal{C}_\mathcal{T}^r$ is supposed to be used not only to steer the plant $\mathcal{P}$ into the desired final state z_F, but also into other desired final states given in the future. Keeping this in mind, the controllability of the common model $\mathcal{A}_\cap$ of the faulty plant becomes a sufficient *and necessary* condition for the reconfigurability of the tracking controller $\mathcal{C}_\mathcal{T}(\mathcal{A}_0)$ with respect to *all possible* final states z_F.

Avoiding illegal transitions. In Section 5.2.1 the reconfigurability with respect to an unambiguous diagnostic result $\mathcal{D}^*$ has been studied, both in the absence and the presence of a set $\mathcal{E}_{ill}$ of illegal transitions. It has been proved that the controllability of the model $\mathcal{A}_{\bar{f}}$ of the faulty plant is a necessary and sufficient condition for the reconfigurability of the nominal tracking controller $\mathcal{C}_\mathcal{T}(\mathcal{A}_0)$ (Theorem 5.5), while for the safe reconfigurability the safe controllability of the automaton $\mathcal{A}_{\bar{f}}$ is required (Theorem 5.2). The same argumentation holds true in the presence of an ambiguous diagnostic result $\mathcal{D}^*$. Therefore, the following result extending Theorem 5.5 can be stated.

Theorem 5.6 (Safe reconfigurability with ambiguous diagnostic result)**.** *A nominal tracking controller $\mathcal{C}_\mathcal{T}(\mathcal{A}_0, \mathcal{E}_{ill})$ is safely reconfigurable with respect to a complete and correct diagnostic result $\mathcal{D}^*$, a set $\mathcal{E}_{ill}$ of illegal transitions and a desired final state z_F if the common model $\mathcal{A}_\cap$ of the faulty plant in (5.27) is safely controllable with respect to the state z_F and the set $\mathcal{E}_{ill}$.*

Proof. The proof is analogous to the one of Theorem 5.5 with the difference that the condition for safe controllability is given in Theorem 3.2 instead of Theorem 3.1 and the reconfigured tracking controller C_T^r is here given by $C_T(A_\cap, \mathcal{E}_{ill})$. The avoidance of illegal transitions by the common model $A_\cap$ of the faulty plant is guaranteed by Theorem 3.4, such that, according to Proposition 5.7, the faulty plant $\mathcal{P} \vDash A_{\tilde{f}}$ also avoids all faulty transitions and reaches the desired final state z_F. □

The above theorem shows that the safe controllability of the common model $A_\cap$ of the faulty plant is a sufficient condition for the safe reconfigurability of the tracking controller based on an ambiguous diagnostic result.

Example 5.2 (cont.) *Reconfiguration of tracking controller for automated warehouse*

The nominal tracking controller $C_T(A_0)$ from Example 3.1 can still be used in the presence of the additional state $A_\#$. Technically, the state $A_\#$ would have to be included into the model A_0 of the faultless plant stored in the nominal trajectory planning unit $T(A_0)$ and into the nominal controller $C(A_0)$. However, since the faulty position $A_\#$ can physically never be reached in the faultless case, it is sufficient to analyze the nominal tracking controller $C_T(A_0)$ from Example 3.1.

The reconfigurability of the nominal tracking controller $C_T(A_0)$ from Example 3.1 with respect to the diagnostic result $\mathcal{D}^*$ in (5.29) and the desired final states $z_F \in \{A, B, C\}$ is analyzed based on Theorem 5.5. That is, the controllability of the common model $A_\cap$ of the faulty plant in Fig. 5.10 with respect to the desired final state z_F has to be checked. In the automaton graph of $A_\cap$ there exists a path from the initial state $z_{\cap 0} = z_{new}$ to every state $z_F \in \{A, B, C\}$ such that an admissible state sequence $Z(0 \ldots k_e + 1) \in \mathcal{Z}_\cap^\infty$ with $z(0) = z_{\cap 0}$ and $z(k_e + 1) = z_F$ for every state $z_F \in \{A, B, C\}$ exists. Consequently, the common model $A_\cap$ of the faulty plant is controllable with respect to every state $z_F \in \{A, B, C\}$. Therefore, according to Theorem 5.5, the nominal tracking controller $C_T(A_0)$ is reconfigurable with respect to the ambiguous diagnostic result $\mathcal{D}^*$ in (5.29) and any desired final state $z_F \in \{A, B, C\}$.

Avoiding illegal transitions. Considering additionally the set $\mathcal{E}_{ill}$ of illegal transitions in (2.54), the safe reconfigurability of the nominal tracking controller $C_T(A_0, \mathcal{E}_{ill})$ from Example 3.1 with respect to the complete and correct, but ambiguous diagnostic result

$$\mathcal{D}^* = \mathcal{Z}^*(1) = \left\{ \begin{pmatrix} B \\ 1 \end{pmatrix}, \begin{pmatrix} C \\ 1 \end{pmatrix}, \begin{pmatrix} B \\ 2 \end{pmatrix}, \begin{pmatrix} C \\ 2 \end{pmatrix} \right\} \tag{5.30}$$

resulting from (4.73), the set $\mathcal{E}_{ill}$ of illegal transitions in (2.54) and every possible desired final state $z_F \in \{A, B, C\}$ is analyzed using Theorem 5.6. That is, the controllability of the legal part $A_{\cap,leg}$ of the common model $A_\cap$ of the faulty plant with respect to the diagnostic result $\mathcal{D}^*$ in (5.30), which is shown in Fig. 5.11, is checked using Theorem 3.2. Compared to the common model $A_\cap$ of the faulty plant in Fig. 5.10, the transition from state B to state A is removed, because it is contained in the set $\mathcal{E}_{ill}$ of illegal transitions in (2.54). Furthermore, due to the different diagnostic result $\mathcal{D}^*$ and the states z_p therein, the artificial initial state z_{new} has a different meaning. Now it combines the states B and C, both, from the automata A_1 and A_2 in Fig. 5.9.

It can be seen that all states A, B and C are reachable from the state $z_{\cap 0} = z_{new}$. Hence, according to Theorem 5.6, the nominal tracking controller $C_T(A_0, \mathcal{E}_{ill})$ is safely reconfigurable with respect to the diagnostic result $\mathcal{D}^*$ in (5.30) and any desired final state $z_F \in \{A, B, C\}$. □

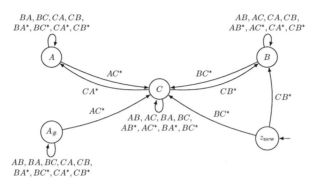

Figure 5.11: Legal part $\mathcal{A}_{\cap,\mathrm{leg}}$ of common model $\mathcal{A}_\cap$ of the faulty plant with respect to $\mathcal{D}^*$ in (5.30).

5.3.4 Reconfiguration of the tracking controller by complete redefinition

In this section a solution to the reconfiguration problems in Problem 5.1 and Problem 5.2 is proposed. It relies on the construction of a new tracking controller $\mathcal{C}_T(\mathcal{A}_\cap)$ based on the common model $\mathcal{A}_\cap$ of the faulty plant. The reconfiguration unit $\mathcal{R}$ knows the complete and correct, but possibly ambiguous, diagnostic result $\mathcal{D}^*$, the set $\{E_{zf}, E_{vf}, E_{wf}, (f \in \mathcal{F})\}$ of error relations and the nominal tracking controller $\mathcal{C}_T(\mathcal{A}_0)$.

In order to compute the common model $\mathcal{A}_\cap$ of the faulty plant based on (5.27), the models $\mathcal{A}_f$ with $z_{f0} = z_\mathrm{p}$, $\begin{pmatrix} z_\mathrm{p} & f \end{pmatrix}^\top \in \mathcal{D}^*$ of the faulty plant have to be known. They can be obtained by applying the error relations E_{zf}, E_{vf} and E_{wf} to the model $\mathcal{A}_0$ of the faultless plant according to (2.32)–(2.35) and (2.37) or (2.38) and setting the respective initial state to $z_{\bar{f}0} = z_\mathrm{p}$. The following proposition shows that the reconfigured tracking controller $\mathcal{C}_T^\mathrm{r} = \mathcal{C}_T(\mathcal{A}_\cap)$ is a solution to the reconfiguration problem in Problem 5.1.

Proposition 5.8 (Reconfiguration based on ambiguous diagnostic result by complete redefinition)**.** *Consider a nominal tracking controller $\mathcal{C}_T(\mathcal{A}_0)$ that is reconfigurable with respect to the complete and correct diagnostic result $\mathcal{D}^*$ and the desired final state z_F. Let $\mathcal{C}_T^\mathrm{r} = \mathcal{C}_T(\mathcal{A}_\cap)$ be the tracking controller reconfigured based on the common model $\mathcal{A}_\cap$ of the faulty plant with respect to the diagnostic result $\mathcal{D}^*$. Let k_e^r be the time horizon of the reference trajectory generated by the reconfigured trajectory planning unit T^r. Then the faulty plant $\mathcal{P} \models \mathcal{A}_{\bar{f}}$ in the closed-loop system $\mathcal{P} \circ \mathcal{C}_T^\mathrm{r}$ reaches the desired final state z_F at time $k_\mathrm{F} = k_\mathrm{e}^\mathrm{r} + 1$.*

Proof. The reconfigurability of $\mathcal{C}_T(\mathcal{A}_0)$ with respect to $\mathcal{D}^*$ and z_F implies the controllability of the automaton $\mathcal{A}_\cap$ with respect to z_F (Theorem 5.5). By Theorem 3.3 a tracking controller $\mathcal{C}_T(\mathcal{A})$ defined based on a controllable automaton $\mathcal{A}$ always steers this automaton $\mathcal{A}$ into the desired final state by applying an appropriate input sequence $V(0\ldots k_e)$ to it. Hence, the tracking controller $C_{\mathcal{T}}(\mathcal{A}_\cap)$ generates an input sequence $V(0\ldots k_e^r)$ such that $\mathcal{A}_\cap$ traverses a state sequence $Z_\cap(0\ldots k_e^r+1)$ with $z_0 = z_{\cap 0}$ and $z(k_e^r+1) = z_F$. By Proposition 5.7 any intersected automaton $\mathcal{A}_f$ with $z_{f0} = z_p$ and $\begin{pmatrix} z_p & f \end{pmatrix}^\top \in \mathcal{D}^*$ therefore also reaches the desired final state z_F at time $k_e^r + 1$.

The correctness of the unambiguous diagnostic result $\mathcal{D}^*$ ensures that $\begin{pmatrix} z_p(k_d) & \bar{f} \end{pmatrix}^\top \in \mathcal{D}^*$. According to Theorem 3.3 the faulty plant $\mathcal{P} \vDash \mathcal{A}_{\bar{f}}$ in the closed-loop system $\mathcal{P} \circ C_{\mathcal{T}}^r$ therefore reaches the desired final state z_F at a finite time k_F. □

The above proposition shows that the reconfiguration of the tracking controller $C_{\mathcal{T}}$ by defining a completely new tracking controller for the common model $\mathcal{A}_\cap$ of the faulty plant in (5.25) is a successful strategy. However, like in the presence of an unambiguous diagnostic result $\mathcal{D}^*$, it is desirable to avoid the complete redefinition of the tracking controller and to rather modify the nominal tracking controller $C_{\mathcal{T}}(\mathcal{A}_0)$ as little as possible.

Avoiding illegal transitions. When during the complete redefinition of the tracking controller additionally a set $\mathcal{E}_{ill}$ of illegal transitions is taken into account, the resulting reconfigured tracking controller solves the safe reconfiguration problem in Problem 5.2 as shown in the following proposition.

Proposition 5.9 (Safe reconfiguration based on ambiguous diagnostic result by complete redefinition). *Consider a nominal tracking controller $C_{\mathcal{T}}(\mathcal{A}_0, \mathcal{E}_{ill})$ that is safely reconfigurable with respect to the complete and correct diagnostic result $\mathcal{D}^*$ and the desired final state z_F. Let $C_{\mathcal{T}}^r = C_{\mathcal{T}}(\mathcal{A}_\cap, \mathcal{E}_{ill})$ be the tracking controller reconfigured based on the common model $\mathcal{A}_\cap$ of the faulty plant with respect to the diagnostic result $\mathcal{D}^*$. Let k_e^r be the time horizon of the reference trajectory generated by the reconfigured trajectory planning unit T^r. Then the faulty plant $\mathcal{P}$ in the closed-loop system $\mathcal{P} \circ C_{\mathcal{T}}^r$ reaches the desired final state z_F at time $k_F = k_e^r + 1$ while not executing any illegal transitions $(z_p(k), z_p(k+1)) \in \mathcal{E}_{ill}$, $(k_r \leq k < k_F)$.*

Proof. The proof follows directly from the proof of Proposition 5.8 and Theorem 3.4 on the avoidance of illegal transitions by a plant controlled by its corresponding tracking controller. □

5.3.5 Partial redefinition of the trajectory planning unit and the controller

It is aimed to modify the reconfiguration method for the reconfiguration of the nominal tracking controller $C_{\mathcal{T}}(\mathcal{A}_0)$ based on an unambiguous diagnostic result $\mathcal{D}^*$ summarized in Algorithm 5.2

such that it can also be applied when the diagnostic result $\mathcal{D}^*$ is ambiguous. Both, for the reconfiguration of the nominal trajectory planning unit $\mathcal{T}(\mathcal{A}_0)$ and the reconfiguration of the nominal controller $\mathcal{C}(\mathcal{A}_0)$, the error relations $E_{z\bar{f}}$ and $E_{v\bar{f}}$ defined in (2.31) and (2.29), respectively, corresponding to the present fault $\bar{f}$ are used. Consequently, as a first step, the common model $\mathcal{A}_\cap$ of the faulty plant in (5.25) has to be expressed as a combination of the model $\mathcal{A}_0$ of the faultless plant and some error relations $E_{z\cap}$ and $E_{v\cap}$. These error relations are defined as follows:

$$
E_{z\cap}(z, v) = \begin{cases} G_0(z, v) & \text{if } \left(\left[v \text{ faultless input} \right] \wedge \left[(z, v) \text{ faultless transition} \right] \right) \\ & \hspace{2cm} \forall \begin{pmatrix} z_p & f \end{pmatrix}^{\mathsf{T}} \in \mathcal{D}^*, \\ z_\# & \text{if } \left[E_{zf}(z, E_{vf}(v)) = z_\#, \quad \forall \begin{pmatrix} z_p & f \end{pmatrix}^{\mathsf{T}} \in \mathcal{D}^* \right] \\ & \hspace{1cm} \vee \left(\left[E_{zf}(z_p, E_{vf}(v)) = z_\#, \quad \forall \begin{pmatrix} z_p & f \end{pmatrix}^{\mathsf{T}} \in \mathcal{D}^* \right] \right. \\ & \hspace{4cm} \left. \wedge \left[z = z_{\cap 0} \right] \right), \\ \text{undefined} & \text{otherwise,} \end{cases}
$$
$$(5.31)$$

$$
E_{v\cap}(v) = v, \quad \forall v \in \mathcal{V}_0 = \mathcal{V}_f, \ (f \in \mathcal{F}). \tag{5.32}
$$

The "usual" error relations E_{zf}, $(f \in \mathcal{F})$ is defined for exactly two cases: either the given transition is changed by the fault $f \in \mathcal{F}$ or it remains unchanged. In contrast, the error relation $E_{z\cap}$ considers all fault candidates $f \in \mathcal{F}^*(k_r)$ included in the diagnostic result $\mathcal{D}^*$. Therefore, a transition remains unchanged only if it remains unchanged by *all* fault candidates $f \in \mathcal{F}^*(k_r)$ included in the diagnostic result $\mathcal{D}^*$. Similarly, a transition is changed only if it is changed in exactly the same way by *all* fault candidates $f \in \mathcal{F}^*(k_r)$ included in the diagnostic result $\mathcal{D}^*$. Consequently, the error relation $E_{z\cap}$ may also lead to an undefined transition. Also note that the domain of the error relation $E_{z\cap}$ now equals not only the state set $\mathcal{Z}_0$ of the model of the faultless plant, but possibly also the artificial initial state $z_{\cap 0} = z_{\text{new}}$.

For the construction of the common model $\mathcal{A}_\cap$ of the faulty plant only identical transitions in the automata $\mathcal{A}_f$, $(f \in \mathcal{F}^*(k_r))$, but not isolated input symbols, are relevant. Therefore, the error relation $E_{v\cap}$ is the identity map.

The following proposition proves that the automaton resulting from the application of these error relations to the model $\mathcal{A}_0$ of the faultless plant according to (2.28) with removed outputs, possibly extended by an artificial initial state z_{new}, equals the common model $\mathcal{A}_\cap$ of the faulty plant in (5.25).

Proposition 5.10 (Error relations corresponding to common model of the faulty plant). *The automaton*

$$
\tilde{\mathcal{A}}_\cap : \begin{cases} \tilde{\mathcal{Z}}_\cap = \mathcal{Z}_0 \cup \{\tilde{z}_{\cap 0}\} & \text{(5.33a)} \\[2mm] \tilde{\mathcal{V}}_\cap = \mathcal{V}_0 & \text{(5.33b)} \\[2mm] \tilde{G}_\cap(z,v) = \begin{cases} E_{z\cap}(z, E_{v\cap}(v)) & \text{if } E_{z\cap}(z, E_{v\cap}(v))!, \\ \text{undefined} & \text{otherwise,} \end{cases} & \text{(5.33c)} \\[4mm] \tilde{z}_{\cap 0} = \begin{cases} z_0 & \text{if } z_p = z_0 \in \mathcal{Z}_f, \ \forall \begin{pmatrix} z_p & f \end{pmatrix}^\top \in \mathcal{D}^*, \\ z_{new} \notin \mathcal{Z}_0 & \text{otherwise,} \end{cases} & \text{(5.33d)} \end{cases}
$$

with $E_{z\cap}$ and $E_{v\cap}$ defined according to (5.31) and (5.32), respectively, equals the common model $\mathcal{A}_\cap$ of the faulty plant with respect to the diagnostic result $\mathcal{D}^*$ defined in (5.27).

Proof. When in (5.25d) the initial states z_{i0} are replaced by the states z_p and the sets $\mathcal{Z}_i$ are replaced by the sets $\mathcal{Z}_f$, where $\begin{pmatrix} z_p & f \end{pmatrix}^\top \in \mathcal{D}^*$ (cf. (5.27)), it is easy to see that the

$$
\bigcup_{(z_p \ f)^\top \in \mathcal{D}^*} \mathcal{Z}_f = \mathcal{Z}_0.
$$

The state sets of $\tilde{\mathcal{A}}_\cap$ and of the common model $\mathcal{A}_\cap$ of the faulty plant are both the union of the state set $\mathcal{Z}_0$ of the faultless plant with the respective initial state. Consequently, also the state sets of $\tilde{\mathcal{A}}_\cap$ and of the common model $\mathcal{A}_\cap$ of the faulty plant are identical:

$$
\tilde{\mathcal{Z}}_\cap = \mathcal{Z}_0 \cup \{z_{\cap 0}\} = \bigcap_{(z_p \ f)^\top \in \mathcal{D}^*} \mathcal{Z}_f \cup \{z_{\cap 0}\} = \mathcal{Z}_\cap.
$$

The input sets of $\tilde{\mathcal{A}}_\cap$ and of the common model $\mathcal{A}_\cap$ of the faulty plant are also the same, because, according to (5.25b) and (5.33b),

$$
\tilde{\mathcal{V}}_\cap = \mathcal{V}_0 = \bigcap_{(z_p \ f)^\top \in \mathcal{D}^*} \mathcal{V}_f = \mathcal{V}_\cap.
$$

Finally it needs to be shown that the state transition functions of the automaton $\tilde{\mathcal{A}}_\cap$ and of the common model $\mathcal{A}_\cap$ of the faulty plant are identical:

$$
\tilde{G}_\cap(z,v), = \begin{cases} E_{z\cap}(z, E_{v\cap}(v)) & \text{if } E_{z\cap}(z, E_{v\cap}(v))!, \\ \text{undefined} & \text{otherwise,} \end{cases}
$$

$$
= \begin{cases} z' & \text{if } \left(E_{\mathsf{z}f}(z, E_{\mathsf{v}f}(v)) = z', \quad \forall \begin{pmatrix} z_{\mathrm{p}} & f \end{pmatrix}^{\mathsf{T}} \in \mathcal{D}^* \right) \\ & \quad \vee \left(\left[E_{\mathsf{z}f}(z_{\mathrm{p}}, E_{\mathsf{v}f}(v)) = z', \quad \forall \begin{pmatrix} z_{\mathrm{p}} & f \end{pmatrix}^{\mathsf{T}} \in \mathcal{D}^* \right] \right. \\ & \qquad\qquad\qquad\qquad\qquad \left. \wedge \left[z = \tilde{z}_{\cap 0} \right] \right), \\ \text{undefined} & \text{otherwise,} \end{cases}
$$

$$
= \begin{cases} z' & \text{if } \left(\left[G_f(z, v) = z', \quad \forall \begin{pmatrix} z_{\mathrm{p}} & f \end{pmatrix}^{\mathsf{T}} \in \mathcal{D}^* \right] \wedge \left[z \neq \tilde{z}_{\cap 0} \right] \right) \\ & \quad \vee \left(\left[G_f(z_{\mathrm{p}}, v) = z', \quad \forall \begin{pmatrix} z_{\mathrm{p}} & f \end{pmatrix}^{\mathsf{T}} \in \mathcal{D}^* \right] \wedge \left[z = \tilde{z}_{\cap 0} \right] \right), \\ \text{undefined} & \text{otherwise,} \end{cases}
$$

$$
= G_{\cap}(z, v), \quad \forall (z, v) \in \tilde{\mathcal{Z}}_{\cap} \times \tilde{\mathcal{V}}_{\cap} = \mathcal{Z}_{\cap} \times \mathcal{V}_{\cap}.
$$

Since all respective elements are identical, the proof is completed. □

The above proposition shows how the common model $\mathcal{A}_{\cap}$ of the faulty plant in (5.27) can be expressed based on the error relations in (5.31) and (5.32) and the model $\mathcal{A}_0$ of the faultless plant. Except for the possible artificial initial state z_{new} it is computed exactly as the models $\mathcal{A}_f$, $(f \in \mathcal{F})$ of the faulty plant. Therefore, it is reasonable to apply the reconfiguration method summarized in Algorithm 5.2 also when the diagnostic result $\mathcal{D}^*$ is ambiguous.

Line 1 of Algorithm 5.2 requires the modification of the automaton graph $\mathcal{A}_0$ stored in the nominal trajectory planning unit $\mathcal{T}(\mathcal{A}_0)$ according to (5.1). Here, the error relations $E_{\mathsf{z}\cap}$ and $E_{\mathsf{v}\cap}$ in (5.31) and (5.32), respectively, instead of the error relations $E_{\mathsf{z}\bar{f}}$ and $E_{\mathsf{v}\bar{f}}$ of the faulty plant have to be used. Additionally, the initial state of the automaton is replaced by $\tilde{z}_{\cap 0}$ in (5.33d) instead of the state $z_{\mathrm{p}}(k_{\mathrm{d}})$. As a result, slightly extending (5.1), the model $\mathcal{A}_0$ of the faultless plant stored in the nominal trajectory planning unit $\mathcal{T}(\mathcal{A}_0)$ is modified according to

$$
\mathcal{A}_{\cap}^{\mathrm{r}} : \begin{cases} \mathcal{Z}_{\cap}^{\mathrm{r}} = \mathcal{Z}_0 \cup \{z_{\cap 0}^{\mathrm{r}}\}, \\ \mathcal{V}_{\cap}^{\mathrm{r}} = \mathcal{V}_0, \\ \mathcal{W}_{\cap}^{\mathrm{r}} = \mathcal{W}_0, \\ G_{\cap}^{\mathrm{r}}(z, v) = \begin{cases} G_0(z, v), & \text{if } (z, v) \text{ faultless transition,} \\ E_{\mathsf{z}\cap}(z, E_{\mathsf{v}\cap}(v)), & \text{if } \left[(z, v) \text{ faulty transition} \right] \vee \left[z = z_{\cap 0}^{\mathrm{r}} \right], \\ \text{undefined} & \text{otherwise,} \end{cases} \\ H_{\cap}^{\mathrm{r}}(z, v) = \begin{cases} H_0(z, v) & \text{if } G_{\cap}^{\mathrm{r}}(z, v)!, \\ \text{undefined} & \text{otherwise,} \end{cases} \\ z_{\cap 0}^{\mathrm{r}} = \begin{cases} z_0 & \text{if } z_{\mathrm{p}} = z_0 \in \mathcal{Z}_f, \quad \forall \begin{pmatrix} z_{\mathrm{p}} & f \end{pmatrix}^{\mathsf{T}} \in \mathcal{D}^*, \\ z_{\mathrm{new}} \notin \mathcal{Z}_0 & \text{otherwise.} \end{cases} \end{cases} \qquad (5.34)
$$

Algorithm 5.3: Partial redefinition of controller based on ambiguous diagnostic result.

Given: Complete and correct diagnostic result $\mathcal{D}^*$.
        Nominal controller $\mathcal{C}(\mathcal{A}_0) = (\mathcal{Z}_c, \mathcal{V}_c, \mathcal{W}_c, G_c, H_c, z_{c0})$.
        Error relations $E_{z\cap}$ **and** $E_{v\cap}$.

1   $\mathcal{Z}_c^r = \mathcal{Z}_c$.   // State set of reconfigured controller based on (5.20a)
2   $\mathcal{V}_c^r = \mathcal{V}_c$.   // Input set of reconfigured controller based on (5.20b)
3   $\mathcal{W}_c^r = \mathcal{W}_c$.   // Output set of reconfigured controller based on (5.20c)

4   $z_{c0}^r = \begin{cases} z_0 & \text{if } z_p = z_0 \in \mathcal{Z}_f, \ \forall \begin{pmatrix} z_p & f \end{pmatrix}^{\mathsf{T}} \in \mathcal{D}^*, \\ z_{\text{new}} \notin \mathcal{Z}_0 & \text{otherwise.} \end{cases}$ // **Initial state of reconfig.**
                                                          controller based on (5.34)

5   Compute the state transition function G_c^- and the output function H_c^- of the pruned automaton $\mathcal{C}(\mathcal{A}_0)^-$ using (5.12) and (5.13), respectively.
6   Compute the set $\mathcal{E}_{\text{new}}$ of transitions to be newly defined using **(5.35)**.
7   Complete the state transition function and the output function of the pruned automaton by (5.20d) and (5.20e), respectively, in order to obtain the state transition function G_c^r and the output function H_c^r of the reconfigured controller.

Result: Reconfigured controller $\mathcal{C}^r = (\mathcal{Z}_c^r, \mathcal{V}_c^r, \mathcal{W}_c^r, G_c^r, H_c^r, z_{c0}^r)$.

According to Line 2 in Algorithm 5.2, the reconfigured controller $\mathcal{C}^r$ is obtained using Algorithm 5.1. This algorithm needs to be slightly changed as well to make it applicable in the presence of an ambiguous diagnostic result. First of all, it uses the error relations $E_{z\cap}$ and $E_{v\cap}$ in (5.31) and (5.32), respectively, instead of the set $\{E_{zf}, E_{vf}, E_{wf}, (f \in \mathcal{F})\}$ of error relations. In Line 4 of Algorithm 5.1, the initial state of the controller is replaced by the state $\tilde{z}_{\cap 0}$ in (5.33d) instead of the state $z_p(k_d)$. In Lines 5–7 the error relations $E_{z\cap}$ and $E_{v\cap}$ in (5.31) and (5.32), respectively, instead of the error relations $E_{z\bar{f}}$ and $E_{v\bar{f}}$ of the faulty plant have to be used. Additionally, the set $\mathcal{E}_{\text{new}}$ of transitions to be newly defined using (5.19) according to Line 6 is now given by

$$\mathcal{E}_{\text{new}} = \mathcal{E}_{\text{new},-} \cup \mathcal{E}_{\text{new},+} \cup \left\{ (z_c, v_c) \in \mathcal{Z}_c \times \mathcal{V}_c : \left[z_c = z_{\text{new}} \right] \right.$$
$$\left. \wedge \left[(\exists w \in \mathcal{W}_c) \, E_{zf}(z_p, E_{vf}(w)) = v_c, \quad \forall \begin{pmatrix} z_p & f \end{pmatrix}^{\mathsf{T}} \in \mathcal{D}^* \right] \right\}$$
$$(5.35)$$

in order to account for the possibly occurring artificial initial state z_{new}. The resulting modified version of Algorithm 5.1 is given by Algorithm 5.3, where changes are indicated in bold text.

Avoiding illegal transitions. In Section 5.2.3 a method for the reconfiguration of the nominal trajectory planning unit $\mathcal{T}(\mathcal{A}_{\text{leg}})$ considering a set $\mathcal{E}_{\text{ill}}$ of illegal transitions has been presented. It was proposed to modify the automaton $\mathcal{A}_{\text{leg}}$ stored in the trajectory planning unit according to (5.2) such that the automaton $\mathcal{A}_{\text{leg}}^{\text{r}}$ results. Compared to the automaton $\mathcal{A}^{\text{r}}$ in (5.1) resulting from the reconfiguration without a set $\mathcal{E}_{\text{ill}}$ of illegal transitions, the main difference was that only those modified transitions which still do not belong to the set $\mathcal{E}_{\text{ill}}$ of illegal transitions are included in the automaton. Following the same rule, in the case of an ambiguous diagnostic result $\mathcal{D}^*$ the automaton

$$\mathcal{A}_{\cap\text{leg}}^{\text{r}} : \begin{cases} \mathcal{Z}_{\cap\text{leg}}^{\text{r}} = \mathcal{Z}_{\text{leg}} \cup \{z_{\cap\text{leg}0}^{\text{r}}\}, \\ \mathcal{V}_{\cap\text{leg}}^{\text{r}} = \mathcal{V}_{\text{leg}}, \\ \mathcal{W}_{\cap\text{leg}}^{\text{r}} = \mathcal{W}_{\text{leg}}, \\ G_{\cap\text{leg}}^{\text{r}}(z,v) = \begin{cases} G_{\text{leg}}(z,v) & \text{if } \Big[(z,v) \text{ faultless transition}\Big] \wedge \Big[G_{\text{leg}}(z,v)!\Big], \\ E_{z\cap}(z, E_{v\cap}(v)) & \text{if } \Big(\Big[(z,v) \text{ faulty transition}\Big] \\ & \qquad \wedge \Big[(z, E_{z\cap}(z, E_{v\cap}(v))) \notin \mathcal{E}_{\text{ill}}\Big]\Big) \\ & \qquad \vee \Big[z = z_{\cap\text{leg}0}^{\text{r}}\Big], \\ \text{undefined} & \text{otherwise}, \end{cases} \\ H_{\cap\text{leg}}^{\text{r}}(z,v) = \begin{cases} H_{\text{leg}}(z,v) & \text{if } \Big[H_{\text{leg}}(z,v)!\Big] \wedge \Big[G_{\cap\text{leg}}^{\text{r}}(z,v)!\Big], \\ \varepsilon & \text{if } \Big[\neg H_{\text{leg}}(z,v)!\Big] \wedge \Big[G_{\cap\text{leg}}^{\text{r}}(z,v)!\Big], \\ \text{undefined} & \text{otherwise}, \end{cases} \\ z_{\cap\text{leg}0}^{\text{r}} = \begin{cases} z_0 & \text{if } z_p = z_0 \in \mathcal{Z}_f, \quad \forall \begin{pmatrix} z_p & f \end{pmatrix}^{\top} \in \mathcal{D}^*, \\ z_{\text{new}} \notin \mathcal{Z}_0 & \text{otherwise}, \end{cases} \end{cases} \tag{5.36}$$

is stored in the reconfigured trajectory planning unit $T^{\text{r}} = T(\mathcal{A}_{\cap\text{leg}}^{\text{r}})$.

Since the controller $\mathcal{C}$ is independent from the set $\mathcal{E}_{\text{ill}}$ of illegal transitions, its reconfiguration can still be executed using Algorithm 5.3.

Example 5.2 (cont.)   *Reconfiguration of tracking controller for automated warehouse*

The trajectory planning unit $T(\mathcal{A}_0)$ and the controller $\mathcal{C}(\mathcal{A}_0)$ in the nominal tracking controller $\mathcal{C}_T(\mathcal{A}_0)$ from Example 3.1 are reconfigured based on the complete and correct, but ambiguous diagnostic result

$$\mathcal{D}^* = \left\{ \begin{pmatrix} A_\# \\ 1 \end{pmatrix}, \begin{pmatrix} A \\ 2 \end{pmatrix} \right\}$$

in (5.29) and the set $\{E_{zf}, E_{vf}, E_{wf}, (f \in \{1,2\})\}$ of error relations in (2.55a), (2.55b), (2.56a)–(2.56c) and (5.28).

The error relations $E_{z\cap}$ and $E_{v\cap}$ are computed according to (5.31) and (5.32), respectively. Therefore, for example, for the state-input pair $(z, v) = (B, BA^*)$ the error relation $E_{z \cap}(z, v)$ is given by

$$
E_{z \cap}(B, BA^*) = \begin{cases} z' & \text{if } \left(E_{zf}(B, E_{vf}(BA^*)) = z', \quad \forall \begin{pmatrix} z_p & f \end{pmatrix}^\top \in \mathcal{D}^* \right) \\ & \quad \lor \left(\left[E_{zf}(z_p, E_{vf}(BA^*)) = z', \quad \forall \begin{pmatrix} z_p & f \end{pmatrix}^\top \in \mathcal{D}^* \right] \right. \\ & \qquad \left. \land \left[B = z_{\cap 0}^r \right] \right), \\ \text{undefined} & \text{otherwise,} \end{cases}
$$

$$
= \begin{cases} z' & \text{if } \left(E_{z1}(B, E_{v1}(BA^*)) = z' \right) \land \left(E_{z2}(B, E_{v2}(BA^*)) = z' \right), \\ \text{undefined} & \text{otherwise,} \end{cases}
$$

$$
= A.
$$

For example, for the state-input pair $(z, v) = (A, AB)$ the error relation is undefined, because

$$
E_{z1}(A, E_{v1}(AB)) = A_\# \neq E_{z2}(A, E_{v2}(AB)) = A
$$

In total, the error relation

$$
E_{z \cap}(z, v) = \begin{cases} G_0(z, v) & \text{if } (z, v) \in \{(A, AC^*), (A, BA^*), (A, BC^*), (A, CA^*), (A, CB^*), \\ & \qquad (B, AB^*), (B, AC^*), (B, BA^*), (B, BC^*), (B, CA^*), \\ & \qquad (B, CB^*), (C, AB^*), (C, AC^*), (C, BA^*), (C, BC^*), \\ & \qquad (C, CA^*), (C, CB^*), (A_\#, AC^*), (A_\#, BA^*), (A_\#, BC^*), \\ & \qquad (A_\#, CA^*), (A_\#, CB^*)\} \\ A & \text{if } (z, v) \in \{(A, BA), (A, BC), (A, CA), (A, CB)\}, \\ B & \text{if } (z, v) \in \{(B, AB), (B, AC), (B, CA), (B, CB)\}, \\ C & \text{if } (z, v) \in \{(z_{\text{new}}, AC^*), (C, AB), (C, AC), (C, BA), (C, BC)\}, \\ A_\# & \text{if } (z, v) \in \{(A_\#, AB), (A_\#, BA), (A_\#, BC), (A_\#, CA), (A_\#, CB)\}, \\ \text{undefined} & \text{otherwise,} \end{cases}
$$
(5.37)

results, which declares 22 transitions as faultless and leaves them therefore unchanged in the model $\mathcal{A}_0$ of the faultless plant. Additionally, the above error relation changes 17 transitions in the model $\mathcal{A}_0$ of the faultless plant, introduces one new transition starting at the artificial initial state z_{new} and removes 9 transitions.

As stated in (5.32), the error relation $E_{v \cap}$ is given by

$$
E_{v \cap}(v) = v, \quad \forall v \in \{AB, AC, BA, BC, CA, CB, AB^*, AC^*, BA^*, BC^*, CA^*, CB^*\}. \quad (5.38)
$$

Then the automaton graph stored in the nominal trajectory planning unit $\mathcal{T}(\mathcal{A}_0)$ is modified according to (5.34). That is, based on the diagnostic result $\mathcal{D}^*$ in (4.56), the initial state of the automaton is changed to the newly defined state $z_{\cap 0}^r = z_{\text{new}}$ and its transitions are modified based on the error relations $E_{z \cap}$ and $E_{v \cap}$ in (5.37) and (5.38), respectively. Thereby the automaton $\mathcal{A}_\cap^r$ shown in Fig. 5.12 results. Compared to the original automaton $\mathcal{A}_0$ in Fig. 2.5 all transitions from state $z = A$ to state $z' = B$ and all transitions whose input v corresponds to a transport with robot M ($v \in \{AB, AC, \dots, CB\}$) are removed.

If the desired final state is given by $z_F = B$, the reconfigured trajectory planning unit $\mathcal{T}^r = \mathcal{T}(\mathcal{A}_\cap^r)$ generates the reference trajectory

$$
R(k_r \dots k_e^r) = (C, B). \quad (5.39)
$$

Finally, Algorithm 5.3 is used to compute the reconfigured controller C^r. Here, only the differences to

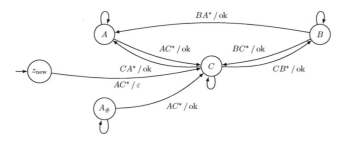

Figure 5.12: Automaton $\mathcal{A}_{\cap}^{r}$ in reconfigured trajectory planning unit $\mathcal{T}^{r}$.

the reconfigured controller $\mathcal{C}^{r}$ in Fig. 5.7 resulting from the reconfiguration based on the unambiguous diagnostic result $\mathcal{D}^{*}$ in (4.56) are discussed. Even though now the error relations $E_{z\cap}$ and $E_{v\cap}$ instead of the error relations E_{z2} and E_{v2} are used, the pruned automaton $\mathcal{C}(\mathcal{A}_{0})^{-}$ resulting from Line 5 in Algorithm 5.3 is the same as before (cf. Fig. 5.6). According to (5.35), the set of transitions to be newly defined is given by

$$\mathcal{E}_{\text{new}} = \Big\{ (A,B),(A,C),(B,A),(B,C),(C,A),(C,B) \Big\} \cup \{(z_{\text{new}},C)\}.$$

That is, every transition between two different states plus the transition starting at the newly defined state z_{new} have to be considered in the next step.

Finally, Line 7 of Algorithm 5.3 is executed such that the reconfigured controller $\mathcal{C}^{r}$ in Fig. 5.13 results. The main difference to the nominal controller $\mathcal{C}(\mathcal{A}_{0})$ in Fig. 3.2 is that for the transport of a parcel always the robot M^{*} instead of the robot M is used. Furthermore note that here the controller does not offer the option to transport a parcel from position A to position B directly.

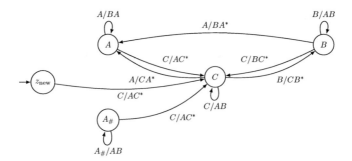

Figure 5.13: Reconfigured controller $\mathcal{C}^{r}$ corresponding to diagnostic result $\mathcal{D}^{*}$ in (5.29).

Avoiding illegal transitions. In the presence of the set $\mathcal{E}_{\text{ill}}$ of illegal transitions in (2.54) the diagnostic result $\mathcal{D}^{*}$ in (5.30) is obtained, which differs from the diagnostic result in the absence of illegal transitions in (5.29). However, since both diagnostic results contain both faults $f \in \{1,2\}$ as fault candidates, only three values of the error relation $E_{z\cap}$ in (5.37) have to be changed, while the error relation $E_{v\cap}$ in (5.38) remains unchanged. The changes occur because of the different meaning of the artificial initial state

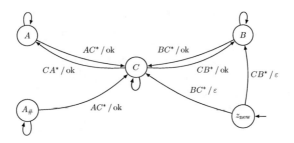

Figure 5.14: Automaton $\mathcal{A}^r_{\cap\text{leg}}$ in reconfigured trajectory planning unit $\mathcal{T}^r$.

z_{new}. Corresponding to this state, the following values for the error relation $E_{z\cap}$ result:

$$E_{z\cap}(z_{\text{new}}, v) = \begin{cases} B & \text{if } v = CB^*, \\ C & \text{if } v = BC^*, \\ \text{undefined}, & \text{otherwise.} \end{cases} \qquad (5.40)$$

The reconfigured trajectory planning unit $\mathcal{T}^r$ contains the automaton $\mathcal{A}^r_{\cap\text{leg}}$ in Fig. 5.14 resulting from the modification of the legal part $\mathcal{A}_{\text{leg}}$ of the model $\mathcal{A}_0$ of the nominal plant in Fig. 2.6 according to (5.36). It is extended by the artificial initial state z_{new}. It can be seen that none of the faulty transitions specified by the error relation $E_{z\cap}$ in (5.37) is contained in the set $\mathcal{E}_{\text{ill}}$ of illegal transitions in (2.54) such that all of them are included into the automaton $\mathcal{A}^r_{\cap\text{leg}}$. When the desired final state is given by $z_F = B$, the reconfigured trajectory planning unit $\mathcal{T}^r = \mathcal{T}(\mathcal{A}^r_{\cap\text{leg}})$ generates the reference trajectory

$$R(k_r \ldots k_e^r) = r(k_r = k_e^r) = B. \qquad (5.41)$$

Finally, the reconfigured controller $\mathcal{C}^r$ is almost the same as the one in Fig. 5.13, where again the artificial initial state z_{new} has a different meaning such that different transitions start from it. The reconfigured controller $\mathcal{C}^r$ corresponding to the diagnostic result $\mathcal{D}^*$ in (5.30) is shown in Fig. 5.15. □

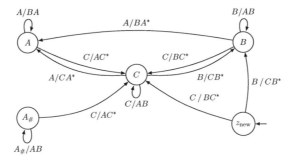

Figure 5.15: Reconfigured controller $\mathcal{C}^r$ corresponding to diagnostic result $\mathcal{D}^*$ in (5.30).

5.3.6 Reconfiguration of the tracking controller by partial redefinition

Combining the method for the reconfiguration of the nominal trajectory planning unit $T(\mathcal{A}_0)$ and the nominal controller $\mathcal{C}(\mathcal{A}_0)$, a reconfiguration method for the nominal tracking controller $\mathcal{C}_T(\mathcal{A}_0)$ based on an ambiguous diagnostic result $\mathcal{D}^*$ results. The reconfiguration method is summarized in Algorithm 5.4. The main difference to Algorithm 5.2 for the reconfiguration based on an unambiguous diagnostic result lies in the computation and use of the error relations $E_{z\cap}$ and $E_{v\cap}$ of the common model $\mathcal{A}_\cap$ of the faulty plant (see bold text).

Analogously to Proposition 5.6, it can be proved that the reconfigured tracking controller resulting from applying Algorithm 5.4 to the nominal tracking controller $\mathcal{C}_T(\mathcal{A}_0)$ is identical to the tracking controller $\mathcal{C}_T(\mathcal{A}_\cap)$ defined for the common model $\mathcal{A}_\cap$ of the faulty plant in (5.27) as follows.

Proposition 5.11 (Identity of tracking controllers). *The tracking controller $\mathcal{C}_T(\mathcal{A}_\cap)$ defined for the common model $\mathcal{A}_\cap$ of the faulty plant in (5.27) is identical to the reconfigured tracking controller $\mathcal{C}_T^r$ resulting from Algorithm 5.4.*

Proof. If no artificial initial state z_{new} is defined, the reconfiguration methods in Algorithm 5.2 and Algorithm 5.4, one based on the model $\mathcal{A}_{\bar{f}}$ and the error relations $E_{z\bar{f}}, E_{v\bar{f}}$ of the faulty plant and one based on the common model $\mathcal{A}_\cap$ of the faulty plant and its error relations $E_{z\cap}, E_{v\cap}$ are exactly the same. Therefore, from Proposition 5.6 it follows that the tracking controllers $\mathcal{C}_T(\mathcal{A}_\cap)$ and $\mathcal{C}_T^r$ are identical if no artificial initial state z_{new} is defined.

If an artificial initial state z_{new} is present, the tracking controllers $\mathcal{C}_T(\mathcal{A}_\cap)$ and $\mathcal{C}_T^r$ are still identical. Any transition starting at the artificial initial state z_{new} is included, both, in the common model $\mathcal{A}_\cap$ of the faulty plant and the automaton $\mathcal{A}_\cap^r$ in (5.34). Consequently, the trajectory planning units $T(\mathcal{A}_\cap)$ and T^r generate the same reference trajectories.

During the construction of the reconfigured controller $\mathcal{C}^r$ based on Algorithm 5.3, the artificial

---

Algorithm 5.4: Reconfiguration of tracking controller with ambiguous diagnostic result.

Given: Complete and correct diagnostic result $\mathcal{D}^*$.
　　　　Nominal tracking controller $\mathcal{C}_T(\mathcal{A}_0) = (T(\mathcal{A}_0), \mathcal{C}(\mathcal{A}_0))$.
　　　　Set $\{E_{zf}, E_{vf}, E_{wf}, (f \in \mathcal{F})\}$ of error relations.

1 **Compute the error relations $E_{z\cap}$ and $E_{v\cap}$ according to (5.31) and (5.32).**
2 Modify the automaton $\mathcal{A}_0$ stored in $T(\mathcal{A}_0)$ according to (5.34) such that the reconfigured trajectory planning unit T^r results.
3 Apply Algorithm 5.3 to $\mathcal{C}(\mathcal{A}_0)$ such that the reconfigured controller $\mathcal{C}^r$ results.

Result: Reconfigured tracking controller $\mathcal{C}_T^r = (T^r, \mathcal{C}^r)$.

---

initial state z_{new} is considered in the definition of the initial state z_{c0}^r in Line 4 and the definition of the set $\mathcal{E}_{new}$ in Line 6. It is easy to see that the initial states of the controller $\mathcal{C}(\mathcal{A}_\cap)$ and of the reconfigured controller $\mathcal{C}^r$ remain identical. The consideration of transitions starting at z_{new} in the set $\mathcal{E}_{new}$ in (5.35) guarantees that the reconfigured controller $\mathcal{C}^r$ includes all transitions starting at its initial state z_{c0}^r that also the controller $\mathcal{C}(\mathcal{A}_\cap)$ defined for the common model $\mathcal{A}_\cap$ of the faulty plant contains. Hence, the controllers $\mathcal{C}(\mathcal{A}_\cap)$ and $\mathcal{C}^r$ also remain identical in the presence of an artificial initial state z_{new}. $\qquad\square$

The above proposition shows that it is not necessary to define a completely new tracking controller $\mathcal{C}_T(\mathcal{A}_\cap)$ for the common model $\mathcal{A}_\cap$ of the faulty plant in (5.27). Rather, an identical reconfigured tracking controller $\mathcal{C}_T^r$ results from a partial redefinition of the nominal tracking controller $\mathcal{C}_T(\mathcal{A}_0)$ according to Algorithm 5.4. Therefore, analogously to Theorem 5.3 on the fulfillment of control aim after the reconfiguration based on an ambiguous diagnostic result, the following theorem can be stated. It shows that the reconfigured tracking controller $\mathcal{C}_T^r$ obtained from Algorithm 5.4 solves the reconfiguration problem in Problem 5.1.

Theorem 5.7 (Fulfillment of control aim after reconfiguration based on ambiguous diagnostic result). *Consider a nominal tracking controller $\mathcal{C}_T(\mathcal{A}_0)$ that is reconfigurable with respect to a complete and correct diagnostic result $\mathcal{D}^*$ and a desired final state z_F. Let $\mathcal{C}_T^r$ be the reconfigured tracking controller resulting from Algorithm 5.4. Let k_e^r be the time horizon of the reference trajectory generated by the reconfigured trajectory planning unit T^r. Then the faulty plant $\mathcal{P} \models \mathcal{A}_f$ in the closed-loop system $\mathcal{P} \circ \mathcal{C}_T^r$ reaches the desired final state z_F at time $k_F = k_e^r + 1$.*

Proof. Since the faulty plant $\mathcal{P}$ in the closed-loop system $\mathcal{P} \circ \mathcal{C}_T(\mathcal{A}_\cap)$ reaches the desired final state z_F at a finite time k_F (Proposition 5.8) if the tracking controller is reconfigurable, and the reconfigured tracking controller $\mathcal{C}_T^r$ is identical to $\mathcal{C}_T(\mathcal{A}_\cap)$ (Proposition 5.11), the proof follows directly. $\qquad\square$

Avoiding illegal transitions. In the presence of a set $\mathcal{E}_{ill}$ of illegal transitions, the reconfiguration based on an ambiguous diagnostic result $\mathcal{D}^*$ is also executed according to Algorithm 5.4. The only difference is that in Line 2 the reconfigured trajectory planning unit T^r is computed by (5.36) instead of (5.34) now. Consequently, in analogy to Theorem 5.7, the following theorem shows that the resulting reconfigured tracking controller $\mathcal{C}_T^r$ solves the safe reconfiguration problem in Problem 5.2.

Theorem 5.8 (Safe fulfillment of control aim after reconfiguration based on ambiguous diagnostic result). *Consider a nominal tracking controller $C_T(A_0, \mathcal{E}_{ill})$ that is safely reconfigurable with respect to a complete and correct diagnostic result $\mathcal{D}^*$, a desired final state z_F and a set $\mathcal{E}_{ill}$ of illegal transitions. Let C_T^r be the reconfigured tracking controller resulting from Algorithm 5.4, where the automaton $A_{\cap leg}^r$ in T^r is computed by (5.36). Let k_e^r be the time horizon of the reference trajectory generated by the reconfigured trajectory planning unit T^r. Then the faulty plant $\mathcal{P} \models A_{\bar{f}}$ in the closed-loop system $\mathcal{P} \circ C_T^r$ reaches the desired final state z_F at time $k_F = k_e^r + 1$ while not executing any illegal transitions $(z_p(k), z_p(k+1)) \in \mathcal{E}_{ill}, (k_r \leq k < k_F)$.*

Proof. If the tracking controller is safely reconfigurable, the faulty plant $\mathcal{P}$ in the closed-loop system $\mathcal{P} \circ C_T(A_\cap, \mathcal{E}_{ill})$ reaches the desired final state z_F at a finite time k_F without executing any illegal transitions $(z_p(k), z_p(k+1)) \in \mathcal{E}_{ill}, (k_r \leq k < k_F)$ (Proposition 5.9). In Proposition 5.11 it has been proved that the tracking controller $C_T(A_\cap)$ and the reconfigured tracking controller C_T^r resulting from Algorithm 5.4 without the consideration of illegal transitions are identical.

The only difference in the reconfigured tracking controller C_T^r obtained now is that the automaton stored in the nominal trajectory planning unit is modified according to (5.36). Following the same argumentation as in the proof on Proposition 5.11, it can be seen that the trajectory planning unit $T(A_\cap, \mathcal{E}_{ill})$ equals the reconfigured trajectory planning unit T^r, which concludes the proof. □

Example 5.2 (cont.)   *Reconfiguration of tracking controller for automated warehouse*

Algorithm 5.4 is used to reconfigure the nominal tracking controller $C_T(A_0)$ from Example 3.1 based on the ambiguous diagnostic result $\mathcal{D}^*$ in (5.29). When Line 1 in Algorithm 5.4 is executed, the error relations $E_{z \cap}$ and $E_{v \cap}$ in (5.37) and (5.38), respectively result. Line 2 of Algorithm 5.4 yields the reconfigured trajectory planning unit T^r in which the automaton $A_\cap^r$ shown in Fig. 5.12 is stored. Finally, from Line 3 of Algorithm 5.4 the reconfigured controller C^r in Fig. 5.13 results.

After the reconfiguration time $k_r = 1$, the reconfigured tracking controller $C_T^r = (T^r, C^r)$ is used to control the faulty plant A_2 in Fig. 5.9b with current state $z_p(k_d = 1) = A$. That is, a parcel lies at position A and the robot M is broken. The trajectory planning unit T^r generates the reference input $r(k_r = 1) = C$ in (5.39) for the controller C^r, which generates the input $v_p(1) = w_c(1) = AC^*$ for the plant and goes to state $z_c(2) = C$. That is, the reconfigured tracking controller C_T^r requests the transport of a parcel from position A to position C using the robot M^*. This transport can be executed by the faulty plant such that the plant goes to state $z_p(2) = C$ and generates the output $w_p(1) = $ ok.

Next, the reference input $r(2) = B$ is given to the reconfigured controller C^r. The reconfigured controller C^r generates input $v_p(2) = w_c(1) = CB^*$ for the plant and goes to state $z_c(3) = B$. That is, the reconfigured tracking controller C_T^r requests the transport of a parcel from position C to position B using the robot M^*. This transport can be executed by the faulty plant such that the plant goes to state $z_p(3) = B$ and generates the output $w_p(2) = $ ok.

Consequently, as stated by Theorem 5.7, the faulty plant reaches the desired final state $z_F = B$ at time $k_F = 3 = k_e^r + 1$. Due to the ambiguity of the diagnostic result $\mathcal{D}^*$ in (5.29), the reconfigured tracking controller C_T^r had to be defined such that it can deal with both, the blocking of the route between position A and position B, and the failure of robot M. If it was known that fault $f = 2$ occurred, the detour over position C could have avoided. Nevertheless, in this example, the early termination of the fault diagnosis at the cost of an ambiguous diagnostic result led to an earlier fulfillment of the control aim compared to the use of the unambiguous diagnostic result as in Example 5.2 ($k_F = 3 < 4$).

Avoiding illegal transitions. Now Algorithm 5.4 is used to safely reconfigure the nominal tracking controller $C_T(\mathcal{A}_0, \mathcal{E}_{ill})$ from Example 3.1 based on the ambiguous diagnostic result $\mathcal{D}^*$ in (5.30) and the set $\mathcal{E}_{ill}$ of illegal transitions in (2.54). When Line 1 in Algorithm 5.4 is executed, the error relations $E_{z\cap}$ and $E_{v\cap}$ in (5.37) combined with (5.40) and (5.38), respectively, result. Line 2 of Algorithm 5.4 yields the reconfigured trajectory planning unit T^r in which the automaton $\mathcal{A}_{\cap leg}^r$ shown in Fig. 5.14 is stored. Finally, from Line 3 of Algorithm 5.4 the reconfigured controller C^r in Fig. 5.15 results.

After the reconfiguration time $k_r = 1$, the reconfigured tracking controller $C_T^r = (T^r, C^r)$ is used to control the faulty plant $\mathcal{A}_2$ with current state $z_p(k_r = 1) = C$. That is, a parcel lies at position C and the robot M is broken. The trajectory planning unit T^r generates the reference input $r(k_r = 1) = B$ in (5.41) for the controller C^r, which generates the input $v_p(1) = w_c(1) = CB^*$ for the plant and goes to state $z_c(2) = B$. That is, the reconfigured tracking controller requests the transport of a parcel from position C to position B using the robot M^*. This transport can be executed by the faulty plant such that the plant goes to state $z_p(2) = B$ and generates the output $w_p(1) = ok$.

Consequently, as stated by Theorem 5.8, the faulty plant reaches the desired final state $z_F = B$ at time $k_F = 2 = k_e^r + 1$ without having used the dangerous route between position A and position B. $\square$

5.3.7 Complexity of the reconfiguration method

Complexity of the reconfigurability analysis. Compared to the complexity of the reconfigurability analysis in presence of an unambiguous diagnostic result $\mathcal{D}^*$ computed in Section 5.2.6, here the construction of the common model $\mathcal{A}_\cap$ of the faulty plant in (5.27) is added (cf. Theorem 5.5). It requires the comparison of at most $n \cdot p$ transitions in all models corresponding to at most F fault candidates included in the ambiguous diagnostic result $\mathcal{D}^*$. The computational complexity of the controllability analysis of $\mathcal{A}_\cap$ is of order $\mathcal{O}(n \cdot p)$. Consequently, the overall computational complexity for the reconfigurability analysis is of order $\mathcal{O}(n \cdot p \cdot F + n \cdot p) = \mathcal{O}(n \cdot p \cdot F)$.

Complexity of the online execution of the reconfiguration method. The reconfiguration of the nominal tracking controller $C_T(\mathcal{A}_0)$ based on an ambiguous diagnostic result $\mathcal{D}^*$ is executed according to Algorithm 5.4. The complexity of the reconfiguration based on an unambiguous diagnostic result was shown to be maximally of order $\mathcal{O}(n \cdot p)$ in Section 5.2.6. Here, only the computation of the error relations $E_{z\cap}$ and $E_{v\cap}$ according to (5.31) and (5.32) is added. The error relation $E_{v\cap}$ declares all inputs as faultless such that the complexity of its computation can be neglected. In the worst case, the computation of the error relation $E_{z\cap}$ requires the comparison of at most $n \cdot p$ transitions in all models corresponding to at most F fault candidates included in the ambiguous diagnostic result $\mathcal{D}^*$. As a result, the computational complexity for the reconfiguration of the tracking controller based on an ambiguous diagnostic result $\mathcal{D}^*$ is of order $\mathcal{O}(n \cdot p \cdot F)$. The number of computations becomes smaller, the less elements are included in the diagnostic result $\mathcal{D}^*$ and the less faulty inputs and transitions are contained in the models $\mathcal{A}_f$, $(f \in \mathcal{F})$ of the faulty plant.

5.3.8 Discussion on reconfiguration with ambiguous diagnostic result

It strongly depends on the "quality" of the diagnostic result $\mathcal{D}^*$, whether the reconfiguration based on an ambiguous diagnostic result $\mathcal{D}^*$ is possible or not. If there are too many states and/or fault candidates included in the diagnostic result $\mathcal{D}^*$, it is likely that the common model $\mathcal{A}_\cap$ of the faulty plant in (5.27) contains only very few transitions or no transitions at all. In this case, there might not exist a path into the desired final state z_F in it, such that the automaton $\mathcal{A}_\cap$ is not controllable and hence the nominal tracking controller $\mathcal{C}_T(\mathcal{A}_0)$ is not reconfigurable (cf. Theorem 5.1).

If the fault candidates included in the diagnostic result $\mathcal{D}^*$ have quite similar effects on the plant $\mathcal{P}$, more transitions will be included into the common model $\mathcal{A}_\cap$ of the faulty plant in (5.27). Then it is more likely for the common model $\mathcal{A}_\cap$ of the faulty plant to be controllable and hence for the nominal tracking controller $\mathcal{C}_T(\mathcal{A}_0)$ to be reconfigurable. This conforms with the intuition that similar faults might share a solution to the reconfiguration problem, while quite different faults usually require different solutions.

As Example 5.2 showed, the plant $\mathcal{P}$ might reach the desired final state z_F at an earlier time compared to the usage of the unambiguous diagnostic result $\mathcal{D}^*$ obtained at the diagnosis time k_d if the reconfiguration is conducted at a reconfiguration time $k_r < k_d$ with an ambiguous diagnostic result $\mathcal{D}^*$. However, the reconfigured tracking controller $\mathcal{C}_T^r$ might be inefficient in the sense that further desired final states z_F are not reached in minimal time. Therefore, there is always a trade-off between a fast reconfiguration and a good performance after reconfiguration.

Further research has to be conducted in order to develop methods, which find the "optimal" reconfiguration time k_r by evaluating the current diagnostic result $\mathcal{D}^*$. For example, the fault diagnosis could be stopped as soon as there exists a path to the desired final state in the common model $\mathcal{A}_\cap$ of the faulty plant.

6 Integrated fault-tolerant control method

In this chapter the methods for tracking control, active fault diagnosis and reconfiguration of the tracking controller described in the previous chapters are combined in order to obtain an integrated fault-tolerant control method. Both, the case that there exist illegal transitions to be avoided by the plant and that there are no such transitions are considered. For both cases, first, the recoverability of the control loop is analyzed. Afterwards the interaction between the previously introduced methods is described and it is proved that the plant in the proposed fault-tolerant control loop always fulfills the control aim of reaching a desired final state if the control loop is recoverable.

6.1 Fault-tolerant control flowchart

The aim of this thesis is to develop a fully integrated fault-tolerant control method. Therefore, in this chapter the tracking controller $\mathcal{C}_{\mathcal{T}}$ presented in Chapter 3, the diagnostic unit $\mathcal{D}$ presented in Chapter 4 and the reconfiguration unit $\mathcal{R}$ presented in Chapter 5 are combined to a single fault-tolerant controller

$$\mathcal{C}_{\mathrm{FTC}} = (\mathcal{C}_{\mathcal{T}}, \mathcal{D}, \mathcal{R}) \tag{6.1}$$

as shown in Fig. 6.1. Depending on whether a set $\mathcal{E}_{\mathrm{ill}}$ of illegal transitions for the plant $\mathcal{P}$ needs to be considered or not, the fault-tolerant controller $\mathcal{C}_{\mathrm{FTC}}$ shall solve the fault-tolerant control problem in Problem 2.1 or the safe fault-tolerant control problem in Problem 2.2, respectively. The aim of this chapter is twofold. On the one hand, it reviews important definitions and conditions from Chapters 3–5. On the other hand, describes and analyzes the interaction of the individually introduced methods.

The following two pictures illustrate the behavior of the fault-tolerant control loop – one showing the dependencies between the components and the other showing the time course of

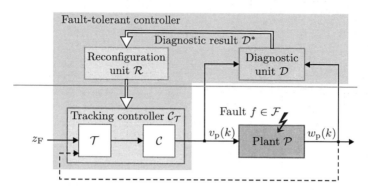

Figure 6.1: Fault-tolerant control loop.

the status of the components.

The flowchart in Fig. 6.2 illustrates the interactions of elements in the fault-tolerant controller $\mathcal{C}_{\text{FTC}} = (\mathcal{C}_\mathcal{T}, \mathcal{D}, \mathcal{R})$ with the plant $\mathcal{P}$. Initially, the nominal tracking controller $\mathcal{C}_\mathcal{T}(\mathcal{A}_0)$ controls the plant $\mathcal{P}$. That is, the trajectory planning unit $\mathcal{T}(\mathcal{A}_0)$ generates a reference trajectory $R(0\ldots k_{\text{e}})$ along which the controller $\mathcal{C}(\mathcal{A}_0)$ steers the plant $\mathcal{P}$ into the desired final state z_{F}. In parallel, the diagnostic unit $\mathcal{D}$ detects whether a fault occurred by evaluating the inputs $v_{\text{p}}(k)$ and outputs $w_{\text{p}}(k)$ of the plant $\mathcal{P}$. If a fault is detected, fault identification is performed in order to obtain a diagnostic result $\mathcal{D}^*$. Once the fault diagnosis is completed, the reconfiguration unit $\mathcal{R}$ uses the diagnostic result $\mathcal{D}^*$ to reconfigure the nominal tracking controller $\mathcal{C}_\mathcal{T}(\mathcal{A}_0)$. Afterwards the reconfigured tracking controller $\mathcal{C}_\mathcal{T}^{\text{r}}$ controls the faulty plant $\mathcal{P}$ until it reaches the desired final state z_{F}.

Figure 6.3 illustrates the behavior of the components in the fault-tolerant control loop on a time bar. Initially, the faultless plant $\mathcal{P} \vDash \mathcal{A}_0$ is controlled by the nominal tracking controller $\mathcal{C}_\mathcal{T}(\mathcal{A}_0)$, while the diagnostic unit $\mathcal{D}$ performs the fault detection. At the fault occurrence time k_{f}, the plant $\mathcal{P}$ becomes faulty, but the fault detection is still continued until the fault detection time k_{f}^*. Then the diagnostic unit $\mathcal{D}$ performs the fault identification. In the meantime the nominal tracking controller $\mathcal{C}_\mathcal{T}(\mathcal{A}_0)$ remains active. At the reconfiguration time k_{r} the fault diagnosis is completed and the reconfiguration unit $\mathcal{R}$ reconfigures the tracking controller $\mathcal{C}_\mathcal{T}$. Afterwards, the reconfigured tracking controller $\mathcal{C}_\mathcal{T}^{\text{r}}$ controls the faulty plant $\mathcal{P} \vDash \mathcal{A}_f$.

Details on the methods realized by the tracking controller $\mathcal{C}_\mathcal{T}$, the diagnostic unit $\mathcal{D}$ and the reconfiguration unit $\mathcal{R}$, including different options on when to call the diagnosis completed, have been presented in the previous chapters and will be shortly reviewed in the following sections.

In the remainder of the chapter it is first analyzed in which cases a fault-tolerant controller

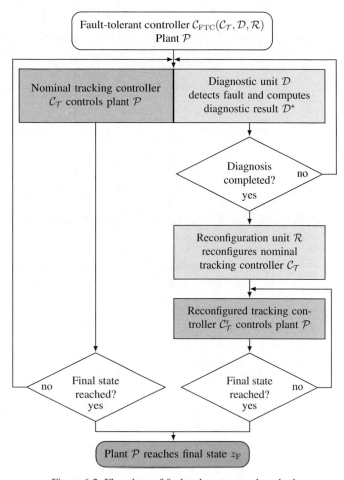

Figure 6.2: Flowchart of fault-tolerant control method.

$\mathcal{C}_{\mathrm{FTC}}$ which solves the fault-tolerant control problem in Problem 2.1 or Problem 2.2, respectively, exists. Afterwards, the definition of the fault-tolerant controller $\mathcal{C}_{\mathrm{FTC}}$ is deduced from the definition of its elements $\mathcal{C}_{\mathcal{T}}$, $\mathcal{D}$, $\mathcal{R}$ from Chapters 3–5. Finally, it is proved that the proposed fault-tolerant controller $\mathcal{C}_{\mathrm{FTC}}$ indeed solves the fault-tolerant control problem in Problem 2.1 or Problem 2.2, respectively, when its existence conditions are fulfilled.

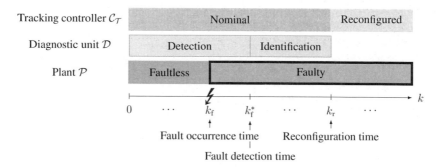

Figure 6.3: Time behavior of elements in fault-tolerant control loop.

6.2 Fault-tolerant control without illegal transitions

In this section the fault-tolerant control problem from Problem 2.1 is considered. That is, the fault-tolerant controller $\mathcal{C}_{\text{FTC}}$ shall steer the plant $\mathcal{P}$ into a desired final state z_{F} despite a possibly occurring fault $f \in \mathcal{F}$.

6.2.1 Recoverability analysis

The recoverability of the closed-loop system $\mathcal{P} \circ \mathcal{C}_{\mathcal{T}}(\mathcal{A}_0)$ consisting of the plant $\mathcal{P}$ and its corresponding nominal tracking controller $\mathcal{C}_{\mathcal{T}}(\mathcal{A}_0)$ from a fault $f \in \mathcal{F}$ refers to the ability to find a fault-tolerant controller $\mathcal{C}_{\text{FTC}}$ that guarantees the fulfillment of the control aim (2.44) by the controlled faulty plant $\mathcal{P} \vDash \mathcal{A}_f$. For the definition of the fault-tolerant controller $\mathcal{C}_{\text{FTC}}$, the model $\mathcal{A}_0$ of the faultless plant in (2.26) and the set $\{E_{\text{z}f}, E_{\text{v}f}, E_{\text{w}f}, (f \in \mathcal{F})\}$ of error relations (cf. (2.29)–(2.31)) can be used. Furthermore, the fault-tolerant controller has access to the input and output sequence $V_{\text{p}}(0 \ldots k_{\text{e}})$ and $W_{\text{p}}(0 \ldots k_{\text{e}})$ of the plant $\mathcal{P}$, respectively. Therefore, recoverability of the closed-loop system is defined as follows.

Definition 6.1 (Recoverability). *The closed-loop system $\mathcal{P} \circ \mathcal{C}_{\mathcal{T}}(\mathcal{A}_0)$ is said to be recoverable from a fault $f \in \mathcal{F}$ with respect to a desired final state z_{F} if it is possible to determine a diagnostic result $\mathcal{D}^*$ with respect to which the nominal tracking controller $\mathcal{C}_{\mathcal{T}}(\mathcal{A}_0)$ is reconfigurable, given the model $\mathcal{A}_0$ of the faultless plant, the set $\{E_{\text{z}f}, E_{\text{v}f}, E_{\text{w}f}, (f \in \mathcal{F})\}$ of error relations as well as the I/O-pair $(V_{\text{p}}(0 \ldots k), W_{\text{p}}(0 \ldots k))$ of the plant $\mathcal{P}$.*

The diagnostic result $\mathcal{D}^*$ is constructed by the diagnostic unit $\mathcal{D}$ at runtime. In general, only a simulation can reveal whether a suitable diagnostic result $\mathcal{D}^*$ can be found and hence, whether or not the closed-loop system $\mathcal{P} \circ \mathcal{C}_{\mathcal{T}}(\mathcal{A}_0)$ is recoverable from a fault $f \in \mathcal{F}$. However, there exists a sufficient condition for the recoverability of the closed-loop system. If the plant $\mathcal{P}$

is diagnosable, that is if its current state $z_\mathrm{p}(k)$ and the present fault $\bar{f} \in \mathcal{F}$ can be uniquely determined, an unambiguous diagnostic result

$$\mathcal{D}^* = \left\{ \begin{pmatrix} z_\mathrm{p}(k_\mathrm{d}) \\ \bar{f} \end{pmatrix} \right\}$$

can be obtained. During the recoverability analysis the state $z_\mathrm{p}(k_\mathrm{d})$ that will be taken by the faulty plant $\mathcal{P} \models \mathcal{A}_{\bar{f}}$ at runtime is unknown. Therefore, reconfigurability with respect to each possible diagnostic result

$$\mathcal{D}^* = \left\{ \begin{pmatrix} z \\ \bar{f} \end{pmatrix} \right\}, \quad z \in \mathcal{Z}_{\bar{f}} \tag{6.2}$$

has to be analyzed. In order to guarantee recoverability from a fault $f \in \mathcal{F}$, the nominal tracking controller $\mathcal{C}_T(\mathcal{A}_0)$ has to be reconfigurable with respect to every diagnostic result $\mathcal{D}^*$ as given in (6.2) and the desired final state z_F (cf. Definition 5.1).

In the following, criteria for the diagnosability of the plant $\mathcal{P}$ and the reconfigurability of the nominal tracking controller $\mathcal{C}_T(\mathcal{A}_0)$ developed in the previous chapters are combined in order to obtain a condition for recoverability of the closed-loop system $\mathcal{P} \circ \mathcal{C}_T(\mathcal{A}_0)$. Note that this condition is only sufficient, but not necessary, because recoverability might be given even if the plant $\mathcal{P}$ is not diagnosable as long as a diagnostic result $\mathcal{D}^*$ with respect to which the tracking controller is reconfigurable can be found. First, the case that passive diagnosis is used is considered

> **Theorem 6.1** (Sufficient condition for recoverability with passive diagnosis). *The closed-loop system $\mathcal{P} \circ \mathcal{C}_T(\mathcal{A}_0)$ is recoverable from a fault $f \in \mathcal{F}$ with respect to the desired final state z_F using passive diagnosis if the following two conditions are fulfilled:*
>
> - *The plant $\mathcal{P}$ in the closed-loop system $\mathcal{P} \circ \mathcal{C}_T(\mathcal{A}_0)$ is diagnosable using passive diagnosis with respect to the fault set $\mathcal{F}$ and the reference trajectory $R(0 \ldots k_\mathrm{e})$ generated by the trajectory planning unit $\mathcal{T}(\mathcal{A}_0)$.*
>
> - *The nominal tracking controller $\mathcal{C}_T(\mathcal{A}_0)$ is reconfigurable with respect to each diagnostic result $\mathcal{D}^*$ as in (6.2) and the desired final state z_F.*

Proof. If the plant $\mathcal{P}$ in the closed-loop system $\mathcal{P} \circ \mathcal{C}_T(\mathcal{A}_0)$ is diagnosable with passive diagnosis with respect to the fault set $\mathcal{F}$ and the reference trajectory $R(0 \ldots k_\mathrm{e})$ generated by the trajectory planning unit $\mathcal{T}(\mathcal{A}_0)$, the present fault $\bar{f}$ and the current state $z_\mathrm{p}(k)$ of the plant

$\mathcal{P}$ can be identified from the models $\{\mathcal{A}_f, (f \in \mathcal{F})\}$ of the faulty plant and the I/O-pair $(V_\mathrm{p}(0 \dots k), W_\mathrm{p}(0 \dots k))$ of the plant, where k_f^* is the fault detection time (Definition 4.5). The models $\{\mathcal{A}_f, (f \in \mathcal{F})\}$ of the faulty plant can be constructed from the model $\mathcal{A}_0$ of the faultless plant and the set $\{E_{zf}, E_{vf}, E_{wf}, (f \in \mathcal{F})\}$ of error relations according to (2.28). Hence, it is possible to determine a diagnostic result $\mathcal{D}^*$ as in (6.2).

If the nominal tracking controller $\mathcal{C}_T(\mathcal{A}_0)$ is reconfigurable with respect to any of these diagnostic results $\mathcal{D}^*$ and the desired final state z_F, the closed-loop system $\mathcal{P} \circ \mathcal{C}_T(\mathcal{A}_0)$ is recoverable from the fault $f \in \mathcal{F}$ with respect to the desired final state z_F (Definition 6.1). $\square$

The above theorem states a rather abstract condition for the recoverability of the closed-loop system $\mathcal{P} \circ \mathcal{C}_T(\mathcal{A}_0)$ from a fault $f \in \mathcal{F}$. In order to evaluate this condition for a given control loop, previously presented conditions for the diagnosability and reconfigurability have to be employed, which is done in the following.

In Theorem 4.1 a necessary and sufficient condition for the diagnosability of the plant $\mathcal{P}$ with passive diagnosis has been stated. First, it requires that all faults $f \in \mathcal{F}$ are detectable. This means that the output sequence $\Phi_f(z_\mathrm{p}(k_\mathrm{f}), V_\mathrm{p}(k_\mathrm{f} \dots k_\mathrm{e}))$ generated by the faulty plant $\mathcal{P} \vDash \mathcal{A}_f$ differs from the output sequence $\Phi_0(z_\mathrm{p}(k_\mathrm{f}), V_\mathrm{p}(k_\mathrm{f} \dots k_\mathrm{e}))$ generated by the faultless plant $\mathcal{P} \vDash \mathcal{A}_0$ before the entire reference trajectory $R(0 \dots k_\mathrm{e})$ has been applied to the controller $\mathcal{C}(\mathcal{A}_0)$ (see Definition 4.3). Second, the reference trajectory $R(k_\mathrm{f}^* \dots k_\mathrm{e})$ generated by the trajectory planning unit $\mathcal{T}(\mathcal{A}_0)$ after the fault detection time $k_\mathrm{f}^* \leq k_\mathrm{e}$ needs to be a preset homing sequence for the overall model $\bar{\mathcal{A}}_\Delta$ of the closed-loop system with respect to the initial uncertainty $\mathcal{Z}^* = \{z_\mathrm{c}(k_\mathrm{f}^*)\} \times \mathcal{Z}_\Delta$, where $z_\mathrm{c}(k_\mathrm{f}^*) = G_\mathrm{c}^\infty(z_{\mathrm{c}0}, R(0 \dots k_\mathrm{f}^* - 1))$. The overall model $\bar{\mathcal{A}}_\Delta$ of the closed-loop system describes the series connection of the nominal controller $\mathcal{C}(\mathcal{A}_0)$ and the overall model $\mathcal{A}_\Delta$ of the faulty plant in (2.42) and is defined in (4.14).

Theorem 5.1 contains a necessary and sufficient condition for the reconfigurability of the nominal tracking controller $\mathcal{C}_T(\mathcal{A}_0)$ with respect to a complete, correct and unambiguous diagnostic result $\mathcal{D}^* = \left\{ \left(z_\mathrm{p}(k_\mathrm{d}) \quad \bar{f} \right)^\top \right\}$ and a desired final state z_F. It states that reconfigurability is given if and only if the automaton $\mathcal{A}_{\bar{f}}$ with initial state $z_{\bar{f}0} = z_\mathrm{p}(k_\mathrm{d})$ is controllable with respect to the state z_F. According to Theorem 3.1 a necessary and sufficient condition for the controllability of the automaton $\mathcal{A}_{\bar{f}}$ is the existence of an admissible state sequence $Z(0 \dots k_\mathrm{e} + 1) \in \mathcal{Z}_{\bar{f}}^\infty$ with $z(0) = z_\mathrm{p}(k_\mathrm{d})$ and $z(k_\mathrm{e} + 1) = z_\mathrm{F}$ for it. By Definition 3.2 such an admissible state sequence $Z(0 \dots k_\mathrm{e} + 1)$ corresponds to a path from the state $z_\mathrm{p}(k_\mathrm{d})$ to the state z_F in the automaton graph of the automaton $\mathcal{A}_{\bar{f}}$.

Integrating the condition for the diagnosability of the plant $\mathcal{P}$ with passive diagnosis and the condition for the reconfigurability of the nominal tracking controller $\mathcal{C}_T(\mathcal{A}_0)$ into Theorem 6.1, the following sufficient condition for the recoverability of the closed-loop system $\mathcal{P} \circ \mathcal{C}_T(\mathcal{A}_0)$ is obtained.

Corollary 6.1 (Sufficient condition for recoverability with passive diagnosis). *The closed-loop system* $\mathcal{P} \circ \mathcal{C}_T(\mathcal{A}_0)$ *is recoverable from a fault* $f \in \mathcal{F}$ *with respect to the desired final state* z_F *using passive diagnosis if all of the following conditions are fulfilled:*

- *(Detectability) There exists a time* $k_f^* \leq k_e$ *such that*

$$\Phi_f(z_p(k_f), V_p(k_f \ldots k_f^*)) \neq \Phi_0(z_p(k_f), V_p(k_f \ldots k_f^*)),$$

  where $V_p(k_f \ldots k_f^*) = \Phi_c(z_c(k_f), R(k_f \ldots k_f^*))$.

- *(Identifiability) The reference trajectory* $R(k_f^* \ldots k_e)$ *after the fault detection time* $k_f^* \leq k_e$ *is a preset homing sequence for the overall model* $\bar{A}_\Delta$ *of the closed-loop system in* (4.14) *with respect to the initial uncertainty* $\mathcal{Z}^* = \{z_c(k_f^*)\} \times \mathcal{Z}_\Delta$, *where*

$$z_c(k_f^*) = G_c^\infty(z_{c0}, R(0 \ldots k_f^* - 1)).$$

- *(Reconfigurability) For every state* $z \in \mathcal{Z}_f$ *there exists a path from the state* z *to the state* z_F *in the automaton graph of* $\mathcal{A}_f$.

Proof. If the first two conditions are fulfilled, the plant $\mathcal{P}$ in the closed-loop system $\mathcal{P} \circ \mathcal{C}_T(\mathcal{A}_0)$ is diagnosable with passive diagnosis with respect to the fault set $\mathcal{F}$ and the reference trajectory $R(0 \ldots k_e)$ generated by the trajectory planning unit $T(\mathcal{A}_0)$ (Theorem 4.1). If additionally the third condition is fulfilled, the nominal tracking controller $\mathcal{C}_T(\mathcal{A}_0)$ is reconfigurable with respect to any of the possible diagnostic results $\mathcal{D}^*$ and the desired final state z_F (Theorems 3.1 and 5.1 and Definition 3.2).

Consequently, according to Theorem 6.1, the closed-loop system $\mathcal{P} \circ \mathcal{C}_T(\mathcal{A}_0)$ is recoverable from the fault $f \in \mathcal{F}$ with respect to the desired final state z_F. □

For the case that active fault diagnosis is used, Theorem 6.1 has to be adjusted as follows.

Theorem 6.2 (Sufficient condition for recoverability with active diagnosis). *The closed-loop system* $\mathcal{P} \circ \mathcal{C}_T(\mathcal{A}_0)$ *is recoverable from a fault* $f \in \mathcal{F}$ *with respect to the desired final state* z_F *using active diagnosis if the following two conditions are fulfilled:*

- *The plant* $\mathcal{P}$ *in the closed-loop system* $\mathcal{P} \circ \mathcal{C}_T(\mathcal{A}_0)$ *is diagnosable using active diagnosis with respect to the fault set* $\mathcal{F}$.

- *The nominal tracking controller* $\mathcal{C}_T(\mathcal{A}_0)$ *is reconfigurable with respect to any diagnostic result* $\mathcal{D}^*$ *as in* (6.2) *and the desired final state* z_F.

Proof. The proof is similar to the one of Theorem 6.1 with the only difference that now active diagnosis is used, that is, Definition 4.6 instead of Definition 4.5 has to be considered. □

Like in the case of passive diagnosis it is possible to deduce conditions from the above theorem that can be directly applied for a given closed-loop system $\mathcal{P} \circ \mathcal{C}_T(\mathcal{A}_0)$. Theorem 4.2 contains a sufficient condition for diagnosability with active diagnosis of the plant $\mathcal{P}$ in the closed-loop system $\mathcal{P} \circ \mathcal{C}_T(\mathcal{A}_0)$. It states that diagnosability is given if all faults $f \in \mathcal{F}$ are detectable and identifiable with active diagnosis. Additionally, Proposition 4.8 states that all faults $f \in \mathcal{F}$ are identifiable if there are no compatible states in the set $\{z_c\} \times \mathcal{Z}_\Delta$ for any $z_c \in \mathcal{Z}_c$. Consequently, from Theorem 6.2 the following corollary can be deduced.

Corollary 6.2 (Sufficient condition for recoverability with active diagnosis). *The closed-loop system* $\mathcal{P} \circ \mathcal{C}_T(\mathcal{A}_0)$ *is recoverable from a fault* $f \in \mathcal{F}$ *with respect to the desired final state* z_F *using active diagnosis if all of the following conditions are fulfilled:*

- *(Detectability) There exists a time* $k_f^* \leq k_e$ *such that*

$$\Phi_f(z_p(k_f), V_p(k_f \dots k_f^*)) \neq \Phi_0(z_p(k_f), V_p(k_f \dots k_f^*)),$$

  where $V_p(k_f \dots k_f^*) = \Phi_c(z_c(k_f), R(k_f \dots k_f^*))$.

- *(Identifiability) There are no compatible states in the set* $\{z_c\} \times \mathcal{Z}_\Delta$ *for any* $z_c \in \mathcal{Z}_c$.

- *(Reconfigurability) For every state* $z \in \mathcal{Z}_f$ *there exists a path from the state* z *to the state* z_F *in the automaton graph of* $\mathcal{A}_f$.

Proof. If the first two conditions are fulfilled, the plant $\mathcal{P}$ in the closed-loop system $\mathcal{P} \circ \mathcal{C}_T(\mathcal{A}_0)$ is diagnosable with active diagnosis with respect to the fault set $\mathcal{F}$ (Theorem 4.2 and Proposition 4.8). If additionally the third condition is fulfilled, the nominal tracking controller $\mathcal{C}_T(\mathcal{A}_0)$ is reconfigurable with respect to each possible diagnostic result $\mathcal{D}^*$ and the desired final state z_F (Theorems 3.1 and 5.1 and Definition 3.2).

Consequently, according to Theorem 6.2, the closed-loop system $\mathcal{P} \circ \mathcal{C}_T(\mathcal{A}_0)$ is recoverable from the fault $f \in \mathcal{F}$ with respect to the desired final state z_F. □

The conditions in Theorem 6.1 and Theorem 6.2 are sufficient but not necessary for recoverability with passive or active diagnosis, respectively. If the plant $\mathcal{P}$ is not diagnosable with the chosen diagnostic method, it might not be possible to identify the present fault $\bar{f} \in \mathcal{F}$ and the current state $z_p(k_d)$ of the plant $\mathcal{P}$ exactly. Hence, it might not be possible to obtain an unambiguous diagnostic result $\mathcal{D}^*$. However, the closed-loop system $\mathcal{P} \circ \mathcal{C}_T(\mathcal{A}_0)$ is recoverable anyhow whenever the tracking controller $\mathcal{C}_T$ is reconfigurable with respect to the obtained *ambiguous* diagnostic result $\mathcal{D}^*$.

Example 6.1 *Fault-tolerant control of automated warehouse*

The recoverability of the automated warehouse example using passive or active diagnosis, respectively, is analyzed. The sufficient condition for recoverability with passive diagnosis in Theorem 6.1 detailed in Corollary 6.1 is not fulfilled, because the reference trajectory $R(k_f^* \ldots k_e) = B$ after the fault detection time $k_f^* = 0$ is no preset homing sequence for the overall model $\bar{A}_\Delta$ of the closed-loop system (cf. Example 4.2). That is, the diagnostic result $\mathcal{D}^*$ will always remain ambiguous.

Nevertheless, the nominal tracking controller $\mathcal{C}_T(\mathcal{A}_0)$ is reconfigurable with respect to the ambiguous diagnostic result

$$\mathcal{D}^* = \left\{ \begin{pmatrix} A \\ 1 \end{pmatrix}, \begin{pmatrix} A \\ 2 \end{pmatrix} \right\}$$

obtained at time $k = 1$ (see common model $\mathcal{A}_\cap$ in Fig. 5.10 of Example 5.2). Therefore, the closed-loop system $\mathcal{P} \circ \mathcal{C}_T(\mathcal{A}_0)$ is recoverable from both faults $f \in \{1, 2\}$ with respect to any desired final state $z_F = \{A, B, C\}$ anyhow. This shows again that the condition in Theorem 6.1 is sufficient, but not necessary for the recoverability of the closed-loop system $\mathcal{P} \circ \mathcal{C}_T(\mathcal{A}_0)$.

The closed-loop system $\mathcal{P} \circ \mathcal{C}_T(\mathcal{A}_0)$ is recoverable from both faults $f \in \{1, 2\}$ with respect to all desired final states $z_F \in \{A, B, C\}$ using active diagnosis, because all conditions in Theorem 6.2 detailed in Corollary 6.2 are fulfilled according to Examples 4.2 and 5.1. $\qquad \square$

6.2.2 Definition of the fault-tolerant controller

The elements of the fault-tolerant controller $\mathcal{C}_{FTC} = (\mathcal{C}_T, \mathcal{D}, \mathcal{R})$, that is, the nominal tracking controller $\mathcal{C}_T(\mathcal{A}_0)$, the diagnostic unit $\mathcal{D}$ and the reconfiguration unit $\mathcal{R}$ have to be defined based on the model $\mathcal{A}_0 = (\mathcal{Z}_0, \mathcal{V}_0, \mathcal{W}_0, G_0, H_0, z_{00})$ of the faultless plant in (2.26) and the set $\{E_{zf}, E_{vf}, E_{wf}, (f \in \mathcal{F})\}$ of error relations in (2.29)–(2.31). Afterwards they have to be combined as shown in Fig. 2.3.

Definition of the tracking controller $\mathcal{C}_T$. The nominal tracking controller $\mathcal{C}_T(\mathcal{A}_0)$ consisting of the nominal trajectory planning unit $\mathcal{T}(\mathcal{A}_0)$ and the nominal controller $\mathcal{C}(\mathcal{A}_0)$ is defined based on the model $\mathcal{A}_0$ of the faultless plant as described in Chapter 3.

In the nominal trajectory planning unit $\mathcal{T}(\mathcal{A}_0)$ a complete and correct graph search algorithm, for example Breadth-first-search, has to be implemented. In addition, only the model of the faultless plant $\mathcal{A}_0$ has to be stored in the nominal trajectory planning unit $\mathcal{T}(\mathcal{A}_0)$.

According to (3.7) the nominal controller $\mathcal{C}(\mathcal{A}_0)$ is given by a deterministic I/O automaton $\mathcal{A}_c = (\mathcal{Z}_c, \mathcal{V}_c, \mathcal{W}_c, G_c, H_c, z_{c0})$. Algorithm 3.1 describes a systematic way to compute the controller $\mathcal{C}(\mathcal{A}_0) \vDash \mathcal{A}_c$ based on the model $\mathcal{A}_0$ of the faultless plant. The elements of the controller $\mathcal{C}(\mathcal{A}_0)$ are given by

$$
\mathcal{C}(\mathcal{A}_0) : \begin{cases} \mathcal{Z}_c = \mathcal{Z}_0, \\ \mathcal{V}_c = \mathcal{Z}_0, \\ \mathcal{W}_c = \mathcal{V}_0, \\ G_c(z_c, v_c) = \begin{cases} v_c & \text{if } \mathcal{V}_{a0}(v_c, z_c) \neq \emptyset, \\ \text{undefined} & \text{otherwise,} \end{cases} \\ H_c(z_c, v_c) = \begin{cases} w_c \in \mathcal{V}_{a0}(v_c, z_c) & \text{if } \mathcal{V}_{a0}(v_c, z_c) \neq \emptyset, \\ \text{undefined} & \text{otherwise,} \end{cases} \\ z_{c0} = z_{00}. \end{cases}
$$

Definition of the diagnostic unit $\mathcal{D}$. The diagnostic unit $\mathcal{D}$ is defined based on the model $\mathcal{A}_0$ of the faultless plant and the overall model $\mathcal{A}_\Delta$ of the faulty plant as described in Chapter 4. Therefore, first, the overall model $\mathcal{A}_\Delta$ of the faulty plant has to be constructed from the set $\{\mathcal{A}_f, (f \in \mathcal{F})\}$ of models of the faulty plant according to (2.42). The models $\mathcal{A}_f, (f \in \mathcal{F})$ can be computed by applying the error relations E_{zf}, E_{vf} and E_{wf}, $(f \in \mathcal{F})$ to the model $\mathcal{A}_0$ of the faultless plant as described by (2.28). As a result, the following overall model of the faulty plant with unknown initial state $z_{\Delta 0}$ is obtained:

$$
\mathcal{A}_\Delta = (\mathcal{Z}_\Delta, \mathcal{V}_\Delta, \mathcal{W}_\Delta, G_\Delta, H_\Delta, z_{\Delta 0}) \tag{6.3}
$$

with

$$
\mathcal{Z}_\Delta = \mathcal{Z}_0 \times \mathcal{F}
$$
$$
\mathcal{V}_\Delta = \mathcal{V}_0
$$
$$
\mathcal{W}_\Delta = \mathcal{W}_0
$$
$$
G_\Delta(z_\Delta, v) = G_\Delta\left(\begin{pmatrix} z \\ f \end{pmatrix}, v \right) = \begin{pmatrix} E_{zf}(z, E_{vf}(v)) \\ f \end{pmatrix}
$$
$$
H_\Delta(z_\Delta, v) = H_\Delta\left(\begin{pmatrix} z \\ f \end{pmatrix}, v \right) = \begin{cases} E_{wf}(H_0(\tilde{z}, \tilde{v})) & \text{if } [z = \tilde{z}] \wedge [E_{vf}(v) = \tilde{v}] \\ & \qquad \wedge [G_f(z, v) = G_0(\tilde{z}, \tilde{v})], \\ \text{manually defined} & \text{otherwise.} \end{cases}
$$

During the modeling process it has to be decided whether some of the conditions in the definition of the output function H_Δ may be dropped due to the independence of the output of the plant from its current state, input symbol or next state, respectively. Thereby, the number of outputs to be defined manually can be reduced.

The diagnostic unit $\mathcal{D}$ needs to perform the following two steps:

1. Fault detection based on the model $\mathcal{A}_0$ of the faultless plant.

2. Fault identification based on the overall model $\mathcal{A}_\Delta$ of the faulty plant.

In both steps, the I/O-pairs $(v_\mathrm{p}(k), w_\mathrm{p}(k))$ of the plant $\mathcal{P}$ are checked for consistency, either with the model $\mathcal{A}_0$ of the faultless plant, or with the states within the overall model $\mathcal{A}_\Delta$ of the faulty plant, respectively.

Figure 6.4 summarizes the active fault diagnosis method reviewed in the following. If passive diagnosis is used, during the fault identification the update step is only performed for the current state estimate $\mathcal{Z}^*(k)$ of the plant. Furthermore, the block labeled "Apply separating sequence [...]" is removed and the input for the nominal controller $\mathcal{C}(\mathcal{A}_0)$ remains to be generated by the nominal trajectory planning unit $\mathcal{T}(\mathcal{A}_0)$.

For **fault detection** (left part of Fig. 6.4), the diagnostic unit $\mathcal{D}$ must analyze the current state estimate $z^*(k)$ of the plant in every time step $k \geq 0$. According to (4.19), the current state

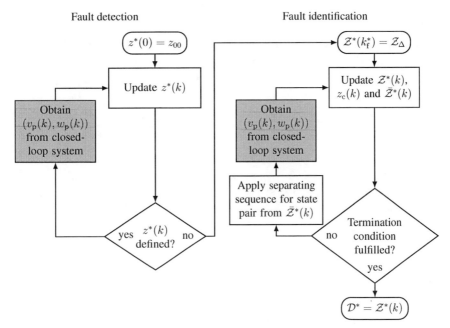

Figure 6.4: Flowchart summarizing active fault diagnosis method (cf. Figs. 4.10 and 4.11).

estimate of the plant is initialized by the initial state of the faultless plant:

$$z^*(0) = z_{00}.$$

The current state estimate of the plant needs to be updated in every time step $k \geq 0$ using (4.21) until a discrepancy between the output $H_0(z^*(k), v_p(k))$ expected based on the model $\mathcal{A}_0$ of the faultless plant and the output $w_p(k)$ generated by the plant $\mathcal{P}$ is observed at the fault detection time k_f^*:

$$z^*(k+1) = \begin{cases} G_0(z^*(k), v_p(k)) & \text{if } w_p(k) = H_0(z^*(k), v_p(k)), \\ \text{undefined} & \text{if } w_p(k) \neq H_0(z^*(k), v_p(k)). \end{cases}$$

When a fault has been detected at the fault detection time k_f^*, the **fault identification** starts. Therefore, the current state estimate of the plant is re-initialized according to (4.28) such that it becomes equal to the state set of the overall model $\mathcal{A}_\Delta$ of the faulty plant in (6.3):

$$\mathcal{Z}^*(k_f^*) = \mathcal{Z}_\Delta = \mathcal{Z}_0 \times \mathcal{F}.$$

Whenever a new I/O-pair $(v_p(k), w_p(k))$ arrives at the diagnostic unit $\mathcal{D}$, the current state estimate of the plant needs to be updated based on (4.29) using the overall model $\mathcal{A}_\Delta$ of the faulty plant in (6.3):

$$\mathcal{Z}^*(k+1) = \Big\{ z'_\Delta \in \mathcal{Z}_\Delta : (\exists z_\Delta \in \mathcal{Z}^*(k)) \\ \Big[z'_\Delta = G_\Delta(z_\Delta, v_p(k)) \Big] \wedge \Big[w_p(k) = H_\Delta(z_\Delta, v_p(k)) \Big] \Big\}.$$

If passive diagnosis shall be used, the inputs $v_p(k)$, $k \geq k_f^*$ for the plant $\mathcal{P}$ after the fault detection time k_f^* are still generated by the nominal tracking controller $\mathcal{C}_T(\mathcal{A}_0)$. Therefore, the diagnostic unit $\mathcal{D}$ merely needs to update and analyze the current state estimate $\mathcal{Z}^*(k)$ of the plant.

In contrast, if active diagnosis shall be used, the diagnostic unit $\mathcal{D}$ additionally needs to generate a diagnostic trajectory $R(k_f^* + 1 \dots k_e)$ as an input for the controller $\mathcal{C}(\mathcal{A}_0)$ after the fault detection time k_f^*. Initially, the overall model $\bar{\mathcal{A}}_\Delta$ of the closed-loop system needs to be computed. According to (4.14), the overall model of the closed-loop system is defined based on the controller $\mathcal{C}(\mathcal{A}_0) \models \mathcal{A}_c$ and the overall model $\mathcal{A}_\Delta$ of the faulty plant by

$$
\bar{\mathcal{A}}_\Delta : \begin{cases}
\bar{\mathcal{Z}}_\Delta = \mathcal{Z}_c \times \mathcal{Z}_\Delta, \\[4pt]
\bar{\mathcal{V}}_\Delta = \mathcal{V}_c, \\[4pt]
\bar{\mathcal{W}}_\Delta = \mathcal{W}_\Delta, \\[6pt]
\bar{G}(\bar{z}_\Delta, \bar{v}_\Delta) = \begin{cases}
\begin{pmatrix} G_c(\bar{z}_{\Delta 1}, \bar{v}_\Delta) \\ G_\Delta(\bar{z}_{\Delta 2}, H_c(\bar{z}_{\Delta 1}, \bar{v}_\Delta)) \end{pmatrix} & \text{if } G_c(\bar{z}_{\Delta 1}, \bar{v}_\Delta)!, \\[12pt]
\text{undefined} & \text{otherwise,}
\end{cases} \\[18pt]
\bar{H}(\bar{z}_\Delta, \bar{v}_\Delta) = \begin{cases}
H_\Delta(\bar{z}_{\Delta 2}, H_c(\bar{z}_{\Delta 1}, \bar{v}_\Delta)) & \text{if } H_c(\bar{z}_{\Delta 1}, \bar{v}_\Delta)!, \\[4pt]
\text{undefined} & \text{otherwise.}
\end{cases}
\end{cases}
$$

Then, the state set $\bar{\mathcal{Z}}_\Delta$ of this overall model of the closed-loop system needs to be partitioned into sets $\mathcal{Z}_i^k$ of k-compatible states ($k = 0, \dots, \bar{n} - 2$, $i = 1, \dots, m_k$). Details on the partitioning procedure can be found in Section 4.3.4.

Based on the overall model $\bar{\mathcal{A}}_\Delta$ of the closed-loop system, the computation of the next part of the diagnostic trajectory $R(k_f^* + 1 \dots k_e)$ can be summarized in four steps.

1. The current state estimate $\bar{\mathcal{Z}}^*(k)$ of the closed-loop system must be known. Therefore, in addition to the current state estimate $\mathcal{Z}^*(k)$ of the plant, the current state $z_c(k)$ of the controller $\mathcal{C}(\mathcal{A}_0)$ needs to be determined in every time step $k \geq k_f^*$. According to (4.34) the current state $z_c(k)$ of the controller $\mathcal{C}(\mathcal{A}_0)$ can be computed iteratively by using the inverted automaton map Φ_c^{-1} of the controller defined in (3.11):

$$
z_c(k+1) = G_c(z_c(k), \Phi_c^{-1}(z_c(k), v_p(k)))
$$
$$
\text{with } \Phi_c^{-1}(z_c(k), v_p(k)) = v_c(k) \Leftrightarrow \Phi_c(z_c(k), v_c(k)) = v_p(k),
$$
$$
z_c(0) = z_{c0} = z_{00}.
$$

Combining the current state estimate $\mathcal{Z}^*(k)$ of the plant and the reconstructed current state $z_c(k)$ of the controller $\mathcal{C}(\mathcal{A}_0)$, the current state estimate of the closed-loop system defined in (4.32) is given by

$$
\bar{\mathcal{Z}}^*(k) = \{z_c(k)\} \times \mathcal{Z}^*(k).
$$

2. The diagnostic unit $\mathcal{D}$ needs to select a pair (z_1, z_2) of states from the current state estimate $\bar{\mathcal{Z}}^*(k)$ of the closed-loop system. The state pair (z_1, z_2) can be selected by chance. Alternatively, if the size of the current state estimate $\mathcal{Z}^*(k)$ of the plant shall be reduced

as fast as possible, the state pair (z_1, z_2) can be selected by using Algorithm 4.3, which analyzes the k-compatibility partitions $\mathcal{Z}_i^k$, $(k = 0, \ldots, \bar{n} - 2,\ i = 1, \ldots, m_k)$ for the "best" state pair according to some heuristic rules.

3. Based on the k-compatibility partitions $\mathcal{Z}_i^k$, $(k = 0, \ldots, \bar{n} - 2,\ i = 1, \ldots, m_k)$, the diagnostic unit $\mathcal{D}$ then has to compute a separating sequence $V_{\mathrm{s}}(0 \ldots k_{\mathrm{e}})$ for the chosen state pair (z_1, z_2). By definition, the separating sequence leads to different outputs of the closed-loop system, depending on whether it is applied in state z_1 or in state z_2:

$$\bar{\Phi}_\Delta(z_1, V_{\mathrm{s}}(0 \ldots k_{\mathrm{e}})) \neq \bar{\Phi}_\Delta(z_2, V_{\mathrm{s}}(0 \ldots k_{\mathrm{e}})).$$

Details on the computation of the separating sequence $V_{\mathrm{s}}(0 \ldots k_{\mathrm{e}})$ based on the k-compatibility partitions $\mathcal{Z}_i^k$, $(k = 0, \ldots, \bar{n} - 2,\ i = 1, \ldots, m_k)$ can be found in Section 4.5.3.

4. Finally, the diagnostic unit $\mathcal{D}$ has to apply the determined separating sequence $V_{\mathrm{s}}(0 \ldots k_{\mathrm{e}})$ in form of the next part of the diagnostic trajectory $R(k_{\mathrm{f}}^* + 1 \ldots k_{\mathrm{e}})$ to the controller $\mathcal{C}(\mathcal{A}_0)$. Afterwards, the process starts over with updating the current state estimate $\mathcal{Z}^*(k)$ of the plant and the current state estimate $\bar{\mathcal{Z}}^*(k)$ of the closed-loop system.

The fault diagnosis is stopped when a previously specified termination condition is fulfilled. Important termination conditions are the unambiguity of the diagnostic result, i.e., $|\mathcal{Z}^*(k_{\mathrm{d}})| = 1$, where k_{d} is called diagnosis time, or solely the identification of the present fault $\bar{f} \in \mathcal{F}$. The termination condition $|\mathcal{Z}^*(k_{\mathrm{d}})| = 1$ is only satisfiable if the plant $\mathcal{P}$ is diagnosable. Other possible termination conditions have been discussed in Section 4.6.2. In order to allow for a later reconfiguration of the nominal tracking controller $\mathcal{C}_{\mathcal{T}}(\mathcal{A}_0)$, it has to be ensured that the fault diagnosis is not stopped too early. Since the validity of the termination condition is application-dependent, the following assumption is made.

Assumption 4 (Validity of the termination condition). *Let $\mathcal{Z}_{\mathrm{F}} \subseteq \mathcal{Z}_0$ be a set of possible desired final states. Then the termination condition for the fault diagnosis is chosen such that if the closed-loop system $\mathcal{P} \circ \mathcal{C}_{\mathcal{T}}(\mathcal{A}_0)$ is recoverable from a fault $f \in \mathcal{F}$ with respect to a desired final state $z_{\mathrm{F}} \in \mathcal{Z}_{\mathrm{F}}$, the nominal tracking controller $\mathcal{C}_{\mathcal{T}}(\mathcal{A}_0)$ is reconfigurable with respect to any possible diagnostic result $\mathcal{D}^*$ obtained with this termination condition and the state z_{F}.*

The choice of the set $\mathcal{Z}_{\mathrm{F}} \subseteq \mathcal{Z}_0$ describes the tolerable loss of functionality of the closed-loop system. If $\mathcal{Z}_{\mathrm{F}} = \mathcal{Z}_0$, the termination condition has to be chosen such that the same result as with a reconfiguration based on an unambiguous diagnostic result $\mathcal{D}^*$ is obtained. In contrast, $\mathcal{Z}_{\mathrm{F}} \subset \mathcal{Z}_0$ means that not all states that are reachable after a reconfiguration based on an unambiguous diagnostic result $\mathcal{D}^*$ have to be reachable such that it might be possible to stop the fault diagnosis early.

The time at which the fault diagnosis is stopped equals the reconfiguration time k_r, because the diagnostic result $\mathcal{D}^*$ is used for the following reconfiguration of the tracking controller $\mathcal{C}_T(\mathcal{A}_0)$ by the reconfiguration unit $\mathcal{R}$. The diagnostic result is defined to be the same as the current state estimate of the plant at this time:

$$\mathcal{D}^* = \mathcal{Z}^*(k_r).$$

Definition of the reconfiguration unit $\mathcal{R}$. The reconfiguration unit $\mathcal{R}$ is defined based on the set $\{E_{zf}, E_{vf}, E_{wf}, (f \in \mathcal{F})\}$ of error relations in (2.29)–(2.31) as described in Chapter 5. When the reconfiguration unit $\mathcal{R}$ receives an unambiguous or ambiguous diagnostic result $\mathcal{D}^*$ from the diagnostic unit $\mathcal{D}$, it has to modify the trajectory planning unit $T(\mathcal{A}_0)$ and the controller $\mathcal{C}(\mathcal{A}_0)$ in the nominal tracking controller $\mathcal{C}_T(\mathcal{A}_0)$ according to Algorithm 5.2 or Algorithm 5.4, respectively. Notice that Algorithm 5.4 describes the reconfiguration for the more general case of an ambiguous diagnostic result $\mathcal{D}^*$ and can also be applied when the diagnostic result $\mathcal{D}^*$ is unambiguous. Therefore, in the following it is only referred to this algorithm, while pointing out simplifications occurring in case of an unambiguous diagnostic result.

Line 1 of Algorithm 5.4. First, the error relations $E_{z\cap}$ and $E_{v\cap}$ for the given diagnostic result $\mathcal{D}^*$ have to be computed according to (5.31) and (5.32), respectively, using the set $\{E_{zf}, E_{vf}, E_{wf}, (f \in \mathcal{F})\}$ of error relations:

$$
E_{z\cap}(z, v) = \begin{cases}
G_0(z, v) & \text{if } \Big[v \text{ faultless input}\Big] \wedge \Big[(z, v) \text{ faultless transition}\Big], \\
& \hspace{3cm} \forall \begin{pmatrix} z_p & f \end{pmatrix}^\top \in \mathcal{D}^*, \\[2mm]
z_\# & \text{if } \Big[E_{zf}(z, E_{vf}(v)) = z_\#, \quad \forall \begin{pmatrix} z_p & f \end{pmatrix}^\top \in \mathcal{D}^*\Big] \\
& \hspace{1cm} \vee \Big(\Big[E_{zf}(z_p, E_{vf}(v)) = z_\#, \quad \forall \begin{pmatrix} z_p & f \end{pmatrix}^\top \in \mathcal{D}^*\Big] \\
& \hspace{5cm} \wedge \Big[z = z_{\cap 0}\Big]\Big), \\[2mm]
\text{undefined} & \text{otherwise,}
\end{cases}
$$

$$E_{v\cap}(v) = v \quad \forall v \in \mathcal{V}_0 = \mathcal{V}_f, \ (f \in \mathcal{F}).$$

If the diagnostic result $\mathcal{D}^*$ is unambiguous, that is, if

$$\mathcal{D}^* = \left\{ \begin{pmatrix} z_p(k_d) \\ \bar{f} \end{pmatrix} \right\},$$

the error relations $E_{z\bar{f}}$ and $E_{v\bar{f}}$ corresponding to the present fault $\bar{f} \in \mathcal{F}$ can be used for the following reconfiguration steps directly. That is, the error relations required for the next steps are given by

$$E_{z\cap}(z,v) = E_{z\bar{f}}(z,v), \quad \forall(z,v) \in \mathcal{Z}_0 \times \mathcal{V}_0$$

and

$$E_{v\cap}(v) = E_{v\bar{f}}(v), \quad \forall v \in \mathcal{V}_0.$$

Line 2 of Algorithm 5.4. Next, the automaton graph $\mathcal{A}_0$ stored in the nominal trajectory planning unit $\mathcal{T}(\mathcal{A}_0)$ needs to be modified based on the diagnostic result $\mathcal{D}^*$ and the error relations $E_{z\cap}$ and $E_{v\cap}$ defined above. According to (5.34) the following automaton results:

$$\mathcal{A}_\cap^r : \begin{cases} \mathcal{Z}_\cap^r = \mathcal{Z}_0 \cup \{z_{\cap 0}^r\}, \\ \mathcal{V}_\cap^r = \mathcal{V}_0, \\ \mathcal{W}_\cap^r = \mathcal{W}_0, \\ G_\cap^r(z,v) = \begin{cases} G_0(z,v) & \text{if } \Big[v \text{ faultless input}\Big] \wedge \Big[(z,v) \text{ faultless transition}\Big], \\ E_{z\cap}(z, E_{v\cap}(v)) & \text{if } \Big[v \text{ faulty input}\Big] \vee \Big[(z,v) \text{ faulty transition}\Big] \\ & \hspace{3cm} \vee \Big[z = z_{\cap 0}^r\Big], \\ \text{undefined} & \text{otherwise,} \end{cases} \\ H_\cap^r(z,v) = \begin{cases} H_0(z,v) & \text{if } \Big[G_\cap^r(z,v)!\Big], \\ \text{undefined} & \text{otherwise,} \end{cases} \\ z_{\cap 0}^r = \begin{cases} z_0 & \text{if } z_p = z_0 \in \mathcal{Z}_f, \quad \forall \big(z_p \;\; f\big)^\top \in \mathcal{D}^*, \\ z_{\text{new}} \notin \mathcal{Z}_0 & \text{otherwise.} \end{cases} \end{cases}$$

In case of an unambiguous diagnostic result $\mathcal{D}^*$, the initial state of the modified automaton $\mathcal{A}_\cap^r$ is always given by

$$z_{\cap 0}^r = z_p(k_d).$$

Line 3 of Algorithm 5.4. Finally, Algorithm 5.3 has to be used to reconfigure the nominal controller $\mathcal{C}(\mathcal{A}_0)$ based on the given diagnostic result $\mathcal{D}^*$ and the corresponding error relations $E_{z\cap}$ and $E_{v\cap}$. That is, faulty transitions are removed from the controller $\mathcal{C}(\mathcal{A}_0)$ and the resulting pruned automaton is completed by transitions from the set $\mathcal{E}_{\text{new}}$ of transitions to be newly defined. According to (5.35) and (5.19) this set is given by

$$\mathcal{E}_{\text{new}} = \Big\{ (z_c, z_c') \in \mathcal{Z}_c \times \mathcal{Z}_c :$$
$$\Big((\exists w \in \mathcal{W}_c) \left(\left[H_c(z_c, z_c') = w \right] \wedge \left[E_{zf}(z_c, E_{vf}(w)) \neq z_c' \right] \right) \Big)$$
$$\vee \Big(\left[(\nexists w \in \mathcal{W}_c) \, H_c(z_c, z_c') = w \right] \wedge \left[(\exists w \in \mathcal{W}_c) \, E_{zf}(z_c, E_{vf}(w)) = z_c' \right] \Big)$$
$$\vee \Big(\left[z_c = z_{\text{new}} \right] \wedge \left[(\exists w \in \mathcal{W}_c) \, E_{zf}(z_p, E_{vf}(w)) = z_c', \quad \forall \begin{pmatrix} z_p & f \end{pmatrix}^\top \in \mathcal{D}^* \right] \Big) \Big\}.$$

The first part of the set $\mathcal{E}_{\text{new}}$ above contains all transitions that are removed from the nominal controller $\mathcal{C}(\mathcal{A}_0)$. The second part contains transitions newly occurring in the faulty plant and the third part takes the possibly occurring artificial initial state z_{new} in case of an ambiguous diagnostic result $\mathcal{D}^*$ into account. As described in (5.20), based on the set $\mathcal{E}_{\text{new}}$ of transitions to be newly defined above, the resulting reconfigured controller $\mathcal{C}^r$ is given by

$$
\mathcal{C}^r : \begin{cases}
\mathcal{Z}_c^r = \mathcal{Z}_c, \\[4pt]
\mathcal{V}_c^r = \mathcal{V}_c, \\[4pt]
\mathcal{W}_c^r = \mathcal{W}_c, \\[4pt]
G_c^r(z_c, v_c) = \begin{cases} G_c(z_c, v_c) & \text{if } \left[H_c(z_c, v_c) \text{ faultless input} \right] \\ & \qquad \wedge \left[(z_c, H_c(z_c, v_c)) \text{ faultless transition} \right], \\ v_c & \text{if } \left[(z_c, v_c) \in \mathcal{E}_{\text{new}} \right] \wedge \left[\mathcal{V}_{a \cap}(v_c, z_c) \neq \emptyset \right], \\ \text{undefined} & \text{otherwise}, \end{cases} \\[4pt]
H_c^r(z_c, v_c) = \begin{cases} H_c(z_c, v_c) & \text{if } \left[H_c(z_c, v_c) \text{ faultless input} \right] \\ & \qquad \wedge \left[(z_c, H_c(z_c, v_c)) \text{ faultless transition} \right], \\ w_c \in \mathcal{V}_{a \cap}(v_c, z_c) & \text{if } \left[(z_c, v_c) \in \mathcal{E}_{\text{new}} \right] \wedge \left[\mathcal{V}_{a \cap}(v_c, z_c) \neq \emptyset \right], \\ \text{undefined} & \text{otherwise}, \end{cases} \\[4pt]
z_{c0}^r = \begin{cases} z_0 & \text{if } z_p = z_0 \in \mathcal{Z}_f, \quad \forall \begin{pmatrix} z_p & f \end{pmatrix}^\top \in \mathcal{D}^*, \\ z_{\text{new}} \notin \mathcal{Z}_0 & \text{otherwise}. \end{cases}
\end{cases}
$$

6.2.3 Fulfillment of the control aim in the fault-tolerant control loop

The following theorem is the main result of the thesis. It states that the proposed fault-tolerant controller $\mathcal{C}_{\text{FTC}} = (\mathcal{C}_T, \mathcal{D}, \mathcal{R})$ solves the fault-tolerant control problem in Problem 2.1.

Theorem 6.3 (Fault-tolerance). *Consider a fault-tolerant control loop with a plant $\mathcal{P}$ and a fault-tolerant controller $\mathcal{C}_{\mathrm{FTC}}$, consisting of a tracking controller $\mathcal{C}_T$, a diagnostic unit $\mathcal{D}$ and a reconfiguration unit $\mathcal{R}$ defined as described in the previous sections. Then the plant $\mathcal{P}$ reaches the desired final state z_F at a finite time $k_\mathrm{F} \geq 0$ for all faults $f \in \mathcal{F}$ from which the closed-loop system $\mathcal{P} \circ \mathcal{C}_T$ is recoverable.*

Proof. If the closed-loop system $\mathcal{P} \circ \mathcal{C}_T$ is recoverable from the fault $f \in \mathcal{F}$, it is possible to determine a diagnostic result $\mathcal{D}^*$ with respect to which the nominal tracking controller $\mathcal{C}_T(\mathcal{A}_0)$ is reconfigurable (Definition 6.1). By Assumption 4, the fault diagnosis is executed until a diagnostic result $\mathcal{D}^*$ with respect to which the nominal tracking controller $\mathcal{C}_T(\mathcal{A}_0)$ is reconfigurable is obtained. The reconfiguration unit $\mathcal{R}$ uses the diagnostic result $\mathcal{D}^*$ in order to modify the nominal tracking controller such that the reconfigured tracking controller $\mathcal{C}_T^{\mathrm{r}}$ results.

The diagnostic result $\mathcal{D}^*$ is always complete and correct (Propositions 4.10 and 4.14). Consequently, the faulty plant $\mathcal{P} \vDash \mathcal{A}_f$ in the closed-loop system $\mathcal{P} \circ \mathcal{C}_T^{\mathrm{r}}$ reaches the desired final state z_F at a finite time k_F (Theorem 5.7). □

The above theorem proves that the proposed fault-tolerant controller $\mathcal{C}_{\mathrm{FTC}} = (\mathcal{C}_T, \mathcal{D}, \mathcal{R})$ solves the fault-tolerant control problem in Problem 2.1 for the most general case of a recoverable closed-loop system $\mathcal{P} \circ \mathcal{C}_T$. Naturally, the same result is also obtained, for the more specific case that additionally the plant $\mathcal{P}$ is diagnosable with respect to the fault set $\mathcal{F}$. In this case, Assumption 4 can be fulfilled by choosing the unambiguity of the diagnostic result $\mathcal{D}^*$ as a termination condition for the fault diagnosis. Then from Theorem 4.3 or Theorem 4.4 on the nature of the diagnostic result $\mathcal{D}^*$ and Theorem 5.3 on the fulfillment of the control aim after a reconfiguration based on an unambiguous diagnostic result, the fulfillment of the control aim (2.44) by the faulty plant in the fault-tolerant control loop follows directly.

Example 6.1 (cont.) *Fault-tolerant control of automated warehouse*

In Examples 3.1, 4.2, 5.1 and 5.2 the behavior of the automated warehouse introduced in Section 2.5 controlled by the fault-tolerant controller $\mathcal{C}_{\mathrm{FTC}} = (\mathcal{C}_T, \mathcal{D}, \mathcal{R})$ has been described in detail. Here, the overall behavior of the fault-tolerant control loop is summarized.

Unambiguous diagnostic result. Figure 6.5 shows the course of all important variables for the case that active diagnosis is used and the fault diagnosis is stopped when an unambiguous diagnostic result $\mathcal{D}^*$ is obtained.

The blue line in the third subplot indicates the desired final state $z_\mathrm{F} = B$, that is, a parcel shall be transported to position B. Initially, the nominal tracking controller $\mathcal{C}_T(\mathcal{A}_0)$ from Example 3.1 controls the plant $\mathcal{P}$. Therefore, the first value of the reference trajectory is $r(0) = B$. The fault $f = 2$ occurs at the fault occurrence time $k_\mathrm{f} = 0$, which means that robot M breaks down. Consequently, the desired transport can not be executed and the parcel remains at position A ($z_\mathrm{p}(1) = A$) such that the fault is immediately detected ($k_\mathrm{f}^* = k_\mathrm{f} = 0$).

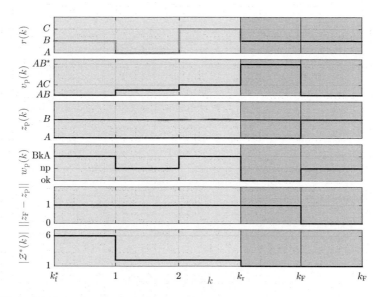

Figure 6.5: Fault-tolerant control of automated warehouse with fault $f = 2$ using active diagnosis.

Afterwards, the active diagnosis starts. That is, the reference trajectory $R(1\dots2) = (A, C)$ is now generated by the diagnostic unit $\mathcal{D}$ (gray line in first subplot) as described in Example 4.2. As a result, the plant $\mathcal{P}$ remains in state $z_\mathrm{p}(3) = A$. The size of the current state estimate $\mathcal{Z}^*(k)$ of the faulty plant is reduced from 6 over 2 at time $k = 1$ to 1 at the diagnosis time $k_\mathrm{d} = 3$. That is, from the output sequence $W_\mathrm{p}(0\dots2) = (\mathrm{BkA}, \mathrm{np}, \mathrm{BkA})$ of the faulty plant $\mathcal{P}$ the present fault $\bar{f} = 2$ and the current state $z_\mathrm{p}(k_\mathrm{d}) = A$ of the plant $\mathcal{P}$ are revealed at the diagnosis time $k_\mathrm{d} = 3$. Based on this diagnostic result

$$\mathcal{D}^* = \left\{ \begin{pmatrix} A \\ 2 \end{pmatrix} \right\},$$

the tracking controller $\mathcal{C}_\mathcal{T}$ is reconfigured at the reconfiguration time $k_\mathrm{r} = k_\mathrm{d} = 3$.

For $k \geq k_\mathrm{r} = 3$, the reconfigured tracking controller $\mathcal{C}_\mathcal{T}^\mathrm{r}$ from Example 5.1 controls the faulty plant $\mathcal{P} \models \mathcal{A}_2$. In order to steer the plant $\mathcal{P}$ into the desired final state $z_\mathrm{F} = B$, the reconfigured trajectory planning unit $\mathcal{T}^\mathrm{r}$ generates the reference trajectory $r(k_\mathrm{r} = k_\mathrm{e}^\mathrm{r} = 3) = B$. As a result, the reconfigured controller $\mathcal{C}^\mathrm{r}$ generates the input $v_\mathrm{p}(k_\mathrm{r} = 3) = AB^*$ for the plant $\mathcal{P}$. That is, a parcel is transported from position A to position B using the robot M^* such that the desired final state is reached at time $k_\mathrm{F} = 4$ ($z_\mathrm{p}(4) = z_\mathrm{F} = B$). The distance to the desired final state $z_\mathrm{F} = B$ decreases by one in the time step after the reconfiguration time $k_\mathrm{r} = 3$. This indicates that after the reconfiguration the shortest path to the desired final state is used.

Ambiguous diagnostic result. Figure 6.6 shows the behavior of the fault-tolerant control loop for the case that passive diagnosis is used and the diagnosis is stopped when the current state estimate $\mathcal{Z}^*(k)$ of the plant contains two elements. That is, the reconfiguration is performed based on an ambiguous diagnostic result now.

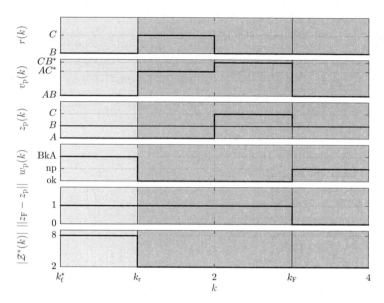

Figure 6.6: Fault-tolerant control of automated warehouse with fault $f = 2$ using passive diagnosis.

Compared to Fig. 6.5 it can be seen that the reconfiguration is performed earlier, namely at the reconfiguration time $k_r = 1 < 3$. At this time the current state estimate $\mathcal{Z}^*(2)$ of the plant still contains two elements, that is, the diagnostic result

$$\mathcal{D}^* = \left\{ \begin{pmatrix} A_\# \\ 1 \end{pmatrix}, \begin{pmatrix} A \\ 2 \end{pmatrix} \right\}$$

is ambiguous.

Starting at the reconfiguration time $k_r = 1$ the reconfigured tracking controller $\mathcal{C}_\mathcal{T}^r$ from Example 5.2 controls the faulty plant $\mathcal{P} \vDash \mathcal{A}_2$. It generates the input sequence $V_p(1 \ldots 2) = (AC^*, CB^*)$ for the plant such that the parcel is transported from position A to position C and then to position B using the robot M^*. Since the distance to the desired final state $z_F = B$ does not decrease in every time step, it can be seen that no shortest path is used. In this example, if it was known that fault $f = 2$ occurred, the parcel could have been transported from position A to position B using the robot M^* directly. However, since the given diagnostic result $\mathcal{D}^*$ is ambiguous, the reconfigured tracking controller $\mathcal{C}_\mathcal{T}^r$ also had to account for the fact that fault $f = 1$ might have occurred, that is, that the direct route between position A and position B might be blocked.

Nevertheless, the desired final state $z_F = B$ is reached at time $k_F = 3$ and therefore even faster than in the previous case of an unambiguous diagnostic result, where $k_F = 4$ was obtained.   □

6.3 Safe fault-tolerant control

This section deals with the safe fault-tolerant control problem in Problem 2.2, where the fault-tolerant controller $\mathcal{C}_{\mathrm{FTC}}$ ensures that the plant $\mathcal{P}$ avoids a set $\mathcal{E}_{\mathrm{ill}}$ of illegal transitions.

6.3.1 Recoverability analysis

The term *safe recoverability* of the closed-loop system $\mathcal{P} \circ \mathcal{C}_{\mathcal{T}}(\mathcal{A}_0, \mathcal{E}_{\mathrm{ill}})$ from a fault $f \in \mathcal{F}$ refers to the ability to find a fault-tolerant controller $\mathcal{C}_{\mathrm{FTC}}$ that guarantees the fulfillment of the safe control aim (2.46) by the controlled faulty plant $\mathcal{P} \vDash \mathcal{A}_f$. That is, additionally to the recoverability of the closed-loop system $\mathcal{P} \circ \mathcal{C}_{\mathcal{T}}(\mathcal{A}_0)$ as defined in Definition 6.1 it has to be guaranteed that the plant $\mathcal{P}$ avoids a given set $\mathcal{E}_{\mathrm{ill}}$ of illegal transitions at all times, starting at the fault occurrence time k_{f} until the desired final state z_{F} is reached at time k_{F}. Therefore, safe recoverability of the closed-loop system is defined as follows.

Definition 6.2 (Safe recoverability). *The closed-loop system $\mathcal{P} \circ \mathcal{C}_{\mathcal{T}}(\mathcal{A}_0, \mathcal{E}_{\mathrm{ill}})$ is said to be safely recoverable from a fault $f \in \mathcal{F}$ with respect to the desired final state z_{F} and a set $\mathcal{E}_{\mathrm{ill}}$ of illegal transitions if it is possible to safely determine a diagnostic result $\mathcal{D}^*$ with respect to which the nominal tracking controller $\mathcal{C}_{\mathcal{T}}(\mathcal{A}_0, \mathcal{E}_{\mathrm{ill}})$ is safely reconfigurable, given the model $\mathcal{A}_0$ of the faultless plant, the set $\{E_{zf}, E_{vf}, E_{wf}, (f \in \mathcal{F})\}$ of error relations and the I/O-pair $(V_{\mathrm{p}}(0 \ldots k), W_{\mathrm{p}}(0 \ldots k))$ of the plant $\mathcal{P}$.*

Regarding conditions for the safe recoverability of a closed-loop system $\mathcal{P} \circ \mathcal{C}_{\mathcal{T}}(\mathcal{A}_0, \mathcal{E}_{\mathrm{ill}})$, the same arguments as in the case without a set $\mathcal{E}_{\mathrm{ill}}$ of illegal transitions discussed in Section 6.2.1 hold true. That is, in general safe recoverability can only be analyzed by means of a simulation and it is only possible to find a sufficient condition for safe recoverability. In order to guarantee the avoidance of illegal transitions, only active diagnosis makes sense. Therefore, the following theorem states a sufficient condition for safe recoverability for the use of active safe diagnosis.

Theorem 6.4 (Sufficient condition for safe recoverability). *The closed-loop system $\mathcal{P} \circ \mathcal{C}_{\mathcal{T}}(\mathcal{A}_0, \mathcal{E}_{\mathrm{ill}})$ is safely recoverable from a fault $f \in \mathcal{F}$ with respect to the desired final state z_{F} and a set $\mathcal{E}_{\mathrm{ill}}$ of illegal transitions if the following two conditions are fulfilled:*

- *The plant $\mathcal{P}$ in the closed-loop system $\mathcal{P} \circ \mathcal{C}_{\mathcal{T}}(\mathcal{A}_0, \mathcal{E}_{\mathrm{ill}})$ is safely diagnosable with respect to the fault set $\mathcal{F}$ and the set $\mathcal{E}_{\mathrm{ill}}$ of illegal transitions.*
- *The nominal tracking controller $\mathcal{C}_{\mathcal{T}}(\mathcal{A}_0, \mathcal{E}_{\mathrm{ill}})$ is safely reconfigurable with respect to any diagnostic result $\mathcal{D}^*$ as in (6.2), the desired final state z_{F} and the set $\mathcal{E}_{\mathrm{ill}}$ of illegal transitions.*

Proof. If the plant $\mathcal{P}$ in the closed-loop system $\mathcal{P} \circ \mathcal{C}_T(\mathcal{A}_0, \mathcal{E}_{\text{ill}})$ is safely diagnosable with respect to the fault set $\mathcal{F}$ and the set $\mathcal{E}_{\text{ill}}$ of illegal transitions, the present fault $\bar{f}$ and the current state $z_p(k)$ of the plant $\mathcal{P}$ can be identified from the models $\{\mathcal{A}_f, \ (f \in \mathcal{F})\}$ of the faulty plant and the I/O-pair $(V_p(k_f^* \ldots k_e), W_p(k_f^* \ldots k_e))$ of the controlled plant. Recall that k_f^* is the fault detection time. At the same time the avoidance of illegal transitions $(z_p(k), z_p(k+1)) \in \mathcal{E}_{\text{ill}}$ by the plant $\mathcal{P}$ is guaranteed for all times $k_f \leq k < k_d$ (Definition 4.10).

The models $\{\mathcal{A}_f, \ (f \in \mathcal{F})\}$ of the faulty plant can be constructed from the model $\mathcal{A}_0$ of the faultless plant and the set $\{E_{zf}, E_{vf}, E_{wf}, \ (f \in \mathcal{F})\}$ of error relations according to (2.28). Hence, it is possible to determine an unambiguous diagnostic result $\mathcal{D}^*$ as in (6.2).

If the nominal tracking controller $\mathcal{C}_T(\mathcal{A}_0, \mathcal{E}_{\text{ill}})$ is safely reconfigurable with respect to any of these diagnostic results $\mathcal{D}^*$, the desired final state z_F and the set $\mathcal{E}_{\text{ill}}$ of illegal transitions, the closed-loop system $\mathcal{P} \circ \mathcal{C}_T(\mathcal{A}_0, \mathcal{E}_{\text{ill}})$ is safely recoverable from the fault $f \in \mathcal{F}$ with respect to the desired final state z_F and the set $\mathcal{E}_{\text{ill}}$ of illegal transitions (Definition 6.1). □

Again it is possible to replace the abstract conditions of safe diagnosability and safe reconfigurability by conditions that can be checked for a given closed-loop system $\mathcal{P} \circ \mathcal{C}_T(\mathcal{A}_0, \mathcal{E}_{\text{ill}})$ directly. Theorem 4.5 states two conditions for safe diagnosability of the plant $\mathcal{P}$.

First, safe detectability has to be given. In Proposition 4.15 two conditions for safe detectability of a fault $f \in \mathcal{F}$ are presented. They state that the output sequence generated by the faulty plant $\mathcal{P} \vDash \mathcal{A}_f$ needs to differ from the output sequence of the faultless plant $\mathcal{P} \vDash \mathcal{A}_0$ given the input sequence $V_p(0 \ldots k_e)$ generated by the nominal tracking controller $\mathcal{C}_T(\mathcal{A}_0, \mathcal{E}_{\text{ill}})$. Furthermore, it has to be guaranteed that between the fault occurrence time k_f and the fault detection time $k_f^* \geq k_f$ the plant $\mathcal{P} \vDash \mathcal{A}_f$ does not execute any illegal transitions.

Secondly, Theorem 4.5 states that for every possible initial uncertainty $\mathcal{Z}^* = \{z_c\} \times \mathcal{Z}_\Delta$ and every possible input $v_c(k_f^*)$ to the nominal controller $\mathcal{C}_T(\mathcal{A}_0, \mathcal{E}_{\text{ill}})$ at the fault detection time k_f^* an adaptive safe homing sequence has to exist which starts with the input symbol $v_c(k_f^*)$. That is, there has to exist a tree that indicates based on the previous outputs which input symbol has to be applied to the closed-loop system $\mathcal{P} \circ \mathcal{C}_T(\mathcal{A}_0, \mathcal{E}_{\text{ill}})$ next in order to guarantee the avoidance of illegal transitions and the identification of the current state $z_p(k)$ of the plant and the fault $f \in \mathcal{F}$ (cf. Definition 4.11).

An adaptive safe homing sequence with respect to a given initial uncertainty $\mathcal{Z}^*$ can be found during the construction of a safe homing tree. Here, only the main steps of the process are described, while details can be found in Sections 4.2.2, 4.2.3 and 4.8.3. The safe homing tree is an unfolding of the overall model $\bar{\mathcal{A}}_\Delta$ of the closed-loop system. Its root vertex contains the initial uncertainty $\mathcal{Z}^*$, from which edges labeled with all possible I/O-pairs $(v, w) \in \bar{\mathcal{V}}_\Delta \times \bar{\mathcal{W}}_\Delta$ emerge. The reached vertices contain the sets of all states z' which are reached from the states z in the preceding vertex given the input symbol v while generating the output symbol w. In the safe homing tree vertices corresponding to the execution of an illegal transition $(z, z') \in \bar{\mathcal{E}}_{\text{ill}}$ by the closed-loop system $\mathcal{P} \circ \mathcal{C}_{\mathcal{T}}(\mathcal{A}_0, \mathcal{E}_{\text{ill}})$ are not included. According to (4.68) the set of illegal transitions for the closed-loop system is given by

$$\bar{\mathcal{E}}_{\text{ill}} = \{(\bar{z}_\Delta, \bar{z}'_\Delta) \in \bar{\mathcal{Z}}_\Delta \times \bar{\mathcal{Z}}_\Delta : (\bar{z}_{\Delta 2}, \bar{z}'_{\Delta 2}) \in \mathcal{E}_{\text{ill}}\} \subset \bar{\mathcal{Z}}_\Delta \times \bar{\mathcal{Z}}_\Delta.$$

New vertices are added to the safe homing tree iteratively. For every new vertex it is checked whether it can contribute to the required adaptive safe homing sequence. If the new vertex only contains a single state, the I/O-sequence leading to this vertex might contribute to the adaptive safe homing sequence. By combining it which previously found I/O-sequences, trees representing candidates for the adaptive safe homing sequence result. The procedure is repeated until an adaptive safe homing sequence is found or no new vertices can be added to the safe homing tree. If no adaptive safe homing sequence can be found, the plant $\mathcal{P}$ is not safely diagnosable. Details on the construction of an adaptive (safe) homing sequence from a (safe) homing tree are given in Algorithm 4.2.

In addition to safe diagnosability, safe reconfigurability of the nominal tracking controller $\mathcal{C}_{\mathcal{T}}(\mathcal{A}_0, \mathcal{E}_{\text{ill}})$ has to be given. A necessary and sufficient condition for safe reconfigurability with respect to an unambiguous diagnostic result $\mathcal{D}^*$ can be found in Theorem 5.2. It states that the model $\mathcal{A}_f$, $(f \in \mathcal{F})$ of the faulty plant with initial state $z_{f0} = z_{\text{p}}(k_{\text{d}})$ has to be safely controllable with respect to the desired final state z_{F} and the set $\mathcal{E}_{\text{ill}}$ of illegal transitions. According to Theorem 3.2 this is given if and only if there exists an admissible state sequence $Z(0 \ldots k_{\text{e}} + 1)$ with $z(0) = z_{\text{p}}(k_{\text{d}})$ and $z(k_{\text{e}} + 1) = z_{\text{F}}$ for the legal part $\mathcal{A}_{f,\text{leg}}$ of the model $\mathcal{A}_f$ of the faulty plant. In Definition 2.4 this legal part $\mathcal{A}_{f,\text{leg}}$ of the automaton $\mathcal{A}_f$ is defined as

$$\mathcal{A}_{f,\text{leg}} : \begin{cases} \mathcal{Z}_{f,\text{leg}} = \mathcal{Z}_f \\ \mathcal{V}_{f,\text{leg}} = \mathcal{V}_f \\ \mathcal{W}_{f,\text{leg}} = \mathcal{W}_f \\ G_{f,\text{leg}}(z, v) = \begin{cases} G_f(z, v) & \text{if } (z, G_f(z, v)) \notin \mathcal{E}_{\text{ill}}, \\ \text{undefined} & \text{otherwise,} \end{cases} \\ H_{f,\text{leg}}(z, v) = \begin{cases} H_f(z, v) & \text{if } (z, G_f(z, v)) \notin \mathcal{E}_{\text{ill}}, \\ \text{undefined} & \text{otherwise,} \end{cases} \\ z_{f,\text{leg}0} = z_{f0} = z_{\text{p}}(k_{\text{d}}). \end{cases}$$

Consequently, the following sufficient condition for safe recoverability of the closed-loop system $\mathcal{P} \circ \mathcal{C}_{\mathcal{T}}(\mathcal{A}_0, \mathcal{E}_{\text{ill}})$ can be stated.

Corollary 6.3 (Sufficient condition for safe recoverability). *Using active safe diagnosis the closed-loop system* $\mathcal{P} \circ \mathcal{C}_T(\mathcal{A}_0, \mathcal{E}_{\text{ill}})$ *is safely recoverable from a fault* $f \in \mathcal{F}$ *with respect to the desired final state* z_F *and a set* $\mathcal{E}_{\text{ill}}$ *of illegal transitions if all of the following conditions are fulfilled:*

- *(Safe detectability) There exists a time* $k_f^* \leq k_e$ *such that*

$$\Phi_f(z_p(k_f), V_p(k_f \ldots k_f^*)) \neq \Phi_0(z_p(k_f), V_p(k_f \ldots k_f^*))$$

  and

$$(G_f^\infty(z_p(k_f), V_p(k_f \ldots k)), G_f^\infty(z_p(k_f), V_p(k_f \ldots k+1)))) \neq \mathcal{E}_{\text{ill}}, \quad \forall k_f \leq k < k_f^*,$$

  where $V_p(k_f \ldots k_f^*) = \Phi_c(z_c(k_f), R(k_f \ldots k_f^*))$.

- *(Safe identifiability) For the overall model* $\bar{A}_\Delta$ *of the closed-loop system in* (4.14) *there exists an adaptive safe homing sequence with first input symbol* v_c *and respect to the set* $\bar{\mathcal{E}}_{\text{ill}}$ *of illegal transitions in* (4.68) *with respect to any initial uncertainty* $\mathcal{Z}^* = \{z_c\} \times \mathcal{Z}_\Delta$, $(z_c \in \mathcal{Z}_c)$ *and every* $v_c \in \{v_c \in \mathcal{V}_c : G_c(z_c, v_c)! \wedge (z_c, v_c) \notin \mathcal{E}_{\text{ill}}\}$.

- *(Safe reconfigurability) For every state* $z \in \mathcal{Z}_f$ *there exists a path from state* z *to state* z_F *in the automaton graph of the legal part* $\mathcal{A}_{f,\text{leg}}$ *of* $\mathcal{A}_f$.

Proof. If the first two conditions are fulfilled, according to Theorem 4.5, the plant $\mathcal{P}$ in the closed-loop system $\mathcal{P} \circ \mathcal{C}_T(\mathcal{A}_0, \mathcal{E}_{\text{ill}})$ is safely diagnosable with respect to the fault set $\mathcal{F}$ and the set $\mathcal{E}_{\text{ill}}$ of illegal transitions. If additionally the third condition is fulfilled, the nominal tracking controller $\mathcal{C}_T(\mathcal{A}_0, \mathcal{E}_{\text{ill}})$ is safely reconfigurable with respect to any of the possible diagnostic results $\mathcal{D}^*$, the desired final state z_F and the set $\mathcal{E}_{\text{ill}}$ of illegal transitions (Theorems 3.2 and 5.2 and Definition 3.2).

Consequently, according to Theorem 6.4, the closed-loop system $\mathcal{P} \circ \mathcal{C}_T(\mathcal{A}_0)$ is safely recoverable from the fault $f \in \mathcal{F}$ with respect to the desired final state z_F and the set $\mathcal{E}_{\text{ill}}$ of illegal transitions. □

Example 6.2   *Safe fault-tolerant control of automated warehouse*

The safe recoverability of the automated warehouse example is analyzed. The sufficient condition for safe recoverability in Theorem 6.4 detailed in Corollary 6.3 is fulfilled for fault $f = 2$ but not for fault $f = 1$. According to Example 4.4 fault $f = 1$ corresponding to the blocking of the route between position A and position B is not safely detectable, because this route is avoided in the presence of illegal transitions anyhow.

Other than that, it has been shown in Example 4.4 and Example 5.1 that both faults $f \in \{1, 2\}$ are safely identifiable and the nominal tracking controller $\mathcal{C}_T(\mathcal{A}_0)$ is safely reconfigurable. □

6.3.2 Definition of the safe fault-tolerant controller

In Section 6.2.2 it has already been summarized how the elements of the fault-tolerant controller $\mathcal{C}_{\text{FTC}} = (\mathcal{C}_T, \mathcal{D}, \mathcal{R})$, that is, the nominal tracking controller $\mathcal{C}_T(\mathcal{A}_0, \mathcal{E}_{\text{ill}})$, the diagnostic unit $\mathcal{D}$ and the reconfiguration unit $\mathcal{R}$ have to be defined. Therefore, in this section only the differences occurring due to the additional consideration of the set $\mathcal{E}_{\text{ill}}$ of illegal transitions are highlighted.

Definition of the tracking controller $\mathcal{C}_T$. In contrast to the case without illegal transitions, in the trajectory planning unit $\mathcal{T}(\mathcal{A}_0, \mathcal{E}_{\text{ill}})$ of the nominal tracking controller $\mathcal{C}_T(\mathcal{A}_0, \mathcal{E}_{\text{ill}})$ now the *legal part* of the model $\mathcal{A}_0$ of the faultless plant has to be stored (cf. (3.19)). According to Definition 2.4 the legal part $\mathcal{A}_{\text{leg}}$ of the automaton $\mathcal{A}_0$ is a subautomaton of $\mathcal{A}_0$ from which all illegal transitions $(z, z') \in \mathcal{E}_{\text{ill}}$ have been removed:

$$
\mathcal{A}_{\text{leg}} : \begin{cases}
\mathcal{Z}_{\text{leg}} = \mathcal{Z}_0 \\
\mathcal{V}_{\text{leg}} = \mathcal{V}_0 \\
\mathcal{W}_{\text{leg}} = \mathcal{W}_0 \\
G_{\text{leg}}(z, v) = \begin{cases} G_0(z, v) & \text{if } (z, G_0(z, v)) \notin \mathcal{E}_{\text{ill}}, \\ \text{undefined} & \text{otherwise}, \end{cases} \\
H_{\text{leg}}(z, v) = \begin{cases} H_0(z, v) & \text{if } (z, G_0(z, v)) \notin \mathcal{E}_{\text{ill}}, \\ \text{undefined} & \text{otherwise}, \end{cases} \\
z_{\text{leg}0} = z_{00}.
\end{cases}
$$

Besides, the definition of the trajectory planning unit $\mathcal{T}(\mathcal{A}_0, \mathcal{E}_{\text{ill}})$ and the controller $\mathcal{C}(\mathcal{A}_0)$ remain the same as in Section 6.2.2.

Definition of the diagnostic unit $\mathcal{D}$. In the diagnostic unit $\mathcal{D}$ the fault detection can be performed in exactly the same way as without the consideration of a set $\mathcal{E}_{\text{ill}}$ of illegal transitions. The basic approach for the fault identification based on the estimation of the current state of the plant remains the same as well. The only change occurs in the generation of the diagnostic trajectory $R(k_f^* + 1 \ldots k_e)$ for the controller $\mathcal{C}(\mathcal{A}_0)$. Instead of generating separating sequences, this diagnostic trajectory is now determined by the adaptive safe homing sequence resulting from the diagnosability analysis. Based on the previous outputs $W_p(k_f^* \ldots k - 1)$ of the plant $\mathcal{P}$ the adaptive safe homing sequence gives a unique next reference input $r(k)$, $(k > k_f^*)$ to be applied to the closed-loop system $\mathcal{P} \circ \mathcal{C}_T(\mathcal{A}_0, \mathcal{E}_{\text{ill}})$. Algorithm 4.1 describes in more detail how to obtain these reference inputs from a given adaptive safe homing sequence.

Definition of the reconfiguration unit $\mathcal{R}$. Like in the case without illegal transitions, the reconfiguration unit $\mathcal{R}$ has to reconfigure the tracking controller $\mathcal{C}_T(\mathcal{A}_0, \mathcal{E}_{\text{ill}})$ both in the presence of an ambiguous or an unambiguous diagnostic result $\mathcal{D}^*$ according to Algorithm 5.4. That is, first the error relations $E_{z\cap}$ and $E_{v\cap}$ have to be computed according to (5.31) and (5.32), respectively. Based on these error relations, the automaton graph stored in the nominal trajectory planning unit $T(\mathcal{A}_0, \mathcal{E}_{\text{ill}})$, which here corresponds to the legal part of the model $\mathcal{A}_0$ of the faulty plant, needs to be modified. Finally, the nominal controller $\mathcal{C}(\mathcal{A}_0)$ has to be reconfigured by removing faulty transitions and adding new transitions, where applicable.

The only difference to the method for the case without a set $\mathcal{E}_{\text{ill}}$ of illegal transitions summarized in Section 6.2.2 is that during the modification of the automaton graph $\mathcal{A}_{\text{leg}}$ stored in the nominal trajectory planning unit $T(\mathcal{A}_0, \mathcal{E}_{\text{ill}})$ illegal transitions $(z, z') \in \mathcal{E}_{\text{ill}}$ have to be excluded. Therefore, according to (5.36), the following automaton results in the reconfigured trajectory planning unit T^{r}:

$$
\mathcal{A}^{\text{r}}_{\cap\text{leg}} : \begin{cases}
\mathcal{Z}^{\text{r}}_{\cap\text{leg}} = \mathcal{Z}_{\text{leg}} \cup \{z^{\text{r}}_{\cap\text{leg}0}\}, \\[4pt]
\mathcal{V}^{\text{r}}_{\cap\text{leg}} = \mathcal{V}_{\text{leg}}, \\[4pt]
\mathcal{W}^{\text{r}}_{\cap\text{leg}} = \mathcal{W}_{\text{leg}}, \\[4pt]
G^{\text{r}}_{\cap\text{leg}}(z, v) = \begin{cases}
G_{\text{leg}}(z, v) & \text{if } \Big[(z, v) \text{ faultless transition}\Big] \wedge \Big[G_{\text{leg}}(z, v)!\Big], \\[4pt]
E_{z\cap}(z, E_{v\cap}(v)) & \text{if } \Big(\Big[(z, v) \text{ faulty transition}\Big] \\
& \qquad \wedge \Big[(z, E_{z\cap}(z, E_{v\cap}(v))) \notin \mathcal{E}_{\text{ill}}\Big]\Big) \\
& \qquad \vee \Big[z = z^{\text{r}}_{\cap\text{leg}0}\Big] \\[4pt]
\text{undefined} & \text{otherwise}, \\
\end{cases} \\[4pt]
H^{\text{r}}_{\cap\text{leg}}(z, v) = \begin{cases}
H_{\text{leg}}(z, v) & \text{if } \Big[H_{\text{leg}}(z, v)!\Big] \wedge \Big[G^{\text{r}}_{\cap\text{leg}}(z, v)!\Big], \\[4pt]
\varepsilon & \text{if } \Big[\neg H_{\text{leg}}(z, v)!\Big] \wedge \Big[G^{\text{r}}_{\cap\text{leg}}(z, v)!\Big], \\[4pt]
\text{undefined} & \text{otherwise}, \\
\end{cases} \\[4pt]
z^{\text{r}}_{\cap\text{leg}0} = \begin{cases}
z_0 & \text{if } z_{\text{p}} = z_0 \in \mathcal{Z}_f, \quad \forall \begin{pmatrix} z_{\text{p}} & f \end{pmatrix}^\top \in \mathcal{D}^*, \\[4pt]
z_{\text{new}} \notin \mathcal{Z}_0 & \text{otherwise}.
\end{cases}
\end{cases}
$$

6.3.3 Safe fulfillment of the control aim in the fault-tolerant control loop

The following theorem states that the proposed fault-tolerant controller $\mathcal{C}_{\text{FTC}} = (\mathcal{C}_T, \mathcal{D}, \mathcal{R})$ solves the safe fault-tolerant control problem in Problem 2.2 if the set $\mathcal{E}_{\text{ill}}$ of illegal transitions is considered during its definition.

Theorem 6.5 (Safe fault-tolerance). *Consider a fault-tolerant control loop with a plant $\mathcal{P}$ and a fault-tolerant controller $\mathcal{C}_{\mathrm{FTC}}$, consisting of a tracking controller $\mathcal{C}_T$, a diagnostic unit $\mathcal{D}$ and a reconfiguration unit $\mathcal{R}$ defined under consideration of a set $\mathcal{E}_{\mathrm{ill}}$ of illegal transitions as described in the previous sections. Then the plant $\mathcal{P}$ reaches the desired final state z_{F} at a finite time $k_{\mathrm{F}} \geq 0$ while not executing any illegal transitions $(z_{\mathrm{p}}(k), z_{\mathrm{p}}(k+1)) \in \mathcal{E}_{\mathrm{ill}}$, $(k_{\mathrm{f}} \leq k < k_{\mathrm{F}})$ for all faults $f \in \mathcal{F}$ from which the closed-loop system $\mathcal{P} \circ \mathcal{C}_T$ is safely recoverable.*

Proof. If the closed-loop system $\mathcal{P} \circ \mathcal{C}_T$ is safely recoverable from the fault $f \in \mathcal{F}$, it is possible to safely determine a diagnostic result $\mathcal{D}^*$ with respect to which the nominal tracking controller $\mathcal{C}_T(\mathcal{A}_0)$ is safely reconfigurable (Definition 6.2). By Assumption 4, the fault diagnosis is executed until a diagnostic result $\mathcal{D}^*$ with respect to which the nominal tracking controller $\mathcal{C}_T(\mathcal{A}_0)$ is safely reconfigurable is obtained. During the fault diagnosis it is guaranteed that the plant $\mathcal{P}$ does not execute any illegal transitions $(z_{\mathrm{p}}(k), z_{\mathrm{p}}(k+1))$, $(k_{\mathrm{f}} \leq k < k_{\mathrm{r}} \leq k_{\mathrm{d}})$ (cf. Theorem 4.5). The reconfiguration unit $\mathcal{R}$ uses the diagnostic result $\mathcal{D}^*$ in order to modify the nominal tracking controller such that the reconfigured tracking controller $\mathcal{C}_T^{\mathrm{r}}$ results.

The diagnostic result $\mathcal{D}^*$ is always complete and correct (Propositions 4.10 and 4.14). Consequently, the faulty plant $\mathcal{P} \vDash \mathcal{A}_f$ in the closed-loop system $\mathcal{P} \circ \mathcal{C}_T^{\mathrm{r}}$ reaches the desired final state z_{F} at a finite time k_{F} while not executing any illegal transitions $(z_{\mathrm{p}}(k), z_{\mathrm{p}}(k+1)) \in \mathcal{E}_{\mathrm{ill}}$, $(k_{\mathrm{r}} \leq k < k_{\mathrm{F}})$ (Theorem 5.8). Thereby it is guaranteed that no illegal transitions are executed at any time $k_{\mathrm{f}} \leq k < k_{\mathrm{F}}$. □

Consider the special case that in addition to the safe recoverability of the closed-loop system $\mathcal{P} \circ \mathcal{C}_T$ safe diagnosability of the plant $\mathcal{P}$ is given and unambiguity of the diagnostic result is chosen as the termination condition for fault diagnosis. Then it follows from Theorem 4.5 on the nature of the diagnostic result $\mathcal{D}^*$ obtained with active safe diagnosis and Theorem 5.4 on the safe fulfillment of the control aim after a reconfiguration based on an unambiguous diagnostic result that the faulty plant $\mathcal{P}$ in the fault-tolerant control loop safely fulfills the control aim (2.46).

Example 6.2 (cont.) *Safe fault-tolerant control of automated warehouse*

In Examples 3.1, 4.4, 5.1 and 5.2 the behavior of the automated warehouse controlled by the fault-tolerant controller $\mathcal{C}_{\mathrm{FTC}} = (\mathcal{C}_T, \mathcal{D}, \mathcal{R})$ under consideration of the set

$$\mathcal{E}_{\mathrm{ill}} = \{(A, B), (B, A)\}$$

of illegal transitions in (2.54) has been described in detail. Here, the overall behavior of the fault-tolerant control loop is summarized (Fig. 6.7). As before, it is considered that fault $f = 2$ occurs at time $k_{\mathrm{f}} = 0$, that is, robot M breaks down before any transport is conducted. The parcel shall be transported to position B, i.e., $z_{\mathrm{F}} = B$, while the direct route between position A and position B has to be avoided.

Initially, the nominal tracking controller $\mathcal{C}_T(\mathcal{A}_0, \mathcal{E}_{ill})$ from Example 3.1 controls the plant $\mathcal{P}$. Therefore, the first value of the reference trajectory is $r(0) = C$. Due to the occurrence of the fault, the desired transport can not be executed and the parcel remains at position A ($z_p(1) = A$). As a result, the fault is immediately detected ($k_f^* = k_f = 0$).

Afterwards, the active safe diagnosis starts. By the I/O-pair $(v_p(k_f^*), w_p(k_f^*)) = (AC, \text{BkA})$ of the plant $\mathcal{P}$ at the fault detection time k_f^*, the current state $z_p(k_d = 1) = A$ and the present fault $\bar{f} = 2$ are already revealed (cf. also adaptive safe homing sequence in Fig. 4.16). Based on this diagnostic result

$$\mathcal{D}^* = \left\{ \begin{pmatrix} A \\ 2 \end{pmatrix} \right\},$$

the tracking controller $\mathcal{C}_T(\mathcal{A}_0, \mathcal{E}_{ill})$ is reconfigured at the reconfiguration time $k_r = k_d = 1$.

For $k \geq k_r = 1$, the reconfigured tracking controller $\mathcal{C}_T^r$ from Example 5.1 controls the faulty plant $\mathcal{P} \models \mathcal{A}_2$. In order to steer the plant $\mathcal{P}$ into the desired final state $z_F = B$, the reconfigured trajectory planning unit $\mathcal{T}^r$ generates the reference trajectory $R(k_r = 1 \ldots k_e^r = 2) = (C, B)$. As a result, the reconfigured controller C^r in Fig. 5.7 generates the input sequence $V_p(1 \ldots 2) = (AC^*, CB^*)$ for the plant $\mathcal{P}$. That is, a parcel is transported from position A to position C and then to position B using the robot M^* such that the desired final state is reached ($z_p(k_F = 3) = z_F = B$).

The illegal transitions between state A and state B are visualized by the red arrows right of the third subplot showing the state sequence $Z_p(0 \ldots k_e)$ of the plant $\mathcal{P}$. It can be seen that the plant $\mathcal{P}$ does not execute any illegal transition. That is, that the dangerous route between position A and position B is avoided by a detour via position C. The distance to the desired final state $z_F = B$ after the reconfiguration time $k_r = 1$ also shows that not the shortest path to the desired final state is used. ☐

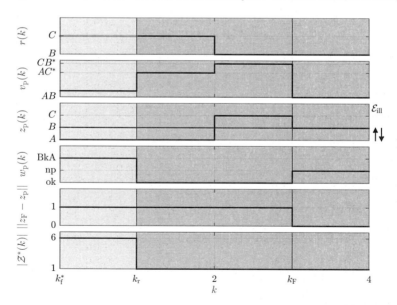

Figure 6.7: Fault-tolerant control of automated warehouse with fault $f = 2$ using active safe diagnosis.

7 Example: Fault-tolerant control of a manufacturing process

In this chapter the developed fault-tolerant control method is applied to the Handling System HANS. In simulations and experiments it is shown that the method is applicable to real-world systems. First, controllability of the Handling System HANS is analyzed and its behavior in the faultless case is discussed. Then, a recoverability analysis is performed and behavior of the fault-tolerant control loop in the faulty case is studied. Various different configurations are taken into account, that is, passive and active diagnosis, unambiguous and ambiguous diagnostic results, as well as the consideration of safety constraints.

7.1 Handling system HANS

This section describes the setup for the experimental evaluation of the developed fault-tolerant control method. Specifically, the Handling System HANS, a lab-sized manufacturing plant at the Institute of Automation and Computer Control of Ruhr-Universität Bochum, as well as the implementation structure are presented.

7.1.1 Experimental setup

The Handling System HANS shown in Fig. 7.1 is a pick-and-place system in which workpieces can be transported to different positions using two pneumatic grippers and a conveyor belt. Workpieces can be heated or cooled in two separate blocks and their temperature can be measured in a third block. A fourth block serves as a storage. Workpieces can be picked up from a buffer rail using the horizontal gripper, which transports them to any position at one of the aforementioned blocks or the conveyor belt. The conveyor belt can only be operated in one direction and moves the workpieces into one out of four slides, depending on the position of the corresponding deflector. From the slides the vertical gripper is able to transport the workpieces

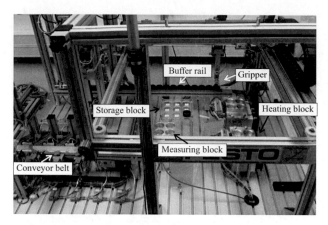

Figure 7.1: Relevant components of Handling System HANS.

to one of five storage positions or into the magazine which is connected to the buffer rail, such that a loop results.

The Handling System HANS is controlled using a Siemens SIMATIC S7-300 PLC on which basic commands like the movement of a gripper to a specified position or the evaluation of sensor signals are already implemented. An Ethernet connection connects the PLC to a PC running MATLAB/Simulink allowing for a data exchange using the protocol UDP. The fault-tolerant controller is realized in MATLAB/Simulink and transmits actuator commands to (respectively receives sensor signals from) the Handling System HANS via the PLC.

7.1.2 Implementation structure

The fault-tolerant controller is implemented in an object-oriented way, where each block in Fig. 6.1 is represented by a MATLAB class object and realized in Simulink as a Level-2 MATLAB S-Function. That is, there is an object of type TrackingController representing the tracking controller C_T, an object of type DiagUnit representing the diagnostic unit $\mathcal{D}$ and an object of type RecUnit representing the reconfiguration unit $\mathcal{R}$. An FtcLogic object coordinates the entire process and routes the correct signals to the plant. The relationship between the objects is illustrated by the UML diagram in Fig. 7.2. Connections with an arrow represent associations between the objects, that is, transmission of signals at runtime. In contrast, connections with a diamond correspond to a composition relationship, which means that one object is contained in another one.

The fault-tolerant controller works with high-level models $\mathcal{A}_f$, $(f \in \mathcal{F} \cup \{0\})$ of the plant that have scalar inputs $v \in \mathcal{V}_f$ and scalar outputs $w \in \mathcal{W}_f$ (see Section 2.3). However, to

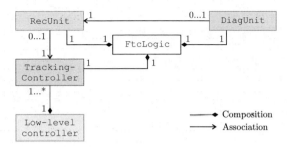

Figure 7.2: Main components of implementation structure.

perform one high-level action in the Handling System HANS, a number of its components have to be activated in the correct order. For example, to transport a workpiece from one position to another, the horizontal gripper has to be moved to the correct position and to be pushed downwards, the compressed air needs to be turned on, then the gripper has to be lifted upwards, to be moved to the target position and has to deposit the workpiece in the same manner. Therefore, a low-level controller converting the high-level commands of the fault-tolerant controller into sequences of vectors of actuator signals and converting sequences of vectors of sensor measurements of the plant into scalar output signals has been developed in [19] and is used in this thesis.

7.2 Models of the Handling System HANS

In this section a model for part of the faultless Handling System HANS is presented. Additionally, several typical fault cases are modeled.

7.2.1 General modeling information

For the experimental evaluation of the fault-tolerant controller $\mathcal{C}_{\mathrm{FTC}}$, a transport process in the Handling System HANS shall be considered. Therefore, different workpiece positions in the Handling System HANS correspond to different states z in the models $\mathcal{A}_f$ $(f \in \mathcal{F} \cup \{0\})$ of the faultless and the faulty plant. In total, the grippers and the conveyor belt can transport workpieces to 33 different discrete positions (see Fig. 7.3). However, in order to keep the models at a presentable size, only 10 of these positions are considered during the experiments. These positions P1, P2, P3, M1, M2, M3, M4, S1, S2 and S3 are marked in green color in Fig. 7.3. Similarly, not all possible transports between these positions are considered. Rather, the number of considered transport options is limited to 15 (see Section 7.2.2 for details).

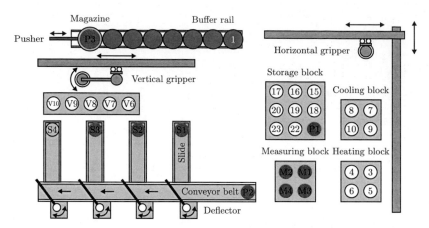

Figure 7.3: Schematic top view of Handling System HANS.

To further limit the complexity of the models, only a single workpiece is considered. That is, the models are only valid if exactly one workpiece occupies one of the 10 aforementioned positions.

7.2.2 Model of the faultless plant

The model $\mathcal{A}_0$ of the faultless Handling System HANS contains 12 states, 15 input symbols and 11 output symbols. As mentioned before, the states $z \in \mathcal{Z}_0$ correspond to the marked positions in Fig. 7.3. According to (2.32) the state set $\mathcal{Z}_0$ of the faultless plant has to include all states that might be reached when some fault $f \in \mathcal{F}$ is present. Therefore, two additional states have to be included into the set $\mathcal{Z}_0$ here. They correspond to two positions on the conveyor belt, where a workpiece is stuck between the first deflector and the first slide or between the second deflector and the second slide, respectively. The description of all 12 states in the state set

$$\mathcal{Z}_0 = \{P1, P2, P3, M1, M2, M3, M4, S1, S2, S3, B1, B2\} \tag{7.1}$$

is summarized in Table 7.1.

The input set

$$\mathcal{V}_0 = \{P1 \rightarrow M1, \ P1 \rightarrow M2, \ P1 \rightarrow M3, \ P1 \rightarrow M4, \ M1 \rightarrow P2, \ M2 \rightarrow P2, \ M3 \rightarrow P2,$$
$$M4 \rightarrow P2, \ P2 \rightarrow S1, \ P2 \rightarrow S2, \ P2 \rightarrow S3, \ S1 \rightarrow P3, \ S2 \rightarrow P3, \ S3 \rightarrow P3\} \tag{7.2}$$

contains 15 input symbols corresponding to the possible transports by the grippers and the

State z	Position of workpiece
P1	On storage block
P2	On conveyor belt
P3	In magazine
M1	Upper right position on measuring block
M2	Upper left position on measuring block
M3	Lower right position on measuring block
M4	Lower left position on measuring block
S1	In first slide
S2	In second slide
S3	In third slide
B1	On conveyor belt before first deflector
B2	On conveyor belt before second deflector

Table 7.1: State set $\mathcal{Z}_0$ of Handling System HANS.

conveyor belt, whose descriptions are given in Table 7.2. The inputs allow for a clockwise transport of workpieces without the possibility to "skip" an element like a block.

The output set

$$\mathcal{W}_0 = \{P1, P2, P3, M1, M2, M3, M4, S1, S2, S3, NP\} \tag{7.3}$$

contains 11 symbols. They correspond to the acknowledgment of the transport of a workpiece to a certain position. Furthermore the output $w = NP$ reflects that the requested action has not been successfully executed. The description of all output symbols is summarized in Table 7.3.

Figure 7.4 shows the model $\mathcal{A}_0$ of the faultless Handling System HANS relating the described states $z \in \mathcal{Z}_0$, input symbols $v \in \mathcal{V}_0$ and output symbols $w \in \mathcal{W}_0$. The labels of the self-loops are omitted for better readability. All self-loops are labeled with all remaining input symbols to make the automaton completely defined (cf. Assumption 2), while the generated output is always $w = NP$. The initial state of the plant is given by

$$z_{00} = P1. \tag{7.4}$$

It can be seen that from position P1 a workpiece can be transported by the horizontal gripper to either one of the four positions M1-M4 on the measuring block. From there the only option is to move the workpiece to position P2 on the conveyor belt. The conveyor belt can transport the workpiece to slide S1, S2 or S3, from where the vertical gripper can lay it into the magazine (P3). In the faultless case the states B1 and B2 corresponding to the blocking of a workpiece in front of a deflector can not be reached.

Input v	Transport of workpiece
P1 $\rightarrow$ M1	Storage block to measuring block
P1 $\rightarrow$ M2	Storage block to measuring block
P1 $\rightarrow$ M3	Storage block to measuring block
P1 $\rightarrow$ M4	Storage block to measuring block
M1 $\rightarrow$ P2	Measuring block to conveyor belt
M2 $\rightarrow$ P2	Measuring block to conveyor belt
M3 $\rightarrow$ P2	Measuring block to conveyor belt
M4 $\rightarrow$ P2	Measuring block to conveyor belt
P2 $\rightarrow$ S1	Conveyor belt to first slide
P2 $\rightarrow$ S2	Conveyor belt to second slide
P2 $\rightarrow$ S3	Conveyor belt to third slide
S1 $\rightarrow$ P3	First slide to magazine
S2 $\rightarrow$ P3	Second slide to magazine
S3 $\rightarrow$ P3	Third slide to magazine
P3 $\rightarrow$ P1	Magazine to storage block

Table 7.2: Input set $\mathcal{V}_0$ of Handling System HANS.

7.2.3 Models of the faulty plant

In Section 2.3 a method for modeling the behavior of the faulty plant by applying error relations $\{E_{zf}, E_{vf}, E_{wf}, \ (f \in \mathcal{F})\}$ to the model $\mathcal{A}_0$ of the faultless plant has been introduced. It allows for a systematic and generic representation of different types of faults. In this example ten different faults

$$f \in \{1, 2, \ldots, 10\} \tag{7.5}$$

are considered, whose descriptions are summarized in Table 7.4. The fault set $\mathcal{F}$ can be partitioned into four kinds of faults:

- Position M1, M2, M3, or M4 on measuring block is blocked ($f \in \{1, 2, 3, 4\}$).

- Position of deflector D1 or D2 toggles ($f \in \{5, 8\}$).

- Deflector D1 or D2 is stuck-closed ($f \in \{6, 9\}$).

- Slide S1 or S2 is blocked ($f \in \{7, 10\}$).

For all faults of the same type the error relations according to Section 2.3.3 are very similar, as will be shown in the following.

Output w	Workpiece has been transported to position
P1	On storage block
P2	On conveyor belt
P3	In magazine
M1	Upper right position on measuring block
M2	Upper left position on measuring block
M3	Lower right position on measuring block
M4	Lower left position on measuring block
S1	In first slide
S2	In second slide
S3	In third slide
NP	Transport failed

Table 7.3: Output set $\mathcal{W}_0$ of Handling System HANS.

Fault f	Description
1	Position M1 on measuring block is blocked
2	Position M2 on measuring block is blocked
3	Position M3 on measuring block is blocked
4	Position M4 on measuring block is blocked
5	Deflector D1 toggles
6	Deflector D1 is stuck-closed
7	Slide S1 is blocked
8	Deflector D2 toggles
9	Deflector D2 is stuck-closed
10	Slide S2 is blocked

Table 7.4: Fault set $\mathcal{F}$ of Handling System HANS.

Blocked position on measuring block. The first type of fault can be seen as a plant fault. Therefore, fault $f = 1$ (position M1 on measuring block is blocked) can be represented by the following error relations:

$$E_{z1}(z, v) = \begin{cases} P1 & \text{if } (z, v) = (P1, P1 \rightarrow M1) \\ G_0(z, v) & \text{otherwise.} \end{cases}$$

$$E_{v1}(v) = v, \qquad \forall v \in \mathcal{V}_0$$

$$E_{w1}(w) = w, \qquad \forall w \in \mathcal{W}_0.$$

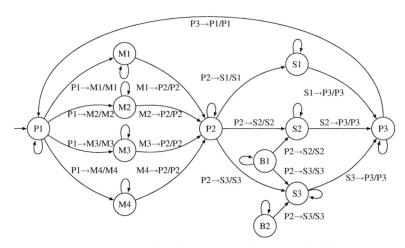

Figure 7.4: Model $\mathcal{A}_0$ of faultless Handling System HANS.

They describe that when a workpiece is at position P1 and a transport from position P1 to position M1 is requested, the workpiece remains at position P1. Faults $f = 2, 3, 4$ can be modeled similarly by replacing M1 by M2, M3 or M4, respectively.

Deflector position toggles. The second type of fault is an actuator fault. Fault $f = 5$ (position of deflector D1 toggles) can be represented by the following error relations:

$$E_{z5}(z, v) = G_0(z, v), \qquad \forall (z, v) \in \mathcal{Z}_0 \times \mathcal{V}_0$$

$$E_{v5}(v) = \begin{cases} \text{P2} \to \text{S1} & \text{if } v \in \{\text{P2} \to \text{S2}, \text{P2} \to \text{S3}\} \\ \text{P2} \to \text{S2} & \text{if } v = \text{P2} \to \text{S1} \\ v & \text{otherwise,} \end{cases}$$

$$E_{w5}(w) = w, \qquad \forall w \in \mathcal{W}_0.$$

In the faultless case a workpiece is transported into slide S1 by closing the first deflector D1. If the position of the first deflector D1 toggles, the workpiece is moved into the second slide S2 instead. Similarly, instead of transporting the workpiece into slide S2 or slide S3, a toggling of deflector D1 leads to a transport into the first slide S1. That is, the fault $f = 5$ interchanges the inputs $v = \text{P2} \to \text{S1}$ and $v = \text{P2} \to \text{S2}$ and replaces the inputs $v = \text{P2} \to \text{S3}$ by the input $v = \text{P2} \to \text{S1}$.

Similarly, the fault $f = 8$ leads to the replacement of the input $v = \text{P2} \rightarrow \text{S2}$ by the input $v = \text{P2} \rightarrow \text{S3}$ and vice versa.

Deflector stuck-closed.   The third type of fault can be seen as an actuator fault as well. Therefore, fault $f = 6$ (deflector D1 is stuck-closed) can be represented by the following error relations:

$$E_{z6}(z,v) = G_0(z,v), \qquad \forall (z,v) \in \mathcal{Z}_0 \times \mathcal{V}_0$$

$$E_{v6}(v) = \begin{cases} \text{P2} \rightarrow \text{S1} & \text{if } v \in \{\text{P2} \rightarrow \text{S2}, \text{P2} \rightarrow \text{S3}\} \\ v & \text{otherwise,} \end{cases}$$

$$E_{w6}(w) = w, \qquad \forall w \in \mathcal{W}_0.$$

In the faultless case, the inputs $v \in \{\text{P2} \rightarrow \text{S2}, \text{P2} \rightarrow \text{S3}\}$ lead to the transport of a workpiece from position P2 into slide S2 or slide S3 by opening the first deflector and closing the second or third deflector, respectively. When fault $f = 6$ is present such that the first deflector can not be opened, the workpiece is transported into slide S1 instead. Consequently, fault $f = 6$ can be modeled as replacing these inputs by the input $v = \text{P2} \rightarrow \text{S1}$.

Fault $f = 9$ can be modeled similarly with the difference that here only input $v = \text{P2} \rightarrow \text{S3}$ is replaced by input $v = \text{P2} \rightarrow \text{S2}$:

$$E_{z9}(z,v) = G_0(z,v), \qquad \forall (z,v) \in \mathcal{Z}_0 \times \mathcal{V}_0$$

$$E_{v9}(v) = \begin{cases} \text{P2} \rightarrow \text{S2} & \text{if } v = \text{P2} \rightarrow \text{S3} \\ v & \text{otherwise,} \end{cases}$$

$$E_{w9}(w) = w, \qquad \forall w \in \mathcal{W}_0.$$

Blocked slide.   The fourth type of fault can be seen as a plant fault. Fault $f = 7$ (slide S1 is blocked) can be represented by the following error relations:

$$E_{z7}(z,v) = \begin{cases} \text{B1} & \text{if } (z,v) = (\text{P2}, \text{P2} \rightarrow \text{S1}), \\ G_0(z,v) & \text{otherwise.} \end{cases} \tag{7.6}$$

$$E_{v7}(v) = v, \qquad \forall v \in \mathcal{V}_0 \tag{7.7}$$

$$E_{w7}(w) = w, \qquad \forall w \in \mathcal{W}_0. \tag{7.8}$$

That is, if a workpiece lies at position $z = \text{P2}$ on the conveyor belt and a transport into slide S1 is requested ($v = \text{P2} \rightarrow \text{S1}$), the workpiece gets stuck on the conveyor belt before the first deflector ($z' = \text{B1}$).

Fault $f = 10$ (slide S2 is blocked) can be modeled similarly by replacing B1 by B2 and P2 → S1 by P2 → S2.

The models $\mathcal{A}_f$, ($f \in \{1, 2, \ldots, 10\}$) of the faulty plant are obtained by applying the error relations $\{E_{zf}, E_{vf}, E_{wf}, (f \in \mathcal{F})\}$ to the model $\mathcal{A}_0$ of the faultless plant according to (2.28). For the Handling System HANS it turns out that the generated output $w(k)$ depends on the current state $z(k)$ and the next state $z(k + 1)$, but not on the input $v(k)$. For example, if deflector D1 is stuck-closed ($f = 6$), and a transport from position P2 into slide S2 is requested ($v = \text{P2} \rightarrow \text{S2}$), the output $w = \text{S1}$ will be generated, because the workpiece ends up in slide S1 instead. That is, even though a different input is applied, the output for the transition from state P2 to state S1 is the same as in the faultless case. This rule is applicable to all other transitions and faults as well. Therefore, the output functions H_f, ($f \in \{1, 2, \ldots, 10\}$) of the faulty plant can be computed by (2.39). Applying (2.39) exemplarily to the transition described above, the following value for the output function corresponding to fault $f = 6$ results:

$$
\begin{aligned}
H_6(\text{P2}, \text{P2} \rightarrow \text{S2}) &= E_{\text{w6}}(H_0(\text{P2}, \tilde{v})), \qquad \text{if } (\exists \tilde{v} \in \mathcal{V}_0) \, G_6(\text{P2}, \text{P2} \rightarrow \text{S2}) = G_0(\text{P2}, \tilde{v}) \\
&= E_{\text{w6}}(H_0(\text{P2}, \text{P2} \rightarrow \text{S1})) \\
&= E_{\text{w6}}(\text{S1}) \\
&= \text{S1}.
\end{aligned}
$$

The resulting automata graphs can be found in Figs. C.1–C.9 in the appendix. Exemplarily, the automaton graph of the automaton $\mathcal{A}_7$ modeling the behavior of the Handling System HANS when slide S1 is blocked is shown in Fig. 7.5. Compared to the model $\mathcal{A}_0$ of the faultless plant in Fig. 7.4 two changes occur. The red bold edge leading from state $z = \text{P2}$ to state $z = \text{B1}$ appears because of the fault, while the transition from state $z = \text{P2}$ to state $z = \text{S1}$ is removed. Furthermore, the initial state z_{70} of the model $\mathcal{A}_7$ is unknown.

7.3 Behavior in the faultless case

This section deals with the control of the Handling System HANS when no fault is present and no safety constraints have to be considered. First, the controllability of the model $\mathcal{A}_0$ of the faultless plant is analyzed. Then, the nominal tracking controller $\mathcal{C}_T$ is defined based on this model. Finally, the behavior of the closed-loop system is considered.

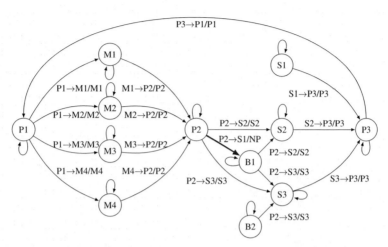

Figure 7.5: Model $\mathcal{A}_7$ of Handling System HANS with blocked slide S1.

7.3.1 Controllability analysis

The controllability of the model $\mathcal{A}_0$ of the faultless plant is analyzed using Theorem 3.1. In the automaton graph of $\mathcal{A}_0$ in Fig. 7.4 it can be seen that from the initial state $z_0 = \text{P1}$ in (7.4) there exists a path to almost every other state $z \in \mathcal{Z}_0$. The only exception are the two states B1 and B2 which correspond to the blocking of a workpiece on the conveyor belt. According to Definition 3.2, these paths are admissible state sequences $Z(0 \ldots k_e)$ with $z(0) = z_0 = \text{P1}$ and $z(k_e + 1) = z$, $z \in \mathcal{Z}_0 \setminus \{\text{B1}, \text{B2}\}$ for the automaton $\mathcal{A}_0$. Consequently, Theorem 3.1 states that the faultless Handling System HANS is controllable with respect to any final state

$$z_F \in \mathcal{Z}_0 \setminus \{\text{B1}, \text{B2}\}. \tag{7.9}$$

7.3.2 Definition of the tracking controller

The trajectory planning unit $\mathcal{T}(\mathcal{A}_0)$ and the controller $\mathcal{C}(\mathcal{A}_0)$ of the tracking controller $\mathcal{C}_T(\mathcal{A}_0)$ are defined according to Section 3.4 and Section 3.3, respectively.

Definition of the trajectory planning unit $\mathcal{T}(\mathcal{A}_0)$. According to (3.18), the model $\mathcal{A}_0$ of the faultless plant shown in Fig. 7.4 with initial state $z_0 = \text{P1}$ is stored in the trajectory planning unit $\mathcal{T}(\mathcal{A}_0)$. Furthermore, a Breadth-first-search algorithm is implemented in order to search for a path in the automaton graph of $\mathcal{A}_0$ at runtime.

Definition of the controller $\mathcal{C}(\mathcal{A}_0)$. The controller $\mathcal{C}(\mathcal{A}_0)$ is defined using Algorithm 3.1. Based on the given model $\mathcal{A}_0$ of the faultless plant in Fig. 7.4, the state set, input set, output set and initial state of the controller $\mathcal{C}(\mathcal{A}_0)$ are initialized as

$$\mathcal{Z}_c = \mathcal{Z}_0 = \{P1, P2, P3, M1, M2, M3, M4, S1, S2, S3, B1, B2\} \tag{7.10}$$

$$\mathcal{V}_c = \mathcal{Z}_0 = \{P1, P2, P3, M1, M2, M3, M4, S1, S2, S3, B1, B2\} \tag{7.11}$$

$$\mathcal{W}_c = \mathcal{V}_0 = \{P1 \rightarrow M1, \ P1 \rightarrow M2, \ P1 \rightarrow M3, \ P1 \rightarrow M4, \ M1 \rightarrow P2, \ M2 \rightarrow P2, \tag{7.12}$$
$$M3 \rightarrow P2, \ M4 \rightarrow P2, \ P2 \rightarrow S1, \ P2 \rightarrow S2, \ P2 \rightarrow S2, \ S1 \rightarrow P3,$$
$$S2 \rightarrow P3, \ S3 \rightarrow P3\}$$

$$z_{c0} = z_{00} = P1, \tag{7.13}$$

respectively.

Then, for all states $z \in \mathcal{Z}_0$ and inputs $v \in \mathcal{V}_0$ of the Handling System HANS Lines 3, 5 and 6 of Algorithm 3.1 are executed. For example, for state $z = P1$ and input $v = P1 \rightarrow M1$ the state transition function and output function of the controller are given by

$$G_c(P1, M1) = M1 \tag{7.14}$$

$$H_c(P1, M1) = P1 \rightarrow M1, \tag{7.15}$$

because

$$z' = G_0(P1, P1 \rightarrow M1) = M1.$$

The resulting controller $\mathcal{C}(\mathcal{A}_0)$ is shown in Fig. 7.6. It can be seen that it has the same structure as the model $\mathcal{A}_0$ of the faultless plant (cf. Proposition 3.2).

7.3.3 Nominal control of the Handling System HANS

The nominal tracking controller $\mathcal{C}_T(\mathcal{A}_0) = (\mathcal{T}(\mathcal{A}_0), \mathcal{C}(\mathcal{A}_0))$ is used to control the faultless Handling System HANS within the setting shown in Fig. 3.1. Let the desired final state for the Handling System HANS be given by

$$z_F = P3. \tag{7.16}$$

That is, it is desired to transport a workpiece to the position P3 in the magazine (cf. Fig. 7.3).

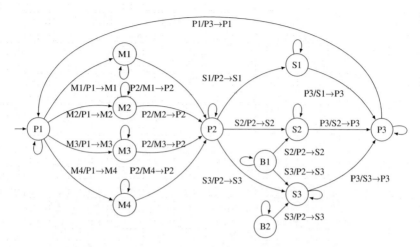

Figure 7.6: Nominal controller $\mathcal{C}(\mathcal{A}_0)$ for Handling System HANS.

The trajectory planning unit $\mathcal{T}(\mathcal{A}_0)$ searches for a path from the initial state $z_0 = \mathrm{P1}$ to the desired final state $z_\mathrm{F} = \mathrm{P3}$ in the automaton graph of the model $\mathcal{A}_0$ of the faultless plant in Fig. 7.4. It can be seen that one shortest path corresponds to the state sequence

$$Z(0\ldots4) = (\mathrm{P1}, \mathrm{M1}, \mathrm{P2}, \mathrm{S1}, \mathrm{P3}).$$

Consequently, the reference trajectory specified by the trajectory planning unit $\mathcal{T}(\mathcal{A}_0)$ is

$$R(0\ldots3) = (\mathrm{M1}, \mathrm{P2}, \mathrm{S1}, \mathrm{P3}). \tag{7.17}$$

In the following, the behavior of the closed-loop system is simulated, where the results are visualized in Fig. 7.7. The first subplot shows the value $r(k)$ of the reference trajectory at every time step, while the second to fourth plot show the input $v_\mathrm{p}(k)$, current state $z_\mathrm{p}(k)$ and output $w_\mathrm{p}(k)$ of the plant $\mathcal{P}$, respectively. The last subplot shows the distance from the current state $z_\mathrm{p}(k)$ to the desired final state $z_\mathrm{F} = \mathrm{P3}$.

The simulation shows that the controller $\mathcal{C}(\mathcal{A}_0)$ in Fig. 7.6 translates the reference trajectory $R(0\ldots3)$ in (7.17) into the input sequence

$$V_\mathrm{p}(0\ldots3) = (\mathrm{P1} \to \mathrm{M1}, \mathrm{M1} \to \mathrm{P2}, \mathrm{P2} \to \mathrm{S1}, \mathrm{S1} \to \mathrm{P3}) \tag{7.18}$$

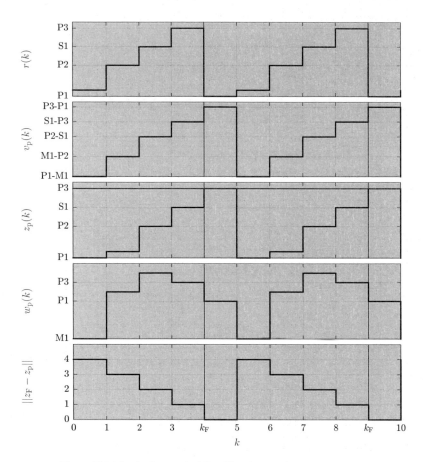

Figure 7.7: Nominal control of Handling System HANS (simulation).

for the plant $\mathcal{P}$. That means, a workpiece is transported from position P1 on the storage block to position M1 on the measuring block by the horizontal gripper and then to position P2 on the conveyor belt. Next, deflector D1 is closed such that the workpiece is transported by the conveyor belt into slide S1. Finally, the vertical gripper puts the workpiece into the magazine.

Consequently, the desired final state $z_F = P3$ is reached by the Handling System HANS at time $k_F = 4$. The resulting output sequence is given by

$$W_p(0\ldots 3) = \Phi_0(P1, (P1 \rightarrow M1, M1 \rightarrow P2, P2 \rightarrow S1, S1 \rightarrow P3)) = (M1, P2, S1, P3).$$
$$(7.19)$$

In order to obtain a cyclic behavior of the Handling System HANS, the desired final state is changed to $z_F = P1$ and then back to $z_F = P3$. As a result the workpiece is transported back to its initial position P1 and the process starts over. The desired final state $z_F = P3$ is reached again at time $k_F = 9$.

Figure 7.8 shows the results of an experiment conducted at the Handling System HANS with the same settings as in the simulation. During the experiment the state $z_p(k)$ of the plant $\mathcal{P}$ can not be measured. However, the course of its input and output signal indicate that the state of the Handling System HANS follows the same trajectory as in the simulation. It can be seen that the first output $w_p(0) = M1$ is generated at time $t = 16s$, when the transport of a workpiece to position M1 is completed. Immediately afterwards the next input $v_p(1) = M1 \to P2$ for the plant $\mathcal{P}$ is generated. That is, in the experiments the output appears to be delayed by one discrete time step with respect to the input. Furthermore, discrete time steps do not map to equidistant time instances. For example, while the transport of the workpiece from position P2 on the conveyor belt into slide S1 at time step $k = 2$ only requires about $8s$, the transport from slide S1 into the magazine at time step $k = 3$ requires almost $50s$.

From the output $w_p(t = 90s) = P3$ it can be seen that the Handling System HANS reaches the desired final state $z_F = P3$ at time $t_F = 90s$ which corresponds to the discrete time step $k_F = 4$.

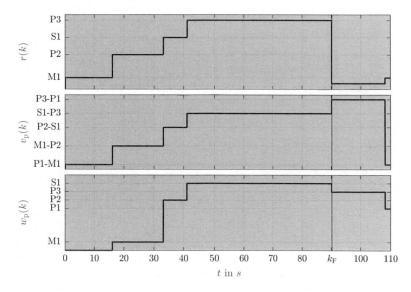

Figure 7.8: Nominal control of Handling System HANS (experiment).

7.4 Recoverability analysis

The recoverability of the closed-loop system $\mathcal{P} \circ \mathcal{C}_T(\mathcal{A}_0)$ is analyzed as described in Section 6.2.1. As mentioned before, there only exist sufficient conditions for the recoverability, but no necessary ones (Theorems 6.1 and 6.2). Both depend on the detectability and identifiability of the faults $f \in \mathcal{F}$ and the reconfigurability of the tracking controller $\mathcal{C}_T(\mathcal{A}_0)$ (cf. Corollaries 6.1 and 6.2).

Detectability. The first step of the analysis reveals that most faults $f \in \mathcal{F}$ are not detectable with respect to the reference trajectory $R(0\ldots 3)$ in (7.17). For faults $f \in \{2, 3, 4, 6, 8, 9, 10\}$ the same output sequence

$$\Phi_f(z_p(k_f), V_p(k_f \ldots k_f^*)) = \Phi_0(z_p(k_f), V_p(k_f \ldots k_f^*)), \; k_f^* \geq k_f \tag{7.20}$$

as in the faultless case will be generated such that (4.10) is never fulfilled. Faults affecting position M2, M3 or M4 will not influence the nominal operation since only position M1 is used. Similarly, faults concerning deflector D2 or slide S2 have no influence, because in nominal operation always slide S1 is used. It is, however, noteworthy that none of these faults leads to the violation of the control aim (2.44) for the desired final state $z_F = $ P3. That is, when one of the above faults is present in the Handling System HANS, the nominal tracking controller $\mathcal{C}_T(\mathcal{A}_0)$ steers the faulty plant $\mathcal{P} \vDash \mathcal{A}_f$ into the desired final state z_F anyhow. If another desired final state $z_F \neq $ P3 is given, the faults might have an impact on the behavior of the plant, but then they also become detectable immediately.

For each of the faults $f \in \{1, 5, 7\}$ under the assumption that $k_f = 0$, a time $k_f^* \leq k_e = 3$ can be found such that (4.10) is fulfilled:

$$\Phi_1(\text{P1}, \text{P1} \rightarrow \text{M1}) = \text{NP}$$
$$\neq \Phi_0(\text{P1}, \text{P1} \rightarrow \text{M1}) = \text{M1} \tag{7.21}$$

$$\Phi_5(\text{P1}, (\text{P1} \rightarrow \text{M1}, \text{M1} \rightarrow \text{P2}, \text{P2} \rightarrow \text{S1})) = (\text{M1}, \text{P2}, \text{S2})$$
$$\neq \Phi_0(\text{P1}, (\text{P1} \rightarrow \text{M1}, \text{M1} \rightarrow \text{P2}, \text{P2} \rightarrow \text{S1})) = (\text{M1}, \text{P2}, \text{S1}) \tag{7.22}$$

$$\Phi_7(\text{P1}, (\text{P1} \rightarrow \text{M1}, \text{M1} \rightarrow \text{P2}, \text{P2} \rightarrow \text{S1})) = (\text{M1}, \text{P2}, \text{NP})$$
$$\neq \Phi_0(\text{P1}, (\text{P1} \rightarrow \text{M1}, \text{M1} \rightarrow \text{P2}, \text{P2} \rightarrow \text{S1})) = (\text{M1}, \text{P2}, \text{S1}) \tag{7.23}$$

That is, these faults are detectable with $k_f^* = 0$ for $f = 1$ and $k_f^* = 2$ for $f \in \{5, 7\}$.

Identifiability. Next, the identifiability of the detectable faults $f \in \{1, 5, 7\}$ is analyzed. First, passive diagnosis is considered. For $f = 1$, the reference trajectory $R(0\ldots 3)$ in (7.17) is no preset homing sequence for the overall model $\bar{A}_\Delta$ of the closed-loop system as defined in (4.14) with respect to the initial uncertainty $\mathcal{Z}^* = \{P1\} \times \mathcal{Z}_\Delta$. A counterexample is

$$G_\Delta \left(\left(\begin{pmatrix} P1 \\ P1 \\ 2 \end{pmatrix} \right), (M1, P2, S1, P3) \right) = \begin{pmatrix} P3 \\ P3 \\ 2 \end{pmatrix}$$

$$\neq G_\Delta \left(\left(\begin{pmatrix} P1 \\ P1 \\ 3 \end{pmatrix} \right), (M1, P2, S1, P3) \right) = \begin{pmatrix} P3 \\ P3 \\ 3 \end{pmatrix},$$

$$\text{but } \Phi_\Delta \left(\left(\begin{pmatrix} P1 \\ P1 \\ 2 \end{pmatrix} \right), (M1, P2, S1, P3) \right) = (M1, P2, S1, P3)$$

$$= \Phi_\Delta \left(\left(\begin{pmatrix} P1 \\ P1 \\ 3 \end{pmatrix} \right), (M1, P2, S1, P3) \right).$$

Neither is the last part $R(2\ldots3)$ of the reference trajectory in (7.17) a preset homing sequence for the overall model $\bar{A}_\Delta$ of the closed-loop system with respect to the initial uncertainty $\mathcal{Z}^* = \{P2\} \times \mathcal{Z}_\Delta$. Consequently, no fault $f \in \mathcal{F}$ is identifiable with passive diagnosis and hence, according to Theorem 4.1, the Handling System HANS is not diagnosable with passive diagnosis.

To analyze the diagnosability of the Handling System HANS with active diagnosis, the compatibility partitions of all state sets $\{z_c\} \times \mathcal{Z}_\Delta$, $z_c \in \mathcal{Z}_c$ are constructed as described in Section 4.3.4. It turns out that there exist multiple compatible state pairs, for example,

$$\left(\begin{pmatrix} S3 \\ B2 \\ 4 \end{pmatrix} \right) \sim \left(\begin{pmatrix} S3 \\ B2 \\ 5 \end{pmatrix} \right). \tag{7.24}$$

That is, by using the nominal tracking controller $\mathcal{C}_T(\mathcal{A}_0)$ it is not possible to distinguish between fault $f = 4$ and fault $f = 5$ if the current state of the controller is $z_c(k_f^*) = S3$ and the current state of the Handling System HANS is $z_p(k_f^*) = B2$ at the fault detection time k_f^*. However, neither when fault $f = 4$ nor when fault $f = 5$ occurs, the plant can ever reach the state $z = B2$ corresponding to the blocking of a workpiece before the second slide (cf. Table 7.4).

Another pair of compatible states is

$$\left(\begin{array}{c} P2 \\ \left(\begin{array}{c} P2 \\ 7 \end{array}\right) \end{array}\right) \sim \left(\begin{array}{c} P2 \\ \left(\begin{array}{c} B1 \\ 7 \end{array}\right) \end{array}\right). \tag{7.25}$$

Both, states $z_p = P2$ and $z_p = B1$ can be reached when fault $f = 7$ occurs, that is, when slide S1 is blocked. However, when the input $v_c = S1$ is given to the controller $C(A_0)$ in Fig. 7.6 in this situation, it will try to move a workpiece from position P2 on the conveyor belt into the first slide. As a result, from both states above a transition into state

$$z' = \left(\begin{array}{c} S1 \\ \left(\begin{array}{c} B1 \\ 7 \end{array}\right) \end{array}\right) \tag{7.26}$$

with output $w_p = NP$ is performed. That is, even though the two states in (7.25) are compatible, a homing sequence can be applied such that the current state is identified.

An analysis of all sets of compatible states reveals that one of both situations always applies. That is, either some of the states within the set are not reachable or a homing sequence can be found besides the compatibility. Hence, in this example it makes sense to drop the simplification of including all states $z \in Z_\Delta$ into the initial uncertainty Z^* (cf. Section 4.4.2). Rather, only physically reachable states should be considered. Then the Handling System HANS is diagnosable with active diagnosis according to Theorem 4.2 even though the sufficient condition for identifiability in Proposition 4.8 is not fulfilled.

Reconfigurability. Finally, the reconfigurability of the tracking controller $C_T(A_0)$ is analyzed. That is, for all automata A_f, $(f \in \mathcal{F})$ sets of reachable states are determined (cf. Theorem 5.1). In general only one or two states become unreachable when a fault $f \in \mathcal{F}$ occurs, while all other states z_F in (7.9) that are reachable in the faultless case remain reachable. For example, if a position on the measuring block is blocked ($f \in \{1, 2, 3, 4\}$), the corresponding state M1, M2, M3 or M4 becomes unreachable in the automaton A_1, A_2, A_3 or A_4, respectively. Table 7.5 shows all reachable states for all automata A_f, $(f \in \mathcal{F})$. That is, all states z_F are displayed with respect to which the tracking controller $C_T(A_0)$ is reconfigurable given that fault $f \in \mathcal{F}$ occurred.

Recoverability. The analysis shows that the closed-loop system $\mathcal{P} \circ C_T(A_0)$ consisting of the Handling System HANS and the tracking controller in Section 7.3.2 is recoverable from all

f	P1	P2	P3	M1	M2	M3	M4	S1	S2	S3	B1	B2
						Reachable states						
1	✓	✓	✓		✓	✓	✓	✓	✓	✓		
2	✓	✓	✓	✓		✓	✓	✓	✓	✓		
3	✓	✓	✓	✓	✓		✓	✓	✓	✓		
4	✓	✓	✓	✓	✓	✓		✓	✓	✓		
5	✓	✓	✓	✓	✓	✓	✓	✓	✓			
6	✓	✓	✓	✓	✓	✓	✓	✓				
7	✓	✓	✓	✓	✓	✓	✓		✓	✓	✓	
8	✓	✓	✓	✓	✓	✓	✓	✓	✓	✓		
9	✓	✓	✓	✓	✓	✓	✓	✓	✓			
10	✓	✓	✓	✓	✓	✓	✓	✓		✓		✓

Table 7.5: Reachable states in automata $\mathcal{A}_f$, $(f \in \mathcal{F})$.

faults $f \in \{1, 2, \ldots, 10\}$. Depending on the present fault $\bar{f}$ the recoverability is limited to the desired final states z_F listed in Table 7.5. Also, it has been shown that active fault diagnosis has to be used in order to identify the present fault $\bar{f}$ and the current state $z_p(k)$ of the faulty Handling System HANS.

7.5 Fault-tolerant control with unambiguous diagnostic result

In this section the behavior of the fault-tolerant control loop is analyzed for the case that the termination condition for the fault diagnosis is an unambiguous diagnostic result. That is, active fault diagnosis is used until the present fault $\bar{f}$ and the current state $z_p(k_d)$ of the Handling System HANS are known. First, the behavior of the fault-tolerant control loop is presented, whereas afterwards the active fault diagnosis and the reconfiguration of the nominal tracking controller $\mathcal{C}_T(\mathcal{A}_0)$ are discussed in detail.

7.5.1 Behavior of the fault-tolerant control loop

Consider the case that fault $\bar{f} = 7$ occurs at the fault occurrence time $k_f = 0$, that is, slide S1 is blocked when the process starts. Figure 7.9 shows the simulated behavior of the fault-tolerant control loop. It can be seen that a fault is detected at the fault detection time $k_f^* = 2$. That is, the fault is detected two time steps after its occurrence. This delay results from the fact that the blocked slide S1 is not used until time $k = 2$ (cf. Fig. 7.7).

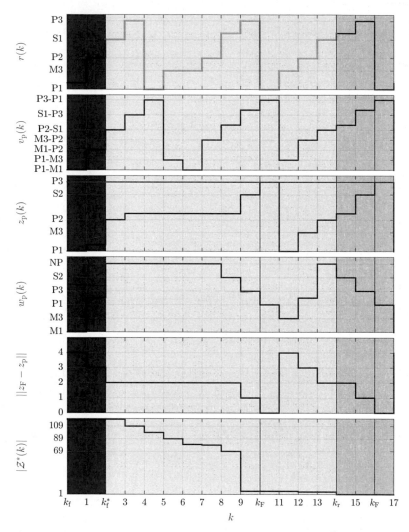

Figure 7.9: Fault-tolerant control of Handling System HANS with fault $\bar{f} = 7$ and unambiguous diagnostic result (simulation).

From the fault detection time until the diagnosis time $k_{\mathrm{d}} = 14$ active diagnosis is used to identify the present fault $\bar{f}$ and the current state $z_{\mathrm{p}}(k_{\mathrm{d}})$ of the plant $\mathcal{P}$. Once the fault diagnosis is completed, the nominal tracking controller $\mathcal{C}_{\mathcal{T}}(\mathcal{A}_0)$ is reconfigured at the reconfiguration time $k_{\mathrm{r}} = k_{\mathrm{d}} = 14$. After the reconfiguration, the reference trajectory

$$R(14\ldots15) = (\mathrm{S2}, \mathrm{P3}) \tag{7.27}$$

is generated by the reconfigured trajectory planning unit $\mathcal{T}^{\mathrm{r}}$. The Handling System HANS reaches the desired final state at time $k_{\mathrm{F}} = 16$. Therefore, as stated in Theorem 6.3, the safe fault-tolerant control problem in Problem 2.1 is solved.

Figure 7.10 shows the results of the corresponding experiment. As expected, the sequences of I/O symbols of the Handling System HANS are the same as in the simulation. A fault is detected at time $t_{\mathrm{f}}^* = 50s$, while it is identified at time $t_{\mathrm{d}} = 262s$. After the reconfiguration, the workpiece arrives at position P3 in the magazine at time $t_{\mathrm{F}} = 327s$.

7.5.2 Fault diagnosis

After the fault detection time $k_{\mathrm{f}}^* = 2$, active diagnosis as described in Chapter 4 is used to identify the present fault $\bar{f}$ and the current state $z_{\mathrm{p}}(k_{\mathrm{d}})$ of the Handling System HANS. That is, the current state estimate $\mathcal{Z}^*(k)$ of the plant is updated in every time step and separating sequences are applied to the plant in order to reduce the size of the current state estimate. Table 7.6 shows the separating sequences computed at different time steps.

Time step k	Separating sequence $V_{\mathrm{S}}(0 \ldots k_{\mathrm{e}})$
3	$(\mathrm{P3}, \mathrm{P1}, \mathrm{M3})$
6	$(\mathrm{M3})$
7	$(\mathrm{P2}, \mathrm{S2})$
9	$(\mathrm{P3}, \mathrm{P1}, \mathrm{M3})$
12	$(\mathrm{P2}, \mathrm{S1})$

Table 7.6: Separating sequences.

For example, at time $k = 3$ the diagnostic reference signal $r(3) = \mathrm{P3}$ is given to the controller $\mathcal{C}(\mathcal{A}_0)$. As a result, the controller requests the Handling System HANS to transport a workpiece from the first slide into the magazine ($v_{\mathrm{p}}(3) = \mathrm{S1} \to \mathrm{P3}$). However, due to the present fault $\bar{f} = 7$, the workpiece lies not in slide S1, but is blocked before it. Therefore, the output $w_{\mathrm{p}}(3) = \mathrm{NP}$ is generated. From this output and the models $\mathcal{A}_f$, ($f \in \mathcal{F}$) of the faulty plant it can be deduced that the workpiece was not in the first slide at time $k = 3$. The application of the remaining separating sequences leads to a further reduction of the current state estimate $\mathcal{Z}^*(k)$

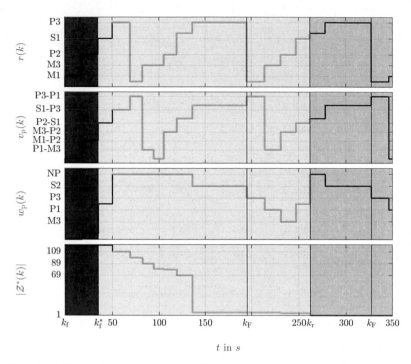

Figure 7.10: Fault-tolerant control of Handling System HANS with fault $\bar{f} = 7$ and unambiguous diagnostic result (experiment).

of the plant. It can be seen that most separating sequences consist of more than one element such that it is not necessary to compute a new separating sequence in every time step.

In Fig. 7.11 the resulting current state estimate $\mathcal{Z}^*(k)$ of the plant is broken down into the set $\mathcal{F}^*(k)$ of fault candidates and the plant states z_p it includes. It can be seen that from time step $k = 9$ the current state $z_\mathrm{p}(k)$ of the Handling System HANS is known, while the set $\mathcal{F}^*(k)$ of fault candidates first becomes a singleton at time step $k = 14$. That is, the unambiguous diagnostic result

$$\mathcal{D}^* = \left\{ \begin{pmatrix} \mathrm{B1} \\ 7 \end{pmatrix} \right\} \tag{7.28}$$

is obtained at the diagnosis time $k_\mathrm{d} = 14$ even though the current state of the Handling System is known unambiguously before. Fault $f = 3$ is excluded from the set of fault candidates at time

$k = 12$, because the workpiece is successfully transported to position M3 on the measuring block. The present fault $\bar{f} = 7$ is not identified before time $k = 14$, because it can only be distinguished from the other remaining fault candidates

$$f \in \mathcal{F}^*(6) \setminus \{7\} = \{1, 2, 4, 9\}$$

when the transport of the workpiece from position P2 on the conveyor belt into the first slide fails.

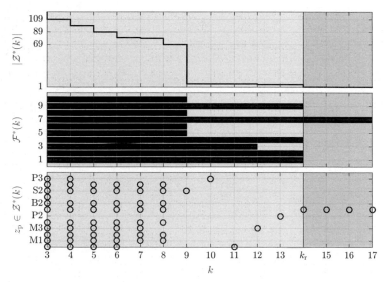

Figure 7.11: Active fault diagnosis of Handling System HANS with fault $\bar{f} = 7$ (simulation).

At time $k = 10 < k_\mathrm{d} = 14$ the plant $\mathcal{P}$ reaches the desired final state $z_\mathrm{F} = $ P3 (cf. Fig. 7.9). That is, a workpiece lies at position P3 in the magazine. However, the process is not stopped at this point, because the termination condition is not fulfilled yet. Rather, active fault diagnosis is continued (with a new workpiece) until the termination condition $|\mathcal{Z}^*(k)| = 1$ is fulfilled at the diagnosis time $k_\mathrm{d} = 14$. The continuation of the fault diagnosis allows for a reconfiguration of the tracking controller such that the transport of another workpiece into the magazine or the achievement of a different desired final state becomes possible.

7.5.3 Reconfiguration of the tracking controller

At the diagnosis time $k_d = 14$, the nominal tracking controller $\mathcal{C}_T(\mathcal{A}_0)$ is reconfigured as described in Chapter 5. That is, the automaton graph stored in the nominal trajectory planning unit $\mathcal{T}(\mathcal{A}_0)$ and transitions in the nominal controller $\mathcal{C}(\mathcal{A}_0)$ are modified according to Algorithm 5.2. The reconfiguration is based on the unambiguous diagnostic result $\mathcal{D}^*$ in (7.28) and the corresponding error relations E_{z7}, E_{v7} and E_{w7} in (7.6)–(7.8).

Reconfiguration of the trajectory planning unit. The automaton graph of the model $\mathcal{A}_0$ of the faultless plant in Fig. 7.4 stored in the trajectory planning unit is modified according to (5.1). That is, faulty transitions are modified by the error relations E_{z7} and E_{v7} corresponding to the identified fault $\bar{f} = 7$. Here, only the transition from state $z = \text{P2}$ to state $z' = \text{S1}$ is affected, while all other transitions and all outputs remain unchanged. Furthermore, the initial state of $\mathcal{A}_0$ is changed to the current state $z_p(k_d) = \text{B1}$ of the faulty plant. The resulting automaton graph $\mathcal{A}_0^r$ stored in the reconfigured trajectory planning unit $\mathcal{T}^r$ is shown in Fig. 7.12. It resembles the model $\mathcal{A}_7$ of the faulty plant in Fig. 7.5, but has a different output at the transition from state $z = \text{P2}$ to state $z' = \text{B1}$. For the desired final state $z_F = \text{P3}$, the reconfigured trajectory planning unit $\mathcal{T}^r$ generates the reference trajectory in (7.27).

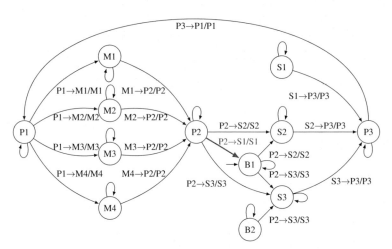

Figure 7.12: Automaton graph $\mathcal{A}_0^r$ in reconfigured trajectory planning unit $\mathcal{T}^r$.

Reconfiguration of the controller. In order to obtain the reconfigured controller $\mathcal{C}^r$, Algorithm 5.1 is applied to the nominal controller $\mathcal{C}(\mathcal{A}_0)$ in Fig. 7.6. The resulting reconfigured controller $\mathcal{C}^r$ is shown in Fig. 7.13. Dashed lines represent transitions that are removed during the reconfiguration because of (5.12) and (5.13). In this example the only pruned transition is the one from state $z = \text{P2}$ to state $z' = \text{S1}$. The set $\mathcal{E}_{\text{new}}$ of transitions to be newly defined is computed using (5.19) as follows:

$$\mathcal{E}_{\text{new}} = \Big\{ (z_c, z_c') \in \mathcal{Z}_c \times \mathcal{Z}_c :$$
$$\Big[(\exists w \in \mathcal{W}_c) \Big(\big[H_c(z_c, z_c') = w \big] \wedge \big[E_{z7}(z_c, E_{v7}(w)) \neq z_c' \big] \Big) \Big]$$
$$\vee \Big[\big((\not\exists w \in \mathcal{W}_c) H_c(z_c, z_c') = w \big) \wedge \big((\exists w \in \mathcal{W}_c) E_{z7}(z_c, E_{v7}(w)) = z_c' \big) \Big] \Big\}$$
$$= \Big\{ \underbrace{(\text{P2}, \text{S1})}_{\mathcal{E}_{\text{new},-}}, \underbrace{(\text{P2}, \text{B1})}_{\mathcal{E}_{\text{new},+}} \Big\}.$$

It contains the removed transition as well as the one transition that newly occurs in the plant $\mathcal{P}$ due to the present fault $\bar{f} = 7$. Finally, (5.20d) and (5.20e) are used to complete the state transition function and the output function, of the reconfigured controller based on the set $\mathcal{E}_{\text{new}}$. When slide S1 is blocked, there is no alternative input which transports a workpiece from position P2 on the conveyor belt into slide S1. Therefore, the first element in the set $\mathcal{E}_{\text{new}}$ does not lead to a new transition in the reconfigured controller $\mathcal{C}^r$. However, corresponding to the newly occurring transition from position P2 on the conveyor belt to the blocked position B1 before slide S1, a new transition is included into the reconfigured controller $\mathcal{C}^r$. In Fig. 7.13 the resulting new transition from state $z = \text{P2}$ to state $z' = \text{B1}$ is marked by a bold green arrow.

7.5.4 Time consumption

The presented fault-tolerant control method proves to be effective, both, in simulations and experiments with the Handling System HANS. The size of the considered models with $n = 12$ states, $p = 15$ input symbols, $q = 11$ output symbols and $F = 10$ faults is manageable with a standard PC with $4\,\text{GB}$ RAM. As expected, the most time consuming task is the computation of the equivalence partition, which takes about $650s$. However, this computation can be performed offline. The time consumption for the definition of the nominal tracking controller $\mathcal{C}_T(\mathcal{A}_0)$ is with less than $0.01s$ negligible.

During the online execution, the computation of the separating sequences during the active diagnosis is most time consuming. In this example the computation times for this task varies between $0.1s$ and $3.8s$. The considered transport process is not time critical such that these delays, which are in the same order as the transport times themselves, are tolerable. The time

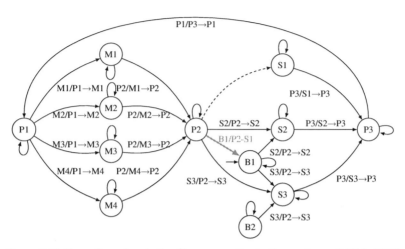

Figure 7.13: Reconfigured controller $\mathcal{C}^r$ corresponding to diagnostic result $\mathcal{D}^*$ in (7.28).

for the reconfiguration of the tracking controller is, just as the time for its definition, very small. In summary, the time consumption for the computations within the fault-tolerant controller does not affect its functionality.

7.6 Fault-tolerant control with ambiguous diagnostic result

In this section the behavior of the fault-tolerant control loop is analyzed within the same setting as before, i.e., slide S1 is blocked when the process is started which means that fault $\bar{f} = 7$ occurs at time $k_f = 0$. The difference to the previously discussed case is that now the termination condition for the fault diagnosis is varied.

7.6.1 Termination of diagnosis with six elements in current state estimate

The current state estimate $\mathcal{Z}^*(k)$ of the plant becomes a singleton at time $k_d = 14$. However, in Section 7.5.2 it has been discussed that the current state of the faulty plant is known already at time $k = 9$. In Fig. 7.11 it can be seen that 6 faults remain in the set $\mathcal{F}^*(9)$ of fault candidates at this time. Therefore, now the termination condition $|\mathcal{Z}^*(k)| \leq 6$ is chosen for the fault diagnosis. A simulation has to reveal whether this termination condition is valid according to Assumption 4. The set $\mathcal{Z}_F$ of possible desired final states that shall still be achievable after a reconfiguration based on an ambiguous diagnostic result $\mathcal{D}^*$ is given by

$$\mathcal{Z}_F = \{P1, P2, P3\}. \tag{7.29}$$

Physically this set describes that after the reconfiguration it shall be possible to transport work-pieces cyclically through the Handling System HANS.

The termination condition $|\mathcal{Z}^*(k)| \leq 6$ is fulfilled at time $k = 9$. The corresponding diagnostic result is given by

$$\mathcal{D}^* = \mathcal{Z}^*(9) = \left\{ \begin{pmatrix} S2 \\ 1 \end{pmatrix}, \begin{pmatrix} S2 \\ 2 \end{pmatrix}, \begin{pmatrix} S2 \\ 3 \end{pmatrix}, \begin{pmatrix} S2 \\ 4 \end{pmatrix}, \begin{pmatrix} S2 \\ 7 \end{pmatrix}, \begin{pmatrix} S2 \\ 9 \end{pmatrix} \right\}. \tag{7.30}$$

That is, it is known that the workpiece lies in slide S2, but it is not known whether one of the four positions M1 − M4 on the measuring block or slide S1 is blocked or deflector D2 toggles. The reconfigurability of the nominal tracking controller $\mathcal{C}_T(\mathcal{A}_0)$ with respect to the diagnostic result $\mathcal{D}^*$ in (7.30) is analyzed according to Theorem 5.5. Therefore, the common model $\mathcal{A}_\cap$ of the faulty plant as defined in (5.27) has to be constructed. The result is shown in Fig. 7.14.

Since only one plant state is contained in the diagnostic result $\mathcal{D}^*$, the initial state of the common model $\mathcal{A}_\cap$ is given by $z_{\cap 0} = $ S2 (cf. (5.25d)). Due to the large number of different faults in the diagnostic result $\mathcal{D}^*$, the common model $\mathcal{A}_\cap$ contains significantly less transitions than the models $\mathcal{A}_f$, $(f \in \mathcal{F})$ of the faulty plant. Most importantly, state P1 becomes absorbing such that only states P3 and state P1 can be reached from the initial state $z_{\cap 0}$ of the common model. Consequently, the common model $\mathcal{A}_\cap$ in Fig. 7.14 is only controllable with respect to the states $z_F \in \{P1, P3, S2\}$ and the tracking controller $\mathcal{C}_T(\mathcal{A}_0)$ is only reconfigurable with respect to these states, but no others. Physically this means that the workpiece can be transported from its current position in the second slide to the magazine and to position P1 on the storage block. However, since any of the positions on the measuring block might be blocked, no further transport is possible afterwards.

For the case of an unambiguous diagnostic result, the closed-loop system $\mathcal{P} \circ \mathcal{C}_T(\mathcal{A}_0)$ has been shown to be recoverable from the fault $\bar{f} = 7$ for desired final states

$$z_F \in \{P1, P2, P3, M1, M2, M3, M4, S2, S3, B1\}$$

(see Section 7.4). With the diagnostic result $\mathcal{D}^*$ in (7.30) obtained due to the termination condition $|\mathcal{Z}^*(k)| \leq 6$ the tracking controller $\mathcal{C}_T(\mathcal{A}_0)$ is only reconfigurable with respect to some of these states. Most importantly, it is *not* reconfigurable with respect to state P2 $\in \mathcal{Z}_F$. Conse-

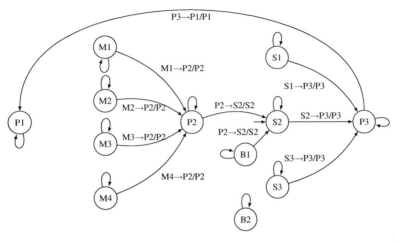

Figure 7.14: Common model $\mathcal{A}_\cap$ of the faulty plant with respect to diagnostic result $\mathcal{D}^*$ in (7.30).

quently Assumption 4 is not fulfilled which means that the chosen termination condition is not valid.

7.6.2 Termination of diagnosis with five elements in current state estimate

In order to fulfill Assumption 4, the termination condition is chosen to $|\mathcal{Z}^*(k)| \leq 5$ now such that the diagnostic result $\mathcal{D}^*$ is narrowed compared to the one in the previous section.

The termination condition $|\mathcal{Z}^*(k)| \leq 5$ is fulfilled at time $k = 12$. The corresponding diagnostic result is given by

$$\mathcal{D}^* = \left\{ \begin{pmatrix} M3 \\ 1 \end{pmatrix}, \begin{pmatrix} M3 \\ 2 \end{pmatrix}, \begin{pmatrix} M3 \\ 4 \end{pmatrix}, \begin{pmatrix} M3 \\ 7 \end{pmatrix}, \begin{pmatrix} M3 \\ 9 \end{pmatrix} \right\} \tag{7.31}$$

Compared to the diagnostic result $\mathcal{D}^*$ in (7.30) the fault $f = 3$ is now removed from the set $\mathcal{F}^*(12)$ of fault candidates. That is, it is known that position M3 on the measuring block is not blocked. The diagnostic unit $\mathcal{D}$ obtains this knowledge by successfully placing a workpiece on position M3. Figure 7.15 shows the common model of the faulty plant with respect to this refined diagnostic result. The additional knowledge leads to an additional transition from state

$z = \mathrm{P1}$ to state $z' = \mathrm{M3}$ in the common model $\mathcal{A}_\cap$ such that state P1 is no longer absorbing. Consequently, the tracking controller $\mathcal{C}_T(\mathcal{A}_0)$ is now reconfigurable with respect to the states $z_F \in \{\mathrm{P1}, \mathrm{M3}, \mathrm{P2}, \mathrm{S2}, \mathrm{P3}\}$ and a complete cycle can be achieved.

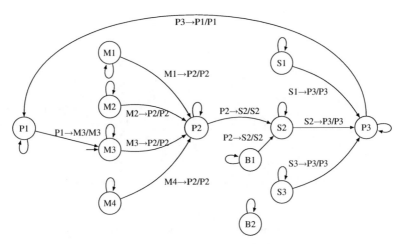

Figure 7.15: Common model $\mathcal{A}_\cap$ of the faulty plant with respect to diagnostic result $\mathcal{D}^*$ in (7.31).

Figure 7.17 shows the behavior of the fault-tolerant control loop when the termination condition $|\mathcal{Z}^*(k)| \leq 5$ is used for the fault diagnosis. At time $k_r = 12$ the fault diagnosis is stopped and the tracking controller $\mathcal{C}_T(\mathcal{A}_0)$ is reconfigured based on the diagnostic result $\mathcal{D}^*$ in (7.31) according to Algorithm 5.4. First, the error relations $E_{z\cap}$ and $E_{v\cap}$ are computed based on (5.31) and (5.32), respectively:

$$E_{z\cap}(z, v) = \begin{cases} G_0(z, v) & \text{if } (z, v) \in \{(\mathrm{P1}, \mathrm{P1} \rightarrow \mathrm{M3}), (\mathrm{M1}, \mathrm{M1} \rightarrow \mathrm{P2}), \dots\} \\ \text{undefined} & \text{if } (z, v) \in \{(\mathrm{P1}, \mathrm{P1} \rightarrow \mathrm{M1}), (\mathrm{P1}, \mathrm{P1} \rightarrow \mathrm{M2}), \dots\} \end{cases} \tag{7.32}$$

$$E_{v\cap}(v) = v, \quad \forall v \in \mathcal{V}_0 \text{ in (7.18).} \tag{7.33}$$

Secondly, the automaton $\mathcal{A}_0$ stored in the trajectory planning unit $\mathcal{T}(\mathcal{A}_0)$ is modified according to (5.34), which here results in exactly the same automaton as the common model $\mathcal{A}_\cap$ of the faulty plant in Fig. 7.15.

Finally, Algorithm 5.3 is applied to the controller $\mathcal{C}(\mathcal{A}_0)$. That is, based on the error relations

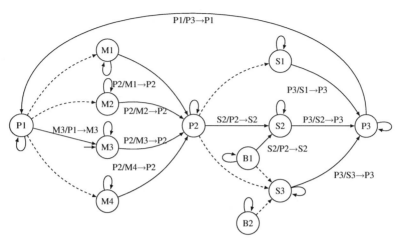

Figure 7.16: Reconfigured controller C^r corresponding to ambiguous diagnostic result $\mathcal{D}^*$ in (7.31).

$E_{z\cap}$ and $E_{v\cap}$ faulty transitions are removed from the controller, the set

$$\mathcal{E}_{\text{new}} = \{(P1, P1 \rightarrow M1), (P1, P1 \rightarrow M2), (P1, P1 \rightarrow M4), (P2, P2 \rightarrow S1),$$
$$(P2, P2 \rightarrow M3), (B1, P2 \rightarrow S3), (B2, P2 \rightarrow S3)\} \qquad (7.34)$$

of transitions to be newly defined is computed and it is found that no new transitions are added to the pruned controller. Thereby, the reconfigured controller C^r in Fig. 7.16 results.

After the reconfiguration the desired final state $z_F = P3$ is reached at time $k_F = 15$, that is, two time steps earlier than when waiting for an unambiguous diagnostic result (cf. Section 7.5.1). At the cost of not being able to use positions M1, M2 and M4 on the measuring block as well as the third slide, the early termination of the fault diagnosis leads to an earlier completing of the specified task by the Handling System HANS in this example.

The results of the corresponding experiment can be seen in Fig. 7.18. It shows that the reconfiguration based on the ambiguous diagnostic result $\mathcal{D}^*$ in (7.31) containing 5 elements is performed at time $t_r = 245s$. The output $w_p(t = 324s) = P3$ indicates that the desired final state $z_F = P3$ is reached at time $t_F = 324s$.

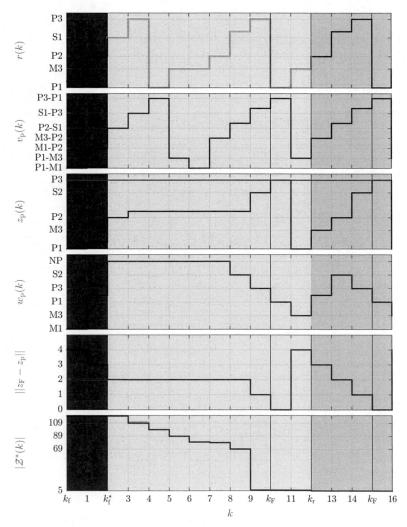

Figure 7.17: Fault-tolerant control of Handling System HANS with fault $\bar{f} = 7$ and termination condition $|\mathcal{Z}^*(k)| = 5$ (simulation).

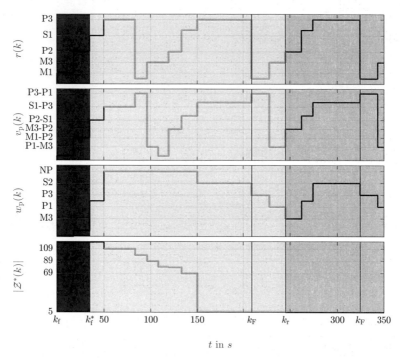

t in s

Figure 7.18: Fault-tolerant control of Handling System HANS with fault $\bar{f} = 7$ and termination condition $|\mathcal{Z}^*(k)| = 5$ (experiment).

7.6.3 Discussion of different termination conditions

The previous sections show that the use of termination conditions for the fault diagnosis different from the unambiguity of the diagnostic result may or may not be beneficial. If the fault diagnosis is stopped too early, the diagnostic result may contain too many different faults. In this case a common reconfiguration solution is not able to restore the desired functionality of the controlled faulty plant. It is application-dependent how much of the functionality is tolerable to be lost in order to allow for an early reconfiguration based on an ambiguous diagnostic result. In this example it is desirable to again achieve a cyclic behavior of the control loop after the reconfiguration, while it is not necessary to have access to all positions in the Handling System HANS. Under these conditions the reconfiguration can be advanced by $\Delta k = 2$ time steps or $\Delta t = 17s$.

7.7 Safe fault-tolerant control

In this section the behavior of the fault-tolerant control loop consisting of the fault-tolerant controller C_{FTC} and the Handling System HANS is analyzed for the case that some safety constraints have to be met by the controlled plant.

7.7.1 Safety constraints

If the heating block in the Handling System HANS is utilized by some other process, workpieces lying at position M1 or position M3 on the measuring block might be heated by the radiated heat as well (see Fig. 7.3). Therefore, the use of positions M1 and M3 shall be avoided. This constraint can be modeled in form of a set $\mathcal{E}_{\mathrm{ill}}$ of illegal transitions. It contains all those transitions $(z, z') \in \mathcal{Z}_0 \times \mathcal{Z}_0$ of the model $\mathcal{A}_0$ of the faultless plant, which end either in state $z' = \mathrm{M1}$ or in state $z' = \mathrm{M3}$. Consequently, the set of illegal transitions for the Handling System HANS is given by

$$\mathcal{E}_{\mathrm{ill}} = \{(\mathrm{P1}, \mathrm{M1}), (\mathrm{P1}, \mathrm{M3})\}. \tag{7.35}$$

7.7.2 Safe controllability analysis

Safe controllability of the Handling System HANS with respect to the set $\mathcal{E}_{\mathrm{ill}}$ of illegal transitions in (7.35) is analyzed based on Theorem 3.2. That is, first, the legal part $\mathcal{A}_{\mathrm{leg}}$ of the model $\mathcal{A}_0$ of the faultless plant in Fig. 7.4 has to be constructed. According to Definition 2.4 the automaton $\mathcal{A}_{\mathrm{leg}}$ is obtained by removing all illegal transitions $(z, z') \in \mathcal{E}_{\mathrm{ill}}$ from the model $\mathcal{A}_0$ of the faultless plant. In this example the state transition function G_{leg} and the output function H_{leg} of the automaton $\mathcal{A}_{\mathrm{leg}}$ are defined for all state-input pairs

$$(z, v) \in (\mathcal{Z}_0 \times \mathcal{V}_0) \setminus \{(\mathrm{P1}, \mathrm{P1} \to \mathrm{M1}), (\mathrm{P1}, \mathrm{P1} \to \mathrm{M3})\},$$

because

$$(\mathrm{P1}, G_0(\mathrm{P1}, \mathrm{P1} \to \mathrm{M1})) = (\mathrm{P1}, \mathrm{M1}) \in \mathcal{E}_{\mathrm{ill}}$$
$$(\mathrm{P1}, G_0(\mathrm{P1}, \mathrm{P1} \to \mathrm{M3})) = (\mathrm{P1}, \mathrm{M3}) \in \mathcal{E}_{\mathrm{ill}}.$$

The resulting automaton $\mathcal{A}_{\mathrm{leg}}$ is shown in Fig. 7.19.

According to Theorem 3.2 a necessary and sufficient condition for the safe controllability of the faultless Handling System HANS with respect to the set $\mathcal{E}_{\mathrm{ill}}$ of illegal transitions in (7.35) is the controllability of the legal part $\mathcal{A}_{\mathrm{leg}}$ of the automaton $\mathcal{A}_0$ in Fig. 7.19. It can be seen that

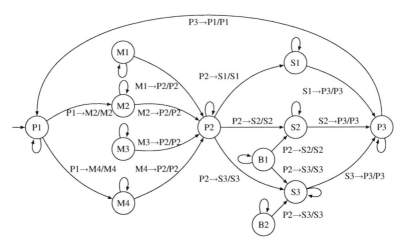

Figure 7.19: Legal part $\mathcal{A}_{\text{leg}}$ of automaton $\mathcal{A}_0$ with respect to set $\mathcal{E}_{\text{ill}}$ of illegal transitions in (7.35).

the states M1, M3, B1 and B2 are not reachable from the initial state $z_0 = \text{P1}$ in the automaton $\mathcal{A}_{\text{leg}}$, but all other states are reachable. That is, the automaton $\mathcal{A}_{\text{leg}}$ is controllable with respect to any final state $z_{\text{F}} \in \mathcal{Z}_0 \setminus \{\text{M1}, \text{M3}, \text{B1}, \text{B2}\}$. Consequently, from Theorem 3.2 it follows that the faultless Handling System HANS is safely controllable with respect to the set $\mathcal{E}_{\text{ill}}$ of illegal transitions in (7.35) and any final state

$$z_{\text{F}} \in \mathcal{Z}_0 \setminus \{\text{M1}, \text{M3}, \text{B1}, \text{B2}\}. \tag{7.36}$$

Compared to the case without illegal transitions, controllability with respect to the final states M1 and M3 is lost, because they can only be reached via illegal transitions.

7.7.3 Nominal safe control of the Handling System HANS

The tracking controller $\mathcal{C}_{\mathcal{T}}(\mathcal{A}_0, \mathcal{E}_{\text{ill}}) = (\mathcal{T}(\mathcal{A}_0, \mathcal{E}_{\text{ill}}), \mathcal{C}(\mathcal{A}_0))$ is defined as described in Chapter 3. That is, the legal part $\mathcal{A}_{\text{leg}}$ of the automaton $\mathcal{A}_0$ in Fig. 7.19 is stored in the trajectory planning unit $\mathcal{T}(\mathcal{A}_0, \mathcal{E}_{\text{ill}})$ according to (3.19). The controller $\mathcal{C}(\mathcal{A}_0)$, however, is the same as the one defined in the case without illegal transitions shown in Fig. 7.6.

When the nominal tracking controller $\mathcal{C}_{\mathcal{T}}(\mathcal{A}_0, \mathcal{E}_{\text{ill}})$ is used to control the faultless Handling System HANS, the results shown in Fig. 7.20 are obtained. It can be seen that the reference trajectory

$$R(0\ldots3) = (\text{M2}, \text{P2}, \text{S1}, \text{P3}) \tag{7.37}$$

is generated and followed by the Handling System HANS. That is, a workpiece is transported from the initial position P1 on the storage block to position M2 on the measuring block and then via the conveyor belt into the first slide S1. Finally, it is placed in the magazine at time $k_\text{F} = 4$, before the process starts over. Compared to the behavior without safety constraints in Fig. 7.7 now position M2 instead of position M1 on the measuring block is used. Thereby the use of the illegal transition $(\text{P1}, \text{P1} \to \text{M1}) \in \mathcal{E}_\text{ill}$ is avoided.

7.7.4 Safe recoverability analysis

The fulfillment of the sufficient condition for safe recoverability in Corollary 6.3 by the closed-loop system $\mathcal{P} \circ \mathcal{C}_\mathcal{T}(\mathcal{A}_0, \mathcal{E}_\text{ill})$ consisting of the tracking controller $\mathcal{C}_\mathcal{T}(\mathcal{A}_0, \mathcal{E}_\text{ill})$ in Section 7.3.2 and the Handling System HANS is analyzed.

Safe detectability. Safe detectability is defined in Definition 4.8, which states that a fault has to induce a deviation from the nominal output sequence and that no illegal transitions are used between the fault occurrence time k_f and the fault detection time k_f^*. Same as in Section 7.4, most faults do not fulfill the first condition here. Only faults $f \in \{2, 5, 7\}$ change the output sequence of the Handling System HANS when the reference trajectory in (7.37) is used. For these faults the fault detection times $k_\text{f}^* = 0$ ($f = 2$) or $k_\text{f}^* = 2$ ($f \in \{5, 7\}$) are obtained if $k_\text{f} = 0$. Note that the second condition for safe detectability in (4.63) is fulfilled for all faults $f \in \mathcal{F}$, for example:

$$(G_1^\infty(\text{P1}, (\text{P1} \to \text{M2})), G_1^\infty(\text{P1}, (\text{P1} \to \text{M2}, \text{M2} \to \text{P2}))) = (\text{M2}, \text{P2}) \notin \mathcal{E}_\text{ill}. \tag{7.38}$$

That is, even if the faults $f \in \{1, 3, 4, 6, 8, 9, 10\}$ are not safely detectable, the safety constraints are not violated. Rather they just have no influence on the Handling System HANS following the current reference trajectory in (7.37).

Safe identifiability. Next, safe identifiability of the Handling System HANS has to be given. Therefore, according to Theorem 4.5, an adaptive safe homing sequence for the overall model $\bar{\mathcal{A}}_\Delta$ of the closed-loop system defined in (4.14) has to exist with respect to the set $\bar{\mathcal{E}}_\text{ill}$ of illegal transitions in (4.68) and any initial uncertainty $\mathcal{Z}^* = \{z_\text{c}\} \times \mathcal{Z}_\Delta$, ($z_\text{c} \in \mathcal{Z}_\text{c}$). These adaptive safe homing sequences have to start with every possible input symbol v_c for which $G_\text{c}(z_\text{c}, v_\text{c})$ is defined.

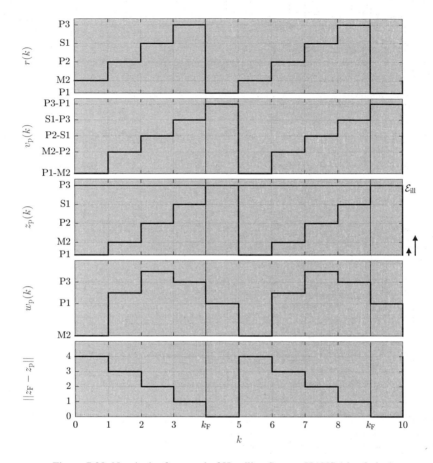

Figure 7.20: Nominal safe control of Handling System HANS (simulation).

In Sections 4.8.3 and 4.8.4 a method for the computation of these adaptive safe homing sequences based on a safe homing tree has been presented. Unfortunately, as described in Section 4.8.6, the computational complexity of this method is exponential in the squared number $n = 12$ of states in the plant model and the number $F = 10$ of faults. This example turns out to be too large to be computationally tractable with the given (inefficient) implementation on the available computer. After $20\,h$ the analysis is still not completed, even though 832 vertices have been added to the safe homing tree. The biggest problem, however, is that about $3.4 \cdot 10^9$ candidate trees for adaptive save homing sequences are stored in the set $\mathcal{S}$ at this time.

Consequently, the presented method is not applicable for the considered example.

Since the rigorous analysis of the safe identifiability fails, some preliminary assumptions are made in order to simplify the problem as follows:

- The fault set $\mathcal{F}$ is reduced to contain only safely detectable faults:

$$\tilde{\mathcal{F}} = \{2, 5, 7\}. \tag{7.39}$$

- The set of states of the overall closed-loop system $\bar{A}_\Delta$ to be considered is reduced by hand based on the fixed fault occurrence time $k_f = 0$:

$$\left\{ \left(\begin{array}{c} z_c(k_f^*) \\ z_p(k_f^*) \\ f \end{array} \right) : z_c(k_f^*) \in \{P1, P2\},\ z_p(k_f^*) \in \mathcal{Z}_0 \setminus \{M1, M3\},\ f \in \tilde{\mathcal{F}} \right\}. \tag{7.40}$$

- Instead of constructing the safe homing tree, the active fault diagnosis method for the case *without* illegal transitions is used, where for the definition of the tracking controller $\mathcal{C}_T$, the *legal part* $\mathcal{A}_{\text{leg}}$ in Fig. 7.19 of the automaton $\mathcal{A}_0$ has to be used. Due to the structure of the model and the kind of transitions included in the set $\mathcal{E}_{\text{ill}}$ of illegal transitions in (7.35) this procedure automatically leads to an avoidance of illegal transitions by the Handling System HANS during the active diagnosis.

  For each of these states active fault diagnosis as described in Section 4.5 is executed. As a result one input sequence per state is obtained. These input sequences can be combined to two adaptive safe homing sequences, one for each possible current state of the controller $\mathcal{C}(\mathcal{A}_{\text{leg}})$ at the fault detection time k_f^*. Part of the adaptive safe homing sequence for $z_c(k_f^*) = P2$ is shown in Fig. 7.21.

By this method all required adaptive safe homing sequences can be found. Therefore, the Handling System HANS is safely diagnosable with active safe diagnosis with respect to set $\tilde{\mathcal{F}}$ of safely detectable faults in (7.39) and the set $\mathcal{E}_{\text{ill}}$ of illegal transitions in (7.35).

Safe reconfigurability. For the safe reconfigurability of the tracking controller $\mathcal{C}_T(\mathcal{A}_0, \mathcal{E}_{\text{ill}})$ the legal parts $\mathcal{A}_{f,\text{leg}}$ of the automata $\mathcal{A}_f$, $(f \in \tilde{\mathcal{F}})$ have to be analyzed (cf. Corollary 6.3). Exemplarily, the legal part of the automaton $\mathcal{A}_7$ is shown in Fig. 7.22, while the legal parts of the automata $\mathcal{A}_2$ and $\mathcal{A}_5$ can be found in the appendix in Figs. C.10 and C.11, respectively. Compared to automaton $\mathcal{A}_7$ the transitions from state $z = P1$ to states $z' = M1$ and $z' = M3$ are removed in the legal part $\mathcal{A}_{7,\text{leg}}$. All reachable states in the automata $\mathcal{A}_{f,\text{leg}}$, $f \in \tilde{\mathcal{F}}$ are listed in

Table 7.7. It shows that the tracking controller $C_T(\mathcal{A}_0, \mathcal{E}_{\text{ill}})$ is recoverable from the faults $f \in \tilde{\mathcal{F}}$ with respect to states

$$z_F \in \{\text{P1}, \text{P2}, \text{P3}, \text{M4}, \text{S2}\} \tag{7.41}$$

and some other desired final states depending on the present fault $\bar{f}$.

					Reachable states							
f	P1	P2	P3	M1	M2	M3	M4	S1	S2	S3	B1	B2
2	✓	✓	✓			✓	✓	✓	✓			
5	✓	✓	✓		✓		✓	✓	✓			
7	✓	✓	✓		✓		✓		✓	✓	✓	

Table 7.7: Reachable states in automata $\mathcal{A}_{f,\text{leg}}$, $f \in \tilde{\mathcal{F}}$.

Safe recoverability. The previous analysis of the safe diagnosability and safe reconfigurability shows that the closed-loop system $\mathcal{P} \circ C_T(\mathcal{A}_0, \mathcal{E}_{\text{ill}})$ consisting of the Handling System HANS and the tracking controller $C_T(\mathcal{A}_0, \mathcal{E}_{\text{ill}})$ in Section 7.7.3 is safely recoverable from the faults $f \in \tilde{\mathcal{F}}$ in (7.39) and the set $\mathcal{E}_{\text{ill}}$ of illegal transitions in (7.35) (Corollary 6.3). Depending on the present fault $\bar{f} \in \tilde{\mathcal{F}}$, the recoverability is limited to a subset of possible desired final states z_F according to Table 7.7.

7.7.5 Behavior of the fault-tolerant control loop

Figure 7.24 shows the behavior of the fault-tolerant control loop for the case that fault $\bar{f} = 7$ occurs, that is, slide S1 is blocked. At the fault detection time $k_f^* = 2$ the fault is detected. Afterwards, active safe fault diagnosis as described in Section 4.8.5 is used for its identification. Since the current state of the controller $C(\mathcal{A}_0)$ at the fault detection time k_f^* is $z_c(k_f^*) = \text{P2}$, the adaptive safe homing sequence in Fig. 7.21 is used to generate inputs for the closed-loop system according to Algorithm 4.1. For example, since the output of the Handling System HANS at the fault detection time k_f^* is $w_p(k_f^* = 2) = \text{NP}$, the next reference input to the controller $C(\mathcal{A}_0)$ is $r(3) = \text{S1}$. Again the output $w_p(3) = \text{NP}$ such that the next reference input is $r(4) = \text{P3}$ and so on. The used path in the adaptive safe homing sequence is marked by bold edges in Fig. 7.21.

The fault diagnosis stops at the diagnosis time $k_d = 22$ when the unambiguous diagnostic result

$$\mathcal{D}^* = \left\{ \begin{pmatrix} \text{M2} \\ 7 \end{pmatrix} \right\} \tag{7.42}$$

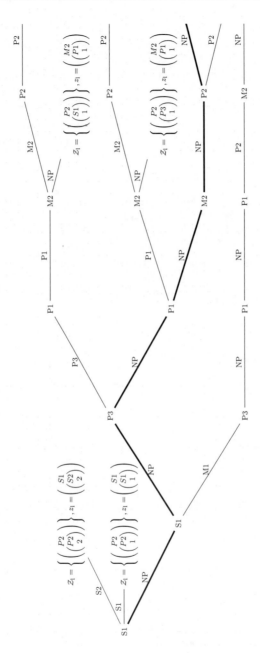

Figure 7.21: Part of adaptive safe homing sequence for Handling System HANS.

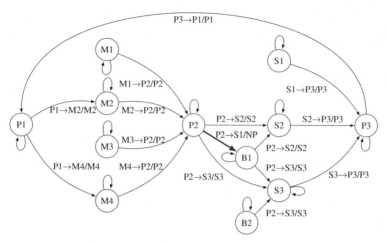

Figure 7.22: Legal part $\mathcal{A}_{7,\mathrm{leg}}$ of model $\mathcal{A}_7$ for Handling System HANS with blocked slide S1.

is obtained. That is, it is known that a workpiece lies at position M2 on the measuring block and slide S1 is blocked. Compared to the fault-tolerant control without illegal transitions, whose results are shown in Fig. 7.9, the fault diagnosis takes much longer here. The reason is that the input symbols for the closed-loop system can not be chosen such as to reduce the size of the current state estimate $\mathcal{Z}^*(k)$ of the plant as fast as possible. Rather, some input symbols are unavailable in some steps, because they might lead to illegal transitions in the Handling System HANS.

Just as in the case without illegal transitions described in the previous section, the reconfiguration of the tracking controller $\mathcal{C}_{\mathcal{T}}(\mathcal{A}_0, \mathcal{E}_{\mathrm{ill}})$ is performed as stated in Section 5.2. The only difference is that now the legal part $\mathcal{A}_{\mathrm{leg}}$ of the model of the faultless Handling System in Fig. 7.19 stored in the trajectory planning unit $\mathcal{T}(\mathcal{A}_0, \mathcal{E}_{\mathrm{ill}})$ is modified according to (5.2). The resulting automaton $\mathcal{A}_{\mathrm{leg}}^{\mathrm{r}}$ is shown in Fig. 7.23. The reconfigured controller $\mathcal{C}^{\mathrm{r}}$ is the same as the one in Fig. 7.13, now with initial state $z_{0c}^{\mathrm{r}} = \mathrm{M2}$.

The desired final state $z_{\mathrm{F}} = \mathrm{P3}$ is reached at time $k_{\mathrm{F}} = 25$. It can be seen that the illegal transitions $\mathcal{E}_{\mathrm{ill}}$ in (7.35) are avoided by the Handling System HANS at all times $0 \leq k \leq k_{\mathrm{F}}$. Therefore, as stated in Theorem 6.5, the safe fault-tolerant control problem in Problem 2.2 is solved.

Figure 7.25 shows the results of the corresponding experiment. A fault is detected at time $t_{\mathrm{f}}^* = 50s$, while it is identified at time $t_{\mathrm{d}} = 344s$. After the reconfiguration, the workpiece

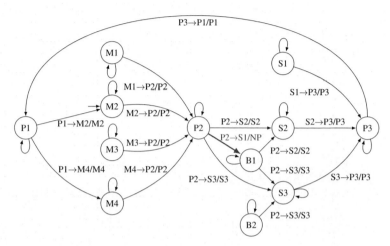

Figure 7.23: Automaton $\mathcal{A}_{\text{leg}}^{\text{r}}$ in reconfigured trajectory planning unit $\mathcal{T}^{\text{r}}$.

arrives at position P3 in the magazine at time $t_F = 423s$.

7.7.6 Time consumption

In Section 7.7.4 it has been discussed that the computational complexity of analyzing the safe diagnosability of the Handling System HANS is too high to be managed in reasonable time. For this example an alternative method for finding the required adaptive safe homing sequences partially relying on simplifications made by hand was used. However, in order to make the presented methods applicable to a larger class of problems, a different approach for finding adaptive safe homing sequences has to be developed.

The example also shows that once the adaptive safe homing sequences have been found, the complexity of the approach becomes very low. The choice of the next input symbol for the closed-loop system based on its previous output is completed within milliseconds. Just as in the case without illegal transitions, the definition and reconfiguration of the tracking controller $\mathcal{C}_{\mathcal{T}}(\mathcal{A}_0, \mathcal{E}_{\text{ill}})$ is accomplished in very little time ($< 0.01s$). That is, if the adaptive safe homing sequences can be found, the safe fault-tolerant control method is applicable to real-world systems.

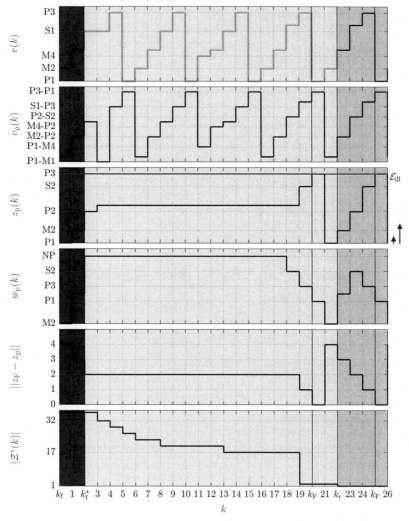

Figure 7.24: Fault-tolerant control of Handling System HANS with fault $\bar{f} = 7$ and $\mathcal{E}_{\mathrm{ill}}$ from (7.35) (simulation).

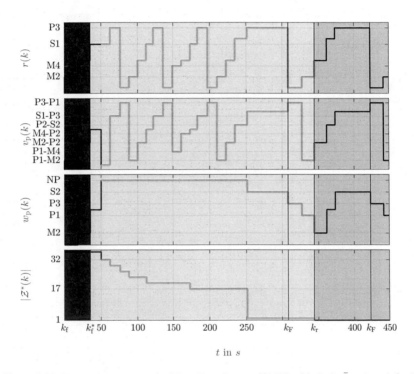

Figure 7.25: Fault-tolerant control of Handling System HANS with fault $\bar{f} = 7$ and $\mathcal{E}_{\mathrm{ill}}$ from (7.35) (experiment).

8 Conclusion

8.1 Summary

The main result of the thesis is a novel method for designing an integrated fault-tolerant controller $\mathcal{C}_{\text{FTC}}$ that realizes the three operating modes of nominal control, fault diagnosis and controller reconfiguration autonomously. Thereby, an active fault-tolerant control loop in which the controlled plant fulfills a given task and respects its safety constraints even in case of a fault was obtained (Theorems 6.3 and 6.5).

For the definition of the fault-tolerant controller $\mathcal{C}_{\text{FTC}}$ the model $\mathcal{A}_0$ of the faultless plant, its safety constraints in form of a set $\mathcal{E}_{\text{ill}}$ of illegal transitions, as well as the **error relations** E_{zf}, E_{vf}, E_{wf}, each of which corresponds to a fault $f \in \mathcal{F}$, have to be known. The use of error relations allows for a generic and simple fault modeling at the cost of limiting the describable behavior to persistent faults. It is assumed that only a single fault occurs.

The newly defined fault-tolerant controller $\mathcal{C}_{\text{FTC}}$ consists of three elements:

- a tracking controller $\mathcal{C}_T$,

- a diagnostic unit $\mathcal{D}$,

- a reconfiguration unit $\mathcal{R}$.

These elements interact with each other in order to put into effect the three operating modes of the fault-tolerant control loop. Therefore, on the one hand for each element new methods had to be developed which guarantee the fulfillment of the individual tasks of control, fault diagnosis and reconfiguration. On the other hand, the interactions between the tracking controller $\mathcal{C}_T$, the diagnostic unit $\mathcal{D}$ and the reconfiguration unit $\mathcal{R}$ had to be defined such that fault-tolerance is achieved.

Within the fault-tolerant controller $\mathcal{C}_{\text{FTC}}$ the **tracking controller** $\mathcal{C}_T$ realizes the nominal operating mode. For the tracking controller $\mathcal{C}_T$, a new structure consisting of a *trajectory planning unit* $\mathcal{T}$ and a *controller* $\mathcal{C}$ has been proposed in Chapter 3 (Fig. 3.1). The definition of the tracking controller follows directly from the model $\mathcal{A}_0$ of the faultless plant and a set $\mathcal{E}_{\text{ill}}$ of illegal transitions. It has been proved that the nominal tracking controller steers the faultless plant into a desired final state, which may be changed at runtime (Theorem 3.3).

For the diagnostic mode a method for **active fault diagnosis** has been proposed in Chapter 4. It allows to systematically influence the faulty plant such that the diagnostic result $\mathcal{D}^*$ is narrowed (Theorem 4.2). The choice of input sequences for the active diagnosis is performed automatically by the diagnostic unit $\mathcal{D}^*$ based on a partitioning of the state set of the plant $\mathcal{P}$. In order to take safety constraints into account, an alternative method for choosing these input sequences based on the evaluation of the behavior of the closed-loop system has been proposed (Theorem 4.5). It has been pointed out that additional effort is required to reduce the computational complexity of this safe fault diagnosis method.

It has been found that the **diagnostic result** $\mathcal{D}^*$ needs to contain not only the present fault $\bar{f}$. Rather, the current state $z_{\mathrm{p}}(k_{\mathrm{d}})$ of the faulty plant has to be known as well, because it is the starting point for the subsequent reconfiguration of the tracking controller $\mathcal{C}_T$. Ideally, the active fault diagnosis is executed until an *unambiguous* diagnostic result $\mathcal{D}^*$ is obtained. Alternatively, it has been shown how to reconfigure the tracking controller $\mathcal{C}_T$ based on an *ambiguous* diagnostic result $\mathcal{D}^*$ as well.

The reconfiguration mode is realized by the **reconfiguration unit** $\mathcal{R}$ presented in Chapter 5. Reconfiguration is obtained by *modifying* the nominal tracking controller $\mathcal{C}_T$ based on the given diagnostic result $\mathcal{D}^*$ and the error relations $\{E_{zf}, E_{vf}, E_{wf}, (f \in \mathcal{F})\}$ (Theorem 5.3). That is, the developed reconfiguration method does *not* rely on a complete redefinition of the tracking controller. The notion of a *common model* of the faulty plant that can be obtained from an ambiguous diagnostic result $\mathcal{D}^*$ has been introduced. Using this common model of the faulty plant, the reconfiguration method has been generalized to the case of an ambiguous diagnostic result $\mathcal{D}^*$ (Theorem 5.7).

It has been shown in Chapter 6 that **recoverability** of the closed-loop system follows from *diagnosability* of the plant $\mathcal{P}$ and *reconfigurability* of the tracking controller $\mathcal{C}_T$. However, due to the possibility of reconfiguration based on an ambiguous diagnostic result, diagnosability is not a necessary condition for recoverability. For the realization of the **integrated fault-tolerant controller** $\mathcal{C}_{\mathrm{FTC}}$, the switching conditions between the different operating modes have been shown to be crucial. The swap from nominal operating mode to diagnostic mode always occurs once a fault is detected. However, when to switch from diagnostic mode to reconfiguration mode depends on the *termination condition* for the fault diagnosis. It has been discussed that a careful choice of this termination condition is essential for the functionality of the fault-tolerant control loop.

The **application example** in Chapter 7 deals with the fault-tolerant control of the Handling System HANS. Both, in simulations and experiments, fault-tolerance could be obtained. Thereby the applicability of the developed fault-tolerant control method was demonstrated at a real system.

8.2 Outlook

The presented fault-tolerant control method for deterministic I/O automata could be extended into different directions. Two examples for such possible extensions are given here.

Different fault classes. The proposed method solves the fault-tolerant control problem under the assumption that only a single fault $f \in \mathcal{F}$ occurs and that the fault is persistent (Assumption 1). The consideration of the presence of time-varying faults or multiple faults at a time is an interesting extension. In this case, a description of the fault had to be incorporated into the plant model. Main problems to be solved when the assumption of a single persistent fault is dropped are, for example, to find new diagnosability conditions depending on the different fault classes. Furthermore, the procedures within the fault-tolerant controller have to be changed in order to be able to react to changes of the present fault or the occurrence of another fault. As a consequence, an clear separation between the three modes of nominal control, fault diagnosis and controller reconfiguration is no longer possible.

Different system class. The fault-tolerant control method has been developed for deterministic I/O automata, but could be extended to deal with nondeterministic I/O automata as well. In [4] it has been shown that the structure of the tracking controller can be used to control nondeterministic plants if a feedback from the plant to the controller is introduced. Based on this framework, the fault diagnosis and reconfiguration methods presented in this thesis can be extended as well. The presence of nondeterminism in the plant immediately questions the existence of an unambiguous diagnostic result. Particularly, the condition to stop the fault diagnosis has to be chosen carefully, because for nondeterministic plants the size of the diagnostic result is usually not monotonically decreasing over time. Regarding the reconfiguration, it has to be analyzed whether the proposed methods for modifying only part of the nominal tracking controller can be transfered.

Bibliography

Contributions of the author

[1] M. Schmidt and J. Lunze. "Active Diagnosis of Deterministic I/O Automata". In: *4th IFAC Workshop on Dependable Control of Discrete Systems*. York, UK, 2013.

[2] M. Schmidt and J. Lunze. "Active Fault Diagnosis of Discrete Event Systems subject to Safety Constraints". In: *2nd Int. Conf. on Control and Fault-tolerant Systems*. Nice, France, 2013.

[3] M. Schmidt and J. Lunze. "A Framework for Active Fault-Tolerant Control of Deterministic I/O Automata". In: *12th IFAC - IEEE Int. Workshop on Discrete Event Systems*. Paris, France, 2014.

[4] M. Schuh and J. Lunze. "Feedback Control of Nondeterministic Input/Output Automata". In: *53rd IEEE Conf. on Decision and Control*. Los Angeles, USA, 2014.

[5] M. Schuh and J. Lunze. "Fault-tolerant control for deterministic discrete event systems with measurable state". In: *2016 American Control Conf.* Boston, USA, 2016.

[6] M. Schuh and J. Lunze. "Fault-tolerant control of deterministic I/O automata using active diagnosis". In: *European Control Conf.* Aalborg, Denmark, 2016.

[7] M. Schuh and J. Lunze. "Fault-tolerant control of deterministic I/O automata with ambiguous diagnostic result". In: *13th IEEE Int. Workshop on Discrete Event Systems*. Xi'an, China, 2016.

[8] M. Schuh and J. Lunze. "Tracking control of deterministic I/O automata". In: *13th IEEE Int. Workshop on Discrete Event Systems*. Xi'an, China, 2016.

[9] M. Schuh, M. Zgorzelski, and J. Lunze. "Experimental Evaluation of an Active Fault-Tolerant Control Method". In: *Control Engineering Practice* 43 (2015), pp. 1–11.

Supervised theses

[10] O. Bayram. "Reglerentwurf für nichtdeterministische Eingangs-/Ausgangsautomaten". Master thesis. Ruhr-Universität Bochum, Lehrstuhl für Automatisierungstechnik und Prozessinformatik, 2016.

[11] N. Ganeshanathan. "Fehlermodellierung für ein virtuelles Pick&Place-System". Bachelor thesis. Ruhr-Universität Bochum, Lehrstuhl für Automatisierungstechnik und Prozessinformatik, 2015.

[12] K. Lamers. "Erprobung und Implementierung einer Methode zur aktiven Fehlerdiagnose ereignisdiskreter Systeme". Bachelor thesis. Ruhr-Universität Bochum, Lehrstuhl für Automatisierungstechnik und Prozessinformatik, 2012.

[13] T. Lieb. "Fehlertolerante Steuerung eines verfahrenstechnischen Prozesses". Bachelor thesis. Ruhr-Universität Bochum, Lehrstuhl für Automatisierungstechnik und Prozessinformatik, 2013.

[14] F. Peters. "Erprobung eines Verfahrens zur aktiven Fehlerdiagnose ereignisdiskreter Systeme am Beispiel einer virtuellen Fertigungszelle". Bachelor thesis. Ruhr-Universität Bochum, Lehrstuhl für Automatisierungstechnik und Prozessinformatik, 2012.

[15] E. Polat. "Implementierung und Erprobung einer Methode zum Steuerungsentwurf für deterministische Eingangs-/Ausgangs-Automaten". Bachelor thesis. Ruhr-Universität Bochum, Lehrstuhl für Automatisierungstechnik und Prozessinformatik, 2014.

[16] M. Quast. "Komponentenorientierte Modellbildung eines virtuellen Pick&Place Systems". Bachelor thesis. Ruhr-Universität Bochum, Lehrstuhl für Automatisierungstechnik und Prozessinformatik, 2013.

[17] T. Rempe. "Modellierung von nichtdeterministischen ereignisdiskreten Prozessen am Handling System HANS". Bachelor thesis. Ruhr-Universität Bochum, Lehrstuhl für Automatisierungstechnik und Prozessinformatik, 2015.

[18] S. Sri Ramanan. "Zustandsbeobachtung am Handling System HANS mit Hilfe von Petrinetzen". Bachelor thesis. Ruhr-Universität Bochum, Lehrstuhl für Automatisierungstechnik und Prozessinformatik, 2016.

[19] M. Zgorzelski. "Fehlertolerante Steuerung eines ereignisdiskreten Prozesses am Handling System HANS". Master thesis. Ruhr-Universität Bochum, Lehrstuhl für Automatisierungstechnik und Prozessinformatik, 2014.

Further literature

[20]   A. Allahham and H. Alia. "Monitoring of a class of timed discrete events systems". In: *Proc. IEEE Int. Conf. on Robotics and Automation*. 2007, pp. 1003–1008.

[21]   K. Altisen, F. Cassez, and S. Tripakis. "Monitoring and fault-diagnosis with digital clocks". In: *6th Int. Conf. on Application of Concurrency to System Design*. 2006.

[22]   K. Andersson, B. Lennartson, and M. Fabian. "Synthesis of Restart States for Manu-facturing Cell Controllers". In: *2nd IFAC Workshop on Dependable Control of Discrete Systems*. Bari, Italy, 2009.

[23]   K. Andersson, B. Lennartson, and M. Fabian. "Restarting Manufacturing Systems; Restart States and Restartability". In: *IEEE Trans. on Automation Science and Engi-neering* 7 (2010), pp. 486–499.

[24]   K. Andersson, B. Lennartson, P. Falkman, and M. Fabian. "Generation of restart states for manufacturing cell controllers". In: *Control Engineering Practice* 19 (2011), pp. 1014–1022.

[25]   F. Baldissera, J. Cury, and J. Raisch. "A Supervisory Control Theory Approach to Con-trol Gene Regulatory Networks". In: *IEEE Trans. on Automatic Control* 61.1 (2016), pp. 18–33.

[26]   S. Balemi, G. J. Hoffmann, P. Gyugyi, H. Wong-Toi, and G. F. Franklin. "Supervisory control of a rapid thermal multiprocessor". In: *IEEE Trans. on Automatic Control* 38 (1993), pp. 1040–1059.

[27]   J.C. Basilio, S.T.S. Lima, S. Lafortune, and M.V. Moreira. "Computation of minimal event bases that ensure diagnosability". In: *Discrete Event Dynamic Systems* 22 (2012), pp. 249–292.

[28]   S. Biswas. "Diagnosability of discrete event systems for temporary failures". In: *Com-puters & Electrical Engineering* 38 (2012), pp. 1534–1549.

[29]   M. Blanke, M. Kinnaert, J. Lunze, and M. Staroswiecki. *Diagnosis and Fault-Tolerant Control*. 3rd ed. Berlin Springer, 2016.

[30]   Y. Brave and M. Heymann. "Stabilization of discrete-event processes". In: *Int. J. of Control* 51.5 (1990), pp. 1101–1117.

[31]   M. Cantarelli and J.-M. Roussel. "Reactive control system design using the supervisory control theory: evaluation of possibilities and limits". In: *9th Int. Workshop On Discrete Event Systems*. Göteborg, Sweden, 2008.

[32] L. Carvalho, J. Basilio, and M. Moreira. "Robust diagnosis of discrete event systems against intermittent loss of observations". In: *Automatica* 48.9 (2012), pp. 2068–2078.

[33] L. Carvalho, J. Basilio, M. Moreira, and L. Clavijo. "Diagnosability of Intermittent Sensor Faults in Discrete Event Systems". In: *2013 American Control Conf.* Chicago, USA, 2013, pp. 929–934.

[34] L. K. Carvalho, M. V. Moreira, and J. C. Basilio. "Generalized Robust Diagnosability of Discrete Event Systems". In: *18th IFAC World Congress*. 2011.

[35] C.G. Cassandras and S. Lafortune. *Introduction to Discrete Event Systems*. Dordrecht, The Netherlands: Kluwer Academic Publishers, 1999.

[36] E. Chanthery and Y. Pencolé. "Monitoring and active diagnosis for discrete-event systems". In: *7th IFAC Symposium on Fault Detection, Supervision and Safety of Technical Processes*. Barcelona, Spain, 2009.

[37] J. Chen and R. Kumar. "Polynomial Test for Stochastic Diagnosability of Discrete Event Systems". In: *8th IEEE Int. Conf. on Automation Science and Engineering*. Seoul, Korea, 2012.

[38] J. Chen and R. Kumar. "Failure Detection Framework for Stochastic Discrete Event Systems With Guaranteed Error Bounds". In: *IEEE Trans. on Automatic Control* 60.6 (2015), pp. 1542–1553.

[39] Y. Chen and G. Provan. "Modeling and diagnosis of timed discrete event systems – a factory automation example". In: *1997 American Control Conf.* 1997.

[40] Z. Chen, F. Lin, C. Wang, L. Y. Wang, and M. Xu. "Active Diagnosability of Discrete Event Systems and its Application to Battery Fault Diagnosis". In: *IEEE Trans. on Control Systems Technology* 22.5 (2014), pp. 1892–1898.

[41] O. Contant, S. Lafortune, and D. Teneketzis. "Diagnosis of intermittent faults". In: *Discrete Event Dynamic Systems* 14.2 (2004), pp. 171–202.

[42] M. Dal Cin. "Verifying Fault-Tolerant Behavior of State Machines". In: *2nd High-Assurance Systems Engineering Workshop*. Washington, USA, 1997, pp. 94–99.

[43] H. Darabi, M. A. Jafari, and A. Buczak. "A Control Switching Theory for Supervisory Control of Discrete Event Systems". In: *IEEE Trans. on Robotics and Automation* 19 (2003), pp. 131–137.

[44] X. Geng and J. Hammer. "Input/Output control of asynchronous sequential machines". In: *IEEE Trans. on Automatic Control* 50.12 (2005), pp. 1956–1970.

[45] A. Girault and É. Rutten. "Automating the addition of fault tolerance with discrete controller synthesis". In: *Formal Methods in System Design* 35.2 (2009), pp. 190–225.

[46]   S. Hashtrudi Zad, R. H. Kwong, and W. M. Wonham. "Fault Diagnosis in Finite-State Automata and Timed Discrete-Event Systems". In: *Topics in Control and its Applications*. Springer, 1999.

[47]   S. Hashtrudi Zad, R. H. Kwong, and W. M. Wonham. "Fault diagnosis in discrete-event systems: framework and model reduction". In: *IEEE Trans. on Automatic Control* 48.7 (2003), pp. 1199–1212.

[48]   S. Hashtrudi Zad, R. H. Kwong, and W. M. Wonham. "Fault diagnosis in discrete-event systems: incorporating timing information". In: *IEEE Trans. on Automatic Control* 50.7 (2005), pp. 1010–1015.

[49]   S. Jiang and R. Kumar. "Diagnosis of dense-time systems using digital-clocks". In: *2006 American Control Conf.* 2006.

[50]   S. Jiang, R. Kumar, and E. Garcia. "Diagnosis of repeated/intermittent failures in discrete event systems". In: *IEEE Trans. on Robotics and Automation* 19 (2003), pp. 310–323.

[51]   S. Jiang, Z. Huang, V. Chandra, and R. Kumar. "A polynomial algorithm for testing diagnosability of discrete-event systems". In: *IEEE Trans. on Automatic Control* 46 (2001), pp. 1348–1321.

[52]   D. Jungnickel. *Graphs, Networks and Algorithms*. 3rd. Berlin: Springer, 2008.

[53]   Z. Kohavi and N.K. Jha. *Switching and finite automata theory*. 3rd. Cambridge University Press, 2010.

[54]   R. Kumar, V. Garg, and S. Marcus. "Language stability and stabilizability of discrete event dynamical system". In: *SIAM J. on Control and Optimization* 31.5 (1993), pp. 1294–1320.

[55]   R. Kumar and S. Takai. "A Framework for Control-Reconfiguration Following Fault-Detection in Discrete Event Systems". In: *Fault Detection, Supervision and Safety of Technical Processes*. 2012, pp. 848–853.

[56]   D. Lee and M. Yannakakis. "Principles and methods of testing finite state machines – a survey". In: *IEEE* 84.8 (1996), pp. 1090–1123.

[57]   G. Lichtenberg and A. Steele. "An approach to fault diagnosis using parallel qualitative observers". In: *Workshop on Discrete Event Systems*. 1996.

[58]   S. Lima, J. Basilio, S. Lafortune, and M. Moreira. "Robust diagnosis of discrete-event systems subject to permanent sensor failures". In: *WODES*. 2010.

[59]   F. Lin. "Robust and adaptive supervisory control of discrete event systems". In: *IEEE Trans. on Systems Automatic Control* 32 (1993), pp. 1848–1852.

[60]   F. Lin. "Diagnosability of discrete-event systems and its applications". In: *Discrete Event Dynamic Systems* 4 (1994), pp. 197–212.

[61]   W.C. Lin, H.E. Garcia, D. Thorsley, and T.S. Yoo. "Sequential window diagnoser for discrete-event systems under unreliable observations". In: *Communication, Control, and Computing, 2009. Allerton 2009. 47th Annual Allerton Conf. on.* IEEE. 2009, pp. 668–675.

[62]   J. Liu and H. Darabi. "Control Reconfiguration of Discrete Event Systems Controllers With Partial Observation". In: *IEEE Trans. on Systems Man and Cybernetics* 34 (2004), pp. 2262–2272.

[63]   J. Lunze. *Ereignisdiskrete Systeme*. München Wien: Oldenbourg Verlag, 2006.

[64]   J. Lunze. "Determination of distinguishing input sequences for the diagnosis of discrete-event systems." In: *2nd IFAC Workshop on Dependable Control of Discrete Systems*. Bari, Italy, 2009.

[65]   J. Lunze. *Automatisierungstechnik*. 3rd. München: Oldenbourg Verlag, 2012.

[66]   J. Lunze. *Künstliche Intelligenz für Ingenieure*. 3rd. De Gruyter Oldenbourg, 2016.

[67]   J. Lunze and J. Schröder. "State observation and diagnosis of discrete-event systems described by stochastic automata". In: *Discrete Event Dynamic Systems* 11 (2001), pp. 319–369.

[68]   J. Lunze and J. Schröder. "Sensor and actuator fault diagnosis of systems with discrete inputs and outputs". In: *IEEE Trans. on Systems, Man and Cybernetics* 34 (2004), pp. 1096–1107.

[69]   J. Lunze and P. Supavatanakul. "Diagnosis of discrete-event systems described by timed automata". In: *IFAC World Congress*. 2002.

[70]   J. Lunze and P. Supavatanakul. "Timed discrete-event method for diagnosis of industrial actuators". In: *IEEE Int. Conf. on Industrial Technology*. 2002.

[71]   J. Lygeros, D.N. Godbole, and M. Broucke. "A fault tolerant control architecture for automated highway systems". In: *IEEE Trans. on Control Systems Technology* 8 (2000), pp. 205–219.

[72]   T. Moor and K. Schmidt. "Fault-tolerant control of Discrete-Event Systems with Lower-Bound Specifications". In: *5th Int. Workshop on Dependable Control of Discrete Systems*. Cancun, Mexico, 2015, pp. 161–166.

[73]   Y. Nke and J. Lunze. "Fault-Tolerant Control of Nondeterministic Input/Output Automata subject to Actuator Faults". In: *WODES*. 2010.

[74]   Y. Nke and J. Lunze. "A fault modeling approach for Input/Output Automata". In: *18th IFAC World Congress*. Milano (Italy), 2011.

[75]   Y. Nke and J. Lunze. "Online control reconfiguration for a faulty manufacturing process". In: *Dependable Control of Discrete Systems*. Saarbrücken, Germany, 2011.

[76]   Y. Nke and J. Lunze. "Control reconfiguration based on unfolding of input/output automata". In: *8th IFAC Symposium on Fault Detection, Supervision and Safety of Technical Processes*. Mexico City, Mexico, 2012.

[77]   Y. Nke and J. Lunze. "Control Design for Nondeterministic Input/Output Automata". In: *European J. of Control* 21 (2015), pp. 1–15.

[78]   A. Nooruldeen and K. Schmidt. "State Attraction Under Language Specification for the Reconfiguration of Discrete Event Systems". In: *IEEE Trans. on Automatic Control* 60.6 (2015), pp. 1630–1634.

[79]   A. Paoli and S. Lafortune. "Safe diagnosability of discrete event systems". In: *42nd IEEE Conf. on Decision and Control*. 2003.

[80]   A. Paoli and S. Lafortune. "Safe diagnosability for fault-tolerant supervision of discrete-event systems". In: *Automatica* 41.8 (2005), pp. 1335–1347.

[81]   A. Paoli, M. Sartini, and S. Lafortune. "Active fault tolerant control of discrete event systems using online diagnostics". In: *Automatica* 47 (2011), pp. 639–649.

[82]   S-J. Park. "Robust Supervisory Control of Uncertain Timed Discrete Event Systems Based on Activity Models and Eligible Time Bounds". In: *IEICE Trans. Fundam. Electron. Commun. Comput. Sci.* E88-A (2005), pp. 782–786.

[83]   S-J. Park and K-H. Cho. "Supervisory control for fault-tolerant scheduling of real-time multiprocessor systems with aperiodic tasks". In: *Int. J. of Control* 82 (2009), pp. 217–227.

[84]   S-J. Park and J-T. Lim. "Fault-tolerant Robust supervisor for DES with model uncertainty and its application to a Workcell". In: *IEEE Trans. on Robotics and Automation* 15 (1999), pp. 386 –391.

[85]   M. Petreczky, R. J. M. Theunissen, R. Su, D. A. van Beek, J. H. van Schuppen, and J. E. Rooda. "Control of Input/Output Discrete-Event Systems". In: *10th European Control Conf.* Budapest, Hungary, 2009.

[86]   G. Provan and Y. Chen. "Model-based diagnosis and control reconfiguration for discrete event systems: an integrated approach". In: *38th IEEE Conf. on Decision and Control*. 1999.

[87] G. Provan and Y-L. Chen. "Model-based Fault Tolerant Control Reconfiguration for Discrete Event Systems". In: *IEEE Int. Conf. on Control Applications*. Anchorage, AK, USA, 2000, pp. 473–478.

[88] P.J.G. Ramadge and W.M. Wonham. "The control of discrete event systems". In: *IEEE* 77.1 (1989), pp. 81–98.

[89] K. R. Rohloff. "Sensor Failure Tolerant Supervisory Control". In: *44th IEEE Conf. on Decision and Control*. 2005.

[90] M. Roth, J.-J. Lesage, and L. Litz. "A residual inspired approach for fault localization in DES". In: *2nd IFAC Workshop on Dependable Control of Discrete Systems*. 2009.

[91] M. Roth, J.J. Lesage, and L. Litz. "The concept of residuals for fault localization in discrete event systems". In: *Control Engineering Practice* 19.9 (2011), pp. 978–988.

[92] M. Roth, J.J. Lesage, L. Litz, et al. "An FDI method for manufacturing systems based on an identified model". In: *13th IFAC Symposium on Information Control Problems in Manufacturing*. 2009.

[93] M. Roth, S. Schneider, J.J. Lesage, and L. Litz. "Fault Detection and Isolation in Manufacturing Systems with an Identified Discrete Event Model". In: *Int. J. of Systems Science* 43.10 (2012), pp. 1826–1841.

[94] J.-M. Roussel and A. Giua. "Designing dependable logic controllers using the supervisory control theory". In: *16th IFAC World Congress*. Praha, Czech Republic, 2005.

[95] A. Saboori and S. Hashtrudi Zad. "Robust nonblocking supervisory control of discrete-event systems under partial observation". In: *Systems and Control Letters* 55 (2006), pp. 839–848.

[96] M. Sampath, S. Lafortune, and D. Teneketzis. "Active Diagnosis of Discrete-Event Systems". In: *IEEE Trans. on Automatic Control* 43.1 (1998), pp. 908–929.

[97] M. Sampath, R. Sengupta, S. Lafortune, K. Sinnamohideen, and D. Teneketzis. "Diagnosability of Discrete-Event Systems". In: *IEEE Trans. on Automatic Control* 40 (1995), pp. 1555–1575.

[98] M. Sampath, R. Sengupta, Lafortune S., K. Sinnamohideen, and D. Teneketzis. "Failure Diagnosis Using Discrete-Event Models". In: *IEEE Trans. on Control Systems Technology* 4 (1996), pp. 105–124.

[99] T. Schlage and J. Lunze. "Data communication reduction in remote diagnosis of discrete-event systems". In: *Fault Detection, Supervision and Safety of Technical Processes*. 2009, pp. 1569–1574.

[100] T. Schlage and J. Lunze. "Remote Diagnosis of Timed I/O-Automata". In: *21st Int. Workshop on the Principles of Diagnosis*. 2010.

[101] S. Schneider, L. Litz, and M. Danancher. "Timed Residuals for Fault Detection and Isolation in Discrete Event Systems". In: *3rd Int. Workshop on Dependable Control of Discrete Systems*. 2011.

[102] S. Shu, Z. Huang, and F. Lin. "Online Sensor Activation for Detectability of Discrete Event Systems". In: *IEEE Trans. on Automation Science and Engineering* 10 (2013), pp. 457–461.

[103] S. Shu and F. Lin. "Detectability of discrete event systems with dynamic event observation". In: *Systems & control letters* 59.1 (2010), pp. 9–17.

[104] S. Shu and F. Lin. "Generalized detectability for discrete event systems". In: *Systems & control letters* 60.5 (2011), pp. 310–317.

[105] S. Shu and F. Lin. "Fault-Tolerant Control for Safety of Discrete-Event Systems". In: *IEEE Trans. on Automation Science and Engineering* 11 (2014), pp. 78–89.

[106] S. Shu, F. Lin, and H. Ying. "Detectability of Discrete Event Systems". In: *IEEE Trans. on Automatic Control* 52.12 (2007), pp. 2356–2359.

[107] S. Shu, F. Lin, H. Ying, and X. Chen. "State estimation and detectability of probabilistic discrete event systems". In: *Automatica* 44.12 (2008), pp. 3054–3060.

[108] Z. Simeu-Abazi, M. Di Mascolo, and M. Knotek. "Fault diagnosis for discrete event systems: Modelling and verification". In: *Reliability Engineering & System Safety* 95.4 (2010), pp. 369–378.

[109] A. N. Sülek and K. Schmidt. "Computation of Fault-tolerant Supervisors for Discrete Event Systems". In: *4th IFAC Workshop on Dependable Control of Discrete Systems*. York, United Kingdom, 2013.

[110] S. Takai. "Verification of robust diagnosability for partially observed discrete event systems". In: *Automatica* 48 (2012), pp. 1913–1919.

[111] S. Takai and T. Ushio. "Reliable decentralized supervisory control of discrete event systems". In: *IEEE Transactions on Systems Man and Cybernetics* 30.5 (2000), pp. 661–667.

[112] D. Thorsley and D. Teneketzis. "Diagnosability of stochastic discrete-event systems". In: *Automatic Control, IEEE Trans. on* 50.4 (2005), pp. 476–492.

[113] D. Thorsley, T.S. Yoo, and H.E. Garcia. "Diagnosability of stochastic discrete-event systems under unreliable observations". In: *American Control Conf., 2008*. IEEE. 2008, pp. 1158–1165.

[114]   Richard F. Tinder. *Asynchronous Sequential Machine Design and Analysis*. Morgan and Claypool Publishers, 2009.

[115]   S. Tripakis. "Fault diagnosis for timed automata". In: *Formal Techniques in Real-Time and Fault-tolerant Systems*. Springer. 2002, pp. 205–221.

[116]   J. van Gorp, A. Giua, M. Defoort, and M. Djemai. "Active diagnosis for a class of switched systems". In: *52nd IEEE Conf. on Decision and Control*. Florence, 2013.

[117]   G. Viana, J. Basilio, and M. Moreira. "Computation of the Maximum Time for Failure Diagnosis of Discrete-Event Systems". In: *American Control Conf.* Chicago, IL, 2015, pp. 396–401.

[118]   X. Wang, I. Chattopadhyay, and A. Ray. "Probabilistic fault diagnosis in discrete event systems". In: *Decision and Control, 2004. CDC. 43rd IEEE Conf. on*. Vol. 5. IEEE. 2004, pp. 4794–4799.

[119]   Q. Wen, R. Kumar, J. Huang, and H. Liu. "Fault-tolerant supervisory control of discrete event systems: formulation and existence results". In: *Dependable Control of Discrete Systems (DCDS)*. Paris, France, 2007.

[120]   Q. Wen, R. Kumar, J. Huang, and H. Liu. "A framework for fault-tolerant control of discrete event systems". In: *IEEE Trans. on Automatic Control* 53.8 (2008), pp. 1839–1849.

[121]   T. Wittmann, J. Richter, and T. Moor. "Fault-Tolerant Control of Discrete Event Systems Based on Fault-Accommodating Models". In: *8th IFAC Symposium on Fault Detection, Supervision and Safety of Technical Processes*. Mexico City, Mexico, 2012.

[122]   T. Yoo and S. Lafortune. "Polynomial-time verification of diagnosability of partially observed discrete-event systems". In: *IEEE Trans. on Automatic Control* 47 (2002), pp. 1491–1495.

[123]   T.S. Yoo and H.E. Garcia. "Event diagnosis of discrete-event systems with uniformly and nonuniformly bounded diagnosis delays". In: *American Control Conf., 2004. 2004*. Vol. 6. IEEE. 2004, pp. 5102–5107.

[124]   T.S. Yoo and H.E. Garcia. "New results on discrete-event counting under reliable and unreliable observation information". In: *Networking, Sensing and Control, 2005. . 2005 IEEE*. IEEE. 2005, pp. 688–693.

[125]   J. Zaytoon and S. Lafortune. "Overview of Fault Diagnosis Methods for Discrete Event Systems". In: *Annual Reviews in Control* (2013), pp. 308–320.

Appendices

A List of symbols

General conventions

- **Scalars** are represented by lower-case italics (i.e., z).

- **Sequences** are represented by upper-case italics (i.e., $Z(0 \ldots k_\mathrm{e}) = (z(0), \ldots, z(k_\mathrm{e})))$.

- **Vectors** are represented by lower-case bold italics (i.e., $\boldsymbol{z}$).

- **Sets** are represented by calligraphic letters (i.e., $\mathcal{Z}$).

Indices and superscripts

$(.)_i$	i-th element of a vector
$(.)_{\#}$	Falsified symbol
$(.)_{\mathrm{sub}}$	Variables related to a subautomaton
$(.)_f$	Variables related to the fault f
$(.)_{\mathrm{p}}$	Variables related to the plant
$(.)_{\mathrm{c}}$	Variables related to the controller
$(.)_{\mathrm{leg}}$	Variables related to the legal part of a plant model
$(.)_{\Delta}$	Variables related to the overall model of the faulty plant
$\overline{(.)}_{\Delta}$	Variables related to the overall model of the closed-loop system
$(.)^{\mathrm{r}}$	Variables related to a reconfigured element
$(.)^{-}$	Variables related to a pruned automaton
$(.)_{\cap}$	Variables related to an intersection automaton

Systems

$\mathcal{P}$	Plant
$\mathcal{C}_{\mathcal{T}}$	Tracking controller
$\mathcal{C}$	Controller

$\mathcal{T}$	Trajectory planning unit
$\mathcal{D}$	Diagnostic unit
$\mathcal{R}$	Reconfiguration unit
$\mathcal{A}$	Deterministic I/O automaton
$\mathcal{A}_\Delta$	Overall model of the faulty plant
$\mathcal{P} \circ \mathcal{C}_\mathcal{T}$	Closed-loop system
$\bar{\mathcal{A}}_\Delta$	Overall model of the closed-loop system
$\mathcal{A}_\cap$	Common model of the faulty plant

Scalars

ε	Empty symbol
k	Time step
k_e	Time horizon of a sequence
k_F	Time at which desired final state is reached
k_f	Fault occurrence time
k_f^*	Fault detection time
k_d	Diagnosis time
k_r	Reconfiguration time
z	State
z_0	Initial state
z'	Next state
z_F	Desired final state
v	Input symbol
w	Output symbol
f	(Arbitrary) Fault
$\bar{f}$	Present fault
r	Reference signal
n	Cardinality of the state set
p	Cardinality of the input set
q	Cardinality of the output set
F	Cardinality of the fault set
m_k	Cardinality of a k-equivalence partition
x	Vertex in a graph
e	Edge in a graph
T	Tree
l	Level in a tree

z_1	State in the leaf of an adaptive homing sequence
z_{new}	Artificial initial state in an intersection automaton

Vectors

z_Δ	State of the overall model of the faulty plant
$\bar{z}_\Delta$	State of the overall model of the closed-loop system

Sequences

$Z(0 \ldots k_e)$	State sequence
$V(0 \ldots k_e)$	Input sequence
$W(0 \ldots k_e)$	Output sequence
$R(0 \ldots k_e)$	Reference trajectory
$V_H(0 \ldots k_e)$	Homing sequence
$V_S(0 \ldots k_e)$	Separating sequence
$V_x(0 \ldots k_e)$	Input sequence corresponding to vertex x in adaptive homing tree
$W_x(0 \ldots k_e)$	Output sequence corresponding to vertex x in adaptive homing tree

Sets

$\emptyset$	Empty set
$\mathcal{D}^*$	Diagnostic result
$\mathcal{Z}$	State set
$\mathcal{V}$	Input set
$\mathcal{W}$	Output set
$\mathcal{F}$	Fault set
$\mathcal{L}$	Behavioral relation
$\mathcal{X}$	Set of vertices
$\mathcal{E}$	Set of edges
$\mathcal{E}_{ill}$	Set of illegal transitions for the plant
$\bar{\mathcal{E}}_{ill}$	Set of illegal transitions for the closed-loop system
$\mathcal{Z}^*$	Initial uncertainty
$\mathcal{Z}^*(k)$	Current state estimate of the plant
$\bar{\mathcal{Z}}^*(k)$	Current state estimate of the closed-loop system

$\mathcal{F}^*(k)$	Set of fault candidates
$\mathcal{Z}_\mathrm{l}$	State set in the leaf of an adaptive homing sequence
$\mathcal{Z}_x$	State set in vertex x in adaptive homing tree
$\mathcal{Z}_x^\mathrm{pre}$	Set of predecessor states of vertex x in adaptive homing tree
$\mathcal{S}$	Set of candidates for an adaptive homing sequence
$\mathcal{Z}_i^k$	Set of k-equivalent states
$\mathcal{E}_\mathrm{new}$	Set of transitions to be newly defined
$\mathcal{Z}_\mathrm{F}$	Set of possible desired final states

Functions

G	State transition function
H	Output function
G^∞	Extended state transition function
Φ	Automaton map
Φ^{-1}	Inverted automaton map
$P(z, \tilde{z})$	Path from state z to state $\tilde{z}$
$\mathcal{V}_\mathrm{a}(.)$	Active input operator
E_{zf}	State error relation
E_{vf}	Input error relation
E_{wf}	Output error relation
E	Error relation operator

Operators

$\times$	Cartesian product
$\sim$	Equivalence
$(.)^\top$	Transpose of a vector
$\wedge$	Logical AND
$\vee$	Logical OR
$\neg$	Logical NOT
$2^{(.)}$	Power set of a set
$|(.)|$	Cardinality of a set
$(.)^\infty$	Set of all sequences with elements from given set
$A \vDash B$	A is modeled by B
$A \overset{!}{=} B$	A shall be equal to B

$A \stackrel{?}{=} B$	Is A equal to B?
$(.)!$	Function is defined for the given arguments
$\mathcal{I}$	Identity map
$\mathcal{O}(.)$	Order of a computational complexity

B Compatibility partitions of the automated warehouse

$z=$	$\mathcal{Z}^0_1$ $\begin{pmatrix}C\\C\\1\end{pmatrix}$	$\mathcal{Z}^0_2$ $\begin{pmatrix}B\\B\\1\end{pmatrix}$	$\mathcal{Z}^0_3$ $\begin{pmatrix}A\\B\\1\end{pmatrix}$	$\mathcal{Z}^0_4$ $\begin{pmatrix}A\\C\\1\end{pmatrix}$	$\begin{pmatrix}B\\C\\1\end{pmatrix}$	$\begin{pmatrix}C\\B\\1\end{pmatrix}$	$\begin{pmatrix}A\\C\\2\end{pmatrix}$	$\begin{pmatrix}B\\C\\2\end{pmatrix}$	$\begin{pmatrix}C\\B\\2\end{pmatrix}$
$v=A$	ok	ok	ok	np	np	np	np	np	np
$v=B$	ok	np	np	np	np	np	np	np	np
$v=C$	np	ok	np	np	np	np	np	np	np

$\mathcal{Z}^0_5$ $\begin{pmatrix}C\\A\\1\end{pmatrix}$	$\begin{pmatrix}C\\A\\2\end{pmatrix}$	$\mathcal{Z}^0_6$ $\begin{pmatrix}A\\A\\1\end{pmatrix}$	$\begin{pmatrix}B\\A\\1\end{pmatrix}$	$\mathcal{Z}^0_7$ $\begin{pmatrix}B\\A\\2\end{pmatrix}$	$\mathcal{Z}^0_8$ $\begin{pmatrix}A\\A\\2\end{pmatrix}$	$\mathcal{Z}^0_9$ $\begin{pmatrix}A\\B\\2\end{pmatrix}$	$\mathcal{Z}^0_{10}$ $\begin{pmatrix}B\\B\\2\end{pmatrix}$	$\mathcal{Z}^0_{11}$ $\begin{pmatrix}C\\C\\2\end{pmatrix}$
np	np	np	np	np	np	BkB	BkB	BkC
np	np	BkA	BkA	BkA	BkA	np	np	BkC
BkA	BkA	ok	np	np	BkA	np	BkB	np

$z=$	$\mathcal{Z}^1_1$ $\begin{pmatrix}C\\C\\1\end{pmatrix}$	$\mathcal{Z}^1_2$ $\begin{pmatrix}B\\B\\1\end{pmatrix}$	$\mathcal{Z}^1_3$ $\begin{pmatrix}A\\B\\1\end{pmatrix}$	$\mathcal{Z}^1_4$ $\begin{pmatrix}C\\B\\1\end{pmatrix}$	$\mathcal{Z}^1_5$ $\begin{pmatrix}A\\C\\1\end{pmatrix}$	$\begin{pmatrix}B\\C\\1\end{pmatrix}$	$\mathcal{Z}^1_6$ $\begin{pmatrix}A\\C\\2\end{pmatrix}$	$\begin{pmatrix}B\\C\\2\end{pmatrix}$	$\mathcal{Z}^1_7$ $\begin{pmatrix}C\\B\\2\end{pmatrix}$
$v=A$	$\mathcal{Z}^0_6$	$\mathcal{Z}^0_6$	$\mathcal{Z}^0_6$	$\mathcal{Z}^0_3$	$\mathcal{Z}^0_4$	$\mathcal{Z}^0_4$	$\mathcal{Z}^0_4$	$\mathcal{Z}^0_4$	$\mathcal{Z}^0_9$
$v=B$	$\mathcal{Z}^0_2$	$\mathcal{Z}^0_2$	$\mathcal{Z}^0_2$	$\mathcal{Z}^0_2$	$\mathcal{Z}^0_4$	$\mathcal{Z}^0_4$	$\mathcal{Z}^0_4$	$\mathcal{Z}^0_4$	$\mathcal{Z}^0_{10}$
$v=C$	$\mathcal{Z}^0_1$	$\mathcal{Z}^0_1$	$\mathcal{Z}^0_4$	$\mathcal{Z}^0_4$	$\mathcal{Z}^0_1$	$\mathcal{Z}^0_1$	$\mathcal{Z}^0_{11}$	$\mathcal{Z}^0_{11}$	$\mathcal{Z}^0_4$

$\mathcal{Z}^1_8$ $\begin{pmatrix}C\\A\\1\end{pmatrix}$	$\mathcal{Z}^1_9$ $\begin{pmatrix}C\\A\\2\end{pmatrix}$	$\mathcal{Z}^1_{10}$ $\begin{pmatrix}A\\A\\1\end{pmatrix}$	$\mathcal{Z}^1_{11}$ $\begin{pmatrix}B\\A\\1\end{pmatrix}$	$\mathcal{Z}^1_{12}$ $\begin{pmatrix}B\\A\\2\end{pmatrix}$	$\mathcal{Z}^1_{13}$ $\begin{pmatrix}A\\A\\2\end{pmatrix}$	$\mathcal{Z}^1_{14}$ $\begin{pmatrix}A\\B\\2\end{pmatrix}$	$\mathcal{Z}^1_{15}$ $\begin{pmatrix}B\\B\\2\end{pmatrix}$	$\mathcal{Z}^1_{16}$ $\begin{pmatrix}C\\C\\2\end{pmatrix}$
$\mathcal{Z}^0_6$	$\mathcal{Z}^0_8$	$\mathcal{Z}^0_6$	$\mathcal{Z}^0_6$	$\mathcal{Z}^0_8$	$\mathcal{Z}^0_8$	$\mathcal{Z}^0_9$	$\mathcal{Z}^0_9$	$\mathcal{Z}^0_4$
$\mathcal{Z}^0_7$	$\mathcal{Z}^0_9$	$\mathcal{Z}^0_7$	$\mathcal{Z}^0_7$	$\mathcal{Z}^0_7$	$\mathcal{Z}^0_7$	$\mathcal{Z}^0_{10}$	$\mathcal{Z}^0_{10}$	$\mathcal{Z}^0_4$
$\mathcal{Z}^0_5$	$\mathcal{Z}^0_5$	$\mathcal{Z}^0_1$	$\mathcal{Z}^0_5$	$\mathcal{Z}^0_5$	$\mathcal{Z}^0_5$	$\mathcal{Z}^0_4$	$\mathcal{Z}^0_4$	$\mathcal{Z}^0_{11}$

C Models of Handling System HANS

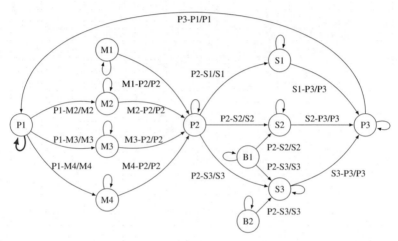

Figure C.1: Model $\mathcal{A}_1$ of Handling System HANS with blocked position M1.

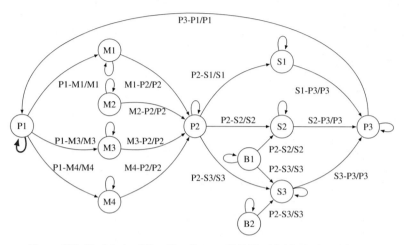

Figure C.2: Model $\mathcal{A}_2$ of Handling System HANS with blocked position M2.

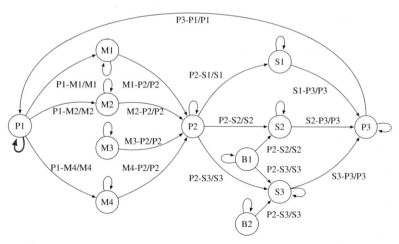

Figure C.3: Model $\mathcal{A}_3$ of Handling System HANS with blocked position M3.

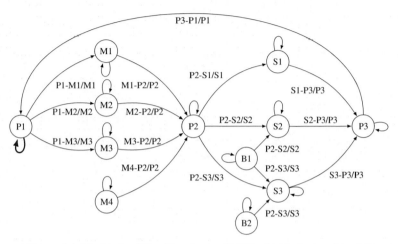

Figure C.4: Model $\mathcal{A}_4$ of Handling System HANS with blocked position M4.

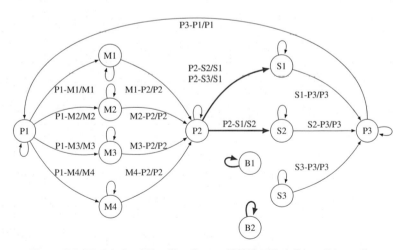

Figure C.5: Model $\mathcal{A}_5$ of Handling System HANS with deflector D1 toggling.

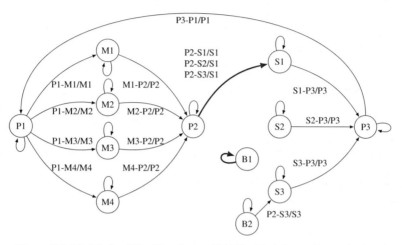

Figure C.6: Model $\mathcal{A}_6$ of Handling System HANS with deflector D1 stuck-closed.

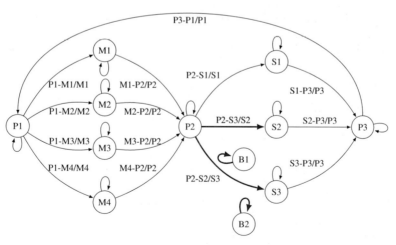

Figure C.7: Model $\mathcal{A}_8$ of Handling System HANS with deflector D2 toggling.

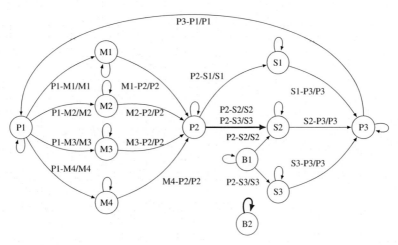

Figure C.8: Model $\mathcal{A}_9$ of Handling System HANS with deflector D2 stuck-closed.

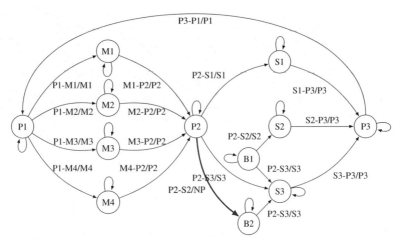

Figure C.9: Model $\mathcal{A}_{10}$ of Handling System HANS with blocked slide S2.

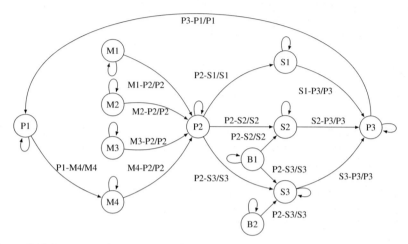

Figure C.10: Legal part $\mathcal{A}_{2,\text{leg}}$ of model $\mathcal{A}_2$ for Handling System HANS with blocked position M2.

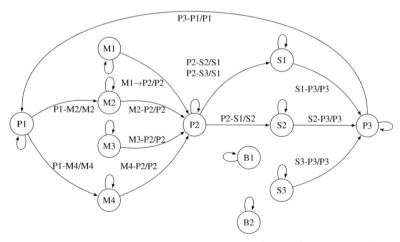

Figure C.11: Legal part $\mathcal{A}_{5,\text{leg}}$ of model $\mathcal{A}_5$ for Handling System HANS with deflector D1 toggling.

Lebenslauf

Persönliche Daten

Name	Melanie Schuh, geb. Schmidt
Geburtsdatum	02.06.1989
Geburtsort	Bochum

Bildungs- und Berufsweg

2011 – 2016	**Ruhr-Universität Bochum**, *Lehrstuhl für Automatisierungstechnik und Prozessinformatik*, Bochum. Wissenschaftliche Mitarbeiterin im Rahmen des TopING-Programms der Fakultät für Elektrotechnik und Informationstechnik
Forschungsthema	*Fehlertolerante Regelung ereignisdiskreter Systeme*
2010 – 2013	**Ruhr-Universität Bochum**, Bochum, Master. Studiengang Elektrotechnik und Informationstechnik, Studienschwerpunkt Automatisierungstechnik
Masterarbeit	*Fault diagnosis in deterministic I/O automata*
2010 – 2011	**Purdue University**, West Lafayette, Indiana, USA. DAAD-gefördertes Auslandsstudium
2010	**ThyssenKrupp Steel Europe AG**, *Direktionsbereich Koordination Metallurgie, Team Anlagentechnik*, Duisburg. Industriepraktikum
2007 – 2010	**Ruhr-Universität Bochum**, Bochum, Bachelor. Studiengang Elektrotechnik und Informationstechnik, Vertiefungsrichtung Informationstechnik
Bachelorarbeit	*Erprobung und Implementierung einer Methode zur fehlertoleranten Steuerung eines ereignisdiskreten verfahrentechnischen Prozesses*
1999 – 2007	**Märkische Schule**, Bochum, Abitur.
1995 – 1999	**Gertrudisschule**, Bochum.